Fiber and
Integrated Optics

NATO ADVANCED STUDY INSTITUTES SERIES

A series of edited volumes comprising multifaceted studies of contemporary scientific issues by some of the best scientific minds in the world, assembled in cooperation with NATO Scientific Affairs Division.

Series B: Physics

RECENT VOLUMES IN THIS SERIES

This series is published by an international board of publishers in conjunction with NATO Scientific Affairs Division

A Life Sciences	Plenum Publishing Corporation
B Physics	London and New York
C Mathematical and Physical Sciences	D. Reidel Publishing Company Dordrecht and Boston
D Behavioral and Social Sciences	Sijthoff International Publishing Company Leiden
E Applied Sciences	Noordhoff International Publishing Leiden

Fiber and
Integrated Optics

Edited by
D.B. Ostrowsky
University of Nice
Nice, France

PLENUM PRESS • NEW YORK AND LONDON
Published in cooperation with NATO Scientific Affairs Division

Library of Congress Cataloging in Publication Data

Nato Advanced Study Institute on Fiber and Integrated Optics, Cargèse, Corsica, 1978.
   Fiber and integrated optics.

   (Nato advanced study institute series: Series B; Physics; v. 41)
   Includes index.
   1. Fiber optics—Congresses. 2. Integrated optics—Congresses. 3. Optical communications—Congresses. 4. Optical wave guides—Congresses. I. Ostrowsky, D. B. II. Title. III. Series.
   TA1800.N37 1978   621.36'92   79-4377
   ISBN-13: 978-1-4613-2939-8   e-ISBN-13: 978-1-4613-2937-4
   DOI: 10.1007/978-1-4613-2937-4

Proceedings of the NATO Advanced Study Institute on Fiber and Integrated Optics, held in Cargèse, France, June 23—July 7, 1978

©1979 Plenum Press, New York
Softcover reprint of the hardcover 1st edition 1979
A Division of Plenum Publishing Corporation
227 West 17th Street, New York, N.Y. 10011

PREFACE

The Advanced Study Institute on Fiber and Integrated Optics was held at Cargese from June 23 to July 7, 1978, at a time when both fields were undergoing a very rapid evolution.

Fiber optics communications systems, in a multimode form, are moving out of laboratories and into practical use, and integrated optics is beginning to produce high performance, single-mode devices. In addition, the spin-off from the technological developments in both fields is beginning to have a growing impact on the general field of experimental physics. The lectures given at Cargese and assembled here illustrate these points and will be of considerable interest to both newcomers and people already in these fields.

The lectures in the first eight chapters of the book deal with fiber and optical communications. The second section, chapters 9-13, is devoted essentially to integrated optics. The third section, chapters 14-17, is devoted to technical seminars and the remaining chapters, 18-22, to national reviews and economic aspects of fiber systems.

On behalf of the organizing committee, which included Drs. Unger, Arnaud, Scheggi, and Daino, I would like to thank the Scientific Affairs Division of NATO, and in particular its director, Dr. T. Kester, for enabling this Advanced Study Institute to be held.

In addition, we would like to offer a very heartfelt thanks to Marie-France Hanseler, who, aided by Aline Medernach and G. Sala, created the memorable atmosphere that pervaded the Institute.

Finally, I offer my personal thanks to my wife, Nicole, whose enthusiasm and aid were irreplaceable.

Nice   D. B. Ostrowsky
October, 1978

CONTENTS

CONTENTS   ix

PROPAGATION IN OPTICAL FIBERS

J. Arnaud

Laboratoire d'Electronique des Microondes
123, rue Albert Thomas
87060 Cédex

1 - INTRODUCTION

Optical fibers have been in use for quite some time
to relay images over short distances. But these optical
guides were not seriously considered for long distance
communication until K.C. Kao[1] discoved in 1968 that the
loss of pure fused silica is below 20 dB/km in the near
infrared. Shortly afterward, workers at Corning Glass[2],
Bell Labs[3], and in Japan[4] managed to fabricate optical
fibers with losses of 20, 2 and very recently, 0.5dB/km.
Because of the very low losses achieved and because of
drastic reductions in pulse broadening obtained by pro-
filing the refractive index distribution, optical fibers
now successfully compete with more conventional techniques for carrying information. Transmission capacities
up to 100 Mbit/sec per fiber over distances of the order
of 10 km without repeater are within the state of the
art. Note that some fiber-optics cables containing as
many as 100 active fibers have been made that have 100
times the capacity quoted above. Compared with twisted
pairs of wire, or coaxial cables, optical fibers present
numerous practical advantages: low size, lack of electromagnetic interference, and, potentially, low cost. The
economic advantage of fiber optics cables, however, will
be fully assessed only during the next 10 years. In this
chapter we discuss mainly multimode circularly symmetric
fibers, because this type of fiber is the most commonly
fabricated today and the most practical in applications.

The basic mechanism for light guidance in fiber optics is total reflection. This condition is most easily understood for meridional rays in step-index fibers, as Fig.1 shows. Such rays are transmitted if they intersect the fiber axis at an angle θ less than the critical angle θ_c

$$|\theta| < \theta_c \tag{1}$$

This critical angle θ_c, and therefore the angular acceptance of the fiber, increases with increasing difference between core and cladding indices.

The basic optical parameters of a fiber are:
the core radius, r_c, usually comprised between 20 and 80μm
- the refractive index on axis, $n_o \simeq 1.5$
- the refractive index of the cladding material n_c
- the relative maximum index variation

$$\Delta = \frac{1}{2}(1-n_c^2/n_o^2) \simeq 1 - n_c/n_o \tag{2}$$

usually between 0.5 and 2 per cent for large capacity fibers. The critical angle $\theta_c = (2\Delta)^{1/2}$ is thus of the order of 0.1 to 0.2 radians. Plastic-clad and lead glass fibers have larger Δ's.
the numerical aperture, NA, defined as the sine of the fiber acceptance angle in air

$$NA = \sin\theta_{ext} = \sqrt{n_o^2-n_c^2} \simeq n_o\theta_c \tag{3}$$

- the V number

$$V = (2\pi r_c/\lambda)\sqrt{2\Delta} \; ; \; \lambda = \lambda_o/n_o \tag{4}$$

where λ_o denotes the free-space wavelength
- the fiber acceptance, N, which is the number of modes that the fiber can transmit. We have approximately

$$N = \frac{1}{2} V^2 \tag{5}$$

In most theoretical calculations, it is assumed that the cladding region extends to infinity. This is often permissible because the optical field decays extremely fast beyond the core region, and has a negligible value at the outer cladding radius. The optical fiber is usually surrounded by a protective plastic jacket (silicone and nylon) that may influence somewhat the optical loss of high-order modes.

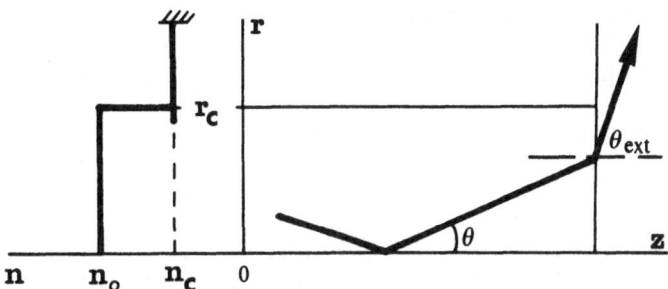

Figure 1 - At left, refractive index profile n(r) of
a step-index fiber. At right, meridional ray trajectory
r = r(z). NA ≡ sinθ$_{ext}$ is the numerical aperture.

Because the loss factor of the cladding material affects
predominantly the higher-order modes (corresponding to
large ray angles θ it tends to reduce the effective
numerical aperture of the fiber. This loss mechanism also
reduces pulse spreading. A rule of thumb is that the
average mode loss is equal to the cladding loss divided
by the fiber V number. Rough core-cladding interfaces
also tend to scatter predominantly high-order modes.
These loss mechanisms are neglected in subsequent calculations.

    There are basically two types of optical sources
that are practical in fiber-optics communication systems: the light emitting diode (LED) and the injection
laser. The LED is characterized by an emissive area on
the order of (50 μm)^2 that radiates light almost isotro-
pically. The coupling of that light to a fiber is me-
diocre, and a loss of more that 15dB is often suffered.
Furthermore, LED's have relatively broad linewidths
Δf, on the order of 9 THz, (or Δλ$_o$ ≃ 40 nm) that cause
pulses to broaden, as we shall see in more detail later.
On the contrary, injection lasers deliver a beam more
easily focused into the fiber core. The linewidth of
lasers is less than 500 GHz. If the control of the laser
modes were perfect, the linewidth of the radiation
would not exceed the reciprocal of the pulse duration,
e.g., 10 GHz for a 100 ps pulse. This condition, however, has not been achieved yet with injection lasers.

    Two types of photodetectors are considered for
fiber-optics communication: the p.i.n. diode and the
avalanche detector. The p.i.n. diode is slightly faster

than the avalanche detector and does not require high
voltages, but it requires 10 times more optical power
than avalanche detectors. At optical frequencies below
300 THz (λ_o >1μm) there are at the present time no low-
noise avalanche detector available. In that frequency
range, combinations of p.i.n. diodes and GaAs field-
effect transistors may prove preferable.

As far as the operating optical frequency is concerned, the trend has been from red light (f≈400 THz,
λ_o ≈0.75μm) to infrared light (f≈230THz, λ_o ≈ 1.3μm)
because the bulk loss of silica and material dispersion
($d^2n/d\lambda_o^2$) are minimal at the longer wavelength.

Let us now survey the content of this chapter. The
most commonly used optical-fiber fabrication technique
(CVD, for chemical vapor deposition) is described in
Section 2, and a few basic measurement techniques in
Section 3. Basic concepts in wave and ray optics are
presented in Section 4. Sections 5 and 6 provide the
mathematical formalism applicable to circularly symmetric optical fibers. In these two sections, we use the
paraxial ray approximation which is simple, and yet, sufficiently accurate for most problems. In section 5, expressions for the spreading of optical pulses are derived. But most optical fibers are imperfect. The most
commonly found defects are random distortions of
the fiber axis, which cause part of the optical power
to be lost. These bends, however, also have the favorable
effect of reducing the rate of pulse spreading, and
thus, increasing the transmission capacity. The
theory presented in section 6 is applicable to fibers
that have arbitrary index profiles and arbitrary (stationary) curvatures. One needs only assume that bending
is weak and circularly symmetric from a statistical
point of view.

The paraxial ray theory used in section 5 and 6
may occasionally be insufficiently accurate. Wave-optics
corrections are therefore briefly considered in Section 7.

2 - FIBER TECHNOLOGY

The basic material is fused silica (SiO_2). The refractive index of silica may be raised by addition of
dopants such as germania (GeO_2) or phosphoric oxide (P_2O_5),
or lowered by addition of boric oxide (B_2O_3) or fluor-
ite. The resulting index change rarely exceeds 1 to 2

per cent, but this is sufficient to ensure guidance of
the optical wave.

Low-loss silica is obtained by a chemical vapor deposition (CVD) technique, which consists of sending
gases (e.g., silicon or germanium tetrachlorides plus
oxygen) into a silica tube heated near the softening
point with a torch. The gases react to form layers of
a mixture of silica and germania. Microwave plasmas are
sometimes used to speed up the chemical reactions. Subsequent heating causes the tube to collapse into a solid rod, called the preform. This preform is about 10mm
in diameter and 400 mm in length. Fibers are pulled
from the preform held vertically with a torch or a CO_2
laser beam. The fiber, about 200µm in diameter, is coated on line with a thin protective layer of silicone
and wound on a drum. The index grading set up in the
preform scales down pretty accurately to that of the
fiber. All subsequent processing is aimed at providing
mechanical and chemical protection to the fibers, and
cabling them together. The number of fibers in a cable
may go from two to more than one hundred. The cable,
about 5mm in diameter,is much smaller and lighter than
coaxial cables of comparable transmission capacity.

The key to the development of fiber optics as a
practical means of transmitting information over long
distances was the reduction of the loss of the bulk material. One part per billion of the transition elements
(iron, cobalt...) and one part per million of the hydroxyl radical (OH^-) increase the optical loss by roughly 1 dB/km.

Thus, extreme purification of the chemical products
used to fabricate the fiber is essential. This problem,
however, appears to have been solved. There remain two
fundamental loss mechanisms: the Rayleigh scattering,
due to fast fluctuations of index and proportional to
the fourth power of the optical frequency, and the vibrations of the SiO_2 or GeO_2 molecules whose fundamental
mode of vibration is in the far infrared, but whose second and third harmonics are troublesome. For pure silica, the minimum loss takes place at an optical frequen
cy of roughly 230 THz(λ_o = 1.3µm). When silica is heavily doped with germania (e.g. with 20 mole per cent
germania), the optimum wavelength is shifted to lower
frequencies, f ≃ 215 THz. There are other sources of
loss which will be discussed later in sections dealing
with propagation.

The number of photons per bit (that is, per elementary time interval for a binary system) required at the optical receiver is 10 for an ideal quantum-limited receiver and roughly 1000 in practical receivers. The total loss that can be tolerated in transmission is thus comprised between 30 and 70 dB depending on the bit rate required (1Mbit/s to 1Gbit/s) and the type of receiver used. In the worst case (30dB),a 2dB/km loss fiber allows a 10 km separation between adjacent repeaters.

3 - MEASUREMENTS

We have selected four types of measurement. A more comprehensive review can be found in Ref.5.

3.1. Measurements on Bulk Samples

Prisms of doped silica can be fabricated with the help of a plasma torch. The refractive index is then measured with the standard minimum deviation method. Because the thermal history of the prism may not be the same as that of the fiber, this technique may not give the absolute value of the index corresponding to a known composition. It is most useful to characterize the dispersive properties of the material, whose knowledge is essential to the determination of the optimum profile. Recent data are given in Ref. 6.

3.2. Profile Measurements

If a beam of light is focused with a spot size of the order of 1µm at the input face of a cleanly cut fiber, the total transmitted power is easily shown to be proportional to $n(x,y)-n_c$, where $n(x,y)$ is the refractive index at the point x,y illuminated by the optical beam, and n_c is the cladding index. Thus the index profile is directly recorded by scanning the fiber core area. The spatial resolution afforded by this technique is limited by wave optics effects to about 2µm
(See Ref.7).

3.3. Backward Scattering

A wealth of information is obtained by turning on an optical source coupled to the fiber and looking at the back-scattered power. This technique gives informa-

tion about the fiber loss as a function of distance.
Defects such as breaks also clearly show up. Note that
one needs access to only one fiber end (Ref.5).

3.4. Power Transfer

The most commonly used technique consists of launching into the fiber an optical signal that has a short
duration, of the order of 100 ps, and looking at the
detected pulse at the other end of the fiber. Alternatively, one may modulate the light sinusoïdally and look
at the baseband frequency response, in amplitude and
phase. The power transfer characteristics depend of
course on the way the light is launched into the fiber
and on the spectral linewidth of the source. Useful information is obtained by using sources at two different
frequencies. An important practical problem, which has
not yet been fully solved, is to predict the behavior
of a sequence of sections of fibers on the basis of
measurements made on each individual section. The field
transfer function of short fibers has also been measured
(Ref.8).

4 - WAVES AND RAYS

The basic concepts in wave and ray propagation are
presented in this section, whose main goal is to derive
(rather than postulate) the paraxial-ray equations that
we shall need. Our viewpoint is that the most fundamental concept in physics is that of time-harmonic plane
waves. The laws of optics (or mechanics) can be derived
from their mathematical properties.

4.1. Time-Harmonic Plane Waves

Let us consider first a homogeneous, time-invariant
lossless medium. The wave equation has solutions of the
form

$$\Psi(x,y,z,t) = g\left(\frac{x}{\lambda_x} + \frac{y}{\lambda_y} + \frac{z}{\lambda_z} - \frac{t}{T}\right) \tag{6}$$

where g is a periodic function of its argument. Without
loss of generality the period of g can be assumed to
be unity.

$$g(u+1) = g(u) \tag{7}$$

For linear media

$$g(u) = \sin(2\pi u) \tag{8}$$

to within amplitude and phase factors.

Next, we define

$$a_x \equiv 1/\lambda_x \; ; \; a_y \equiv 1/\lambda_y \; ; \; a_z \equiv 1/\lambda_z \; ; \; f \equiv 1/T \tag{9a}$$

$$\vec{a} \equiv (a_x, a_y, a_z) \; ; \; |\vec{a}| \equiv a \equiv 1/\lambda \tag{9b}$$

where f is called the frequency, $\vec{a}$ the wave vector and λ the wavelength in the medium. The alternative notation $\omega = 2\pi f$, $\vec{k} = 2\pi\vec{a}$ is less convenient.

The concise notation

$$\Psi(x,y,z,t) = g(\sum_{i=1}^{4} A_i X^i) \tag{10}$$

where

$$X^i = (x,y,z,t)$$

$$A^i = (a_x, a_y, a_z, f)$$

$$A_i = (a_x, a_y, a_z, -f) \tag{11.}$$

is useful to unite space and time.

If we keep the frequency of the wave, or T, a constant but vary the direction of $\vec{a}$, the wavelength λ, and its reciprocal the length a of $\vec{a}$, usually vary. When this happens, the medium is called anisotropic. The tip of the vector $\vec{a}$ describes a surface, the surface of wave vector, that specifies the wavelength in the medium for each spatial direction of propagation. It is described mathematically in implicit form by

$$H(\vec{a},f) = 0 \tag{12}$$

In the special case of isotropic media, the wavelength λ being independent of the direction of propagation of the wave, Eq. (12) is

$$H(\vec{a},f) = a_x^2 + a_y^2 + a_z^2 - a^2(f) = 0 \tag{13}$$

The surface of wave vectors is in that case a sphere of radius a. Note that the phase velocity

$$v = f/a \tag{14}$$

is the velocity of motion of wave crests. In optics, one frequently makes use of the free-space wavelength

$$\lambda_o = c/f \tag{15}$$

where $c = 3 \times 10^8$ m/s is the free-space velocity of light, in place of f, and of a (phase) refractive index

$$n = \lambda_o/\lambda \tag{16}$$

but it is mathematically easier to use f and a than λ_o and n.

A medium is called nondispersive if a is proportional to f or, equivalently, if n and v are independent of f. The dispersion properties of a medium are described by the so-called dispersion curve $f = f(a)$, or equivalently by Sellmeier's law

$$n^2(\lambda_o) = 1 + \sum_{\gamma=1}^{3} A_\gamma (1-\pi_\gamma)^{-1} \; ; \; \pi_\gamma = (\ell_\gamma/\lambda_o)^2 \tag{17}$$

where $\ell_1, \ell_2, \ell_3, A_1, A_2, A_3$ are coefficients that can be found from a best fit to the experimental data. Each term in Eq.(17) corresponds to a molecular resonance of the material. One resonance takes place in the infrared (free-space wavelength ℓ_3) and the other two (ℓ_1, ℓ_2) in the ultraviolet. But Sellmeier's law law gives a good fit to measurements independently of any physical justification. A typical variation of n with λ_o is shown in Figure 2. Note the inflexion point near $\lambda_o = 1.3$ μm.

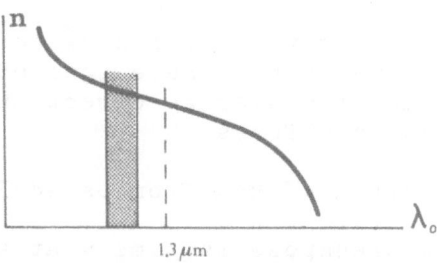

Figure 2 - Typical variation of n as a function of λ_o. The **hatched** area corresponds to visible light.

4.2. Rays Defined from Truncated Time-Harmonic Waves

Let the time-harmonic plane wave considered in the previous section be incident on an aperture whose width $\Delta \ell$ is somewhat larger than the medium wavelength λ. The wave intensity is strong after the aperture only near a straight line that defines a ray. Let that ray trajectory be denoted $\vec{x} = \vec{x}(\sigma)$ or

$$x = x(\sigma) \; ; \; y = y(\sigma) \; ; \; z = z(\sigma) \qquad (18)$$

Our main objective is to show that, in isotropic media, rays have the same direction as the wave vector. However, a more general formalism is first given.

To obtain the direction of the ray, note that, just after the aperture, the incident plane wave is replaced by a spectrum of plane waves with slightly different wave vectors. A geometric construction readily shows that the wavelengths of these waves coincide (and thus the waves reinforce each other) along the normal to the surface of wave vector $H(\vec{a}, f) = 0$. Thus, the direction $d\vec{x}/d\sigma$ of a ray is always given by

$$\frac{d\vec{x}}{d\sigma} = \frac{\partial H(\vec{a}, f)}{\partial \vec{a}} \qquad (19)$$

a relation known as the first Hamilton equation (hence the H notation used). For isotropic media, we obtain, introducing the expression given for H in Eq.(13) into Eq.(19), the relations

$$\frac{dx}{d\sigma} = 2a_x \; ; \; \frac{dy}{d\sigma} = 2a_y \; ; \; \frac{dz}{d\sigma} = 2a_z \qquad (20a)$$

or

$$dx/dz = a_x/a_z \; ; \; dy/dz = a_y/a_z \qquad (20b)$$

which show that the ray direction $d\vec{x}/d\sigma$ is that of the wave vector $\vec{a}$. Note that we have ignored the spreading of the beam around its average direction and thus neglected diffraction effects.

4.3. Group Velocity Defined from Gated Plane Waves

Let us now transpose in time what we have done above in space. The time-harmonic plane wave propagates along the z axis and is normally incident on a gate that opens up during a time somewhat larger than the

wave period T. The time-harmonic wave thus acquires a
temporal spectrum whose various components remain in
phase and reinforce each other along a particular direction in space-time that defines the group velocity u
of the wave packet. A reasoning similar to the one made
in the preceding section shows that

$$\frac{dt}{d\sigma} = - \frac{\partial H(a,f)}{\partial f} \tag{21}$$

Thus, for isotropic media, using Eq.(13) and the last
expression in Eq.(20a) with $a_z = a$, we obtain

$$\frac{dt}{dz} = \frac{da}{df} = u^{-1} \tag{22}$$

4.4. Refraction at Plane Interfaces

Let us consider next a stratified medium, invariant
under a translation along the z axis, with a = $1/\lambda$ a
function of x only, and a plane incident wave. Clearly
a_z must be invariant because of the continuity of the
field, which has an $\exp(2\pi i a_z z)$ dependence on z. Thus
differentiating a_x with respect to z in Eq.(13) (where
a now depends on x but whose dependence on f can be
ignored) we obtain, using Eq.(20b)

$$2a_x \frac{da_x}{dz} = 2a\frac{da}{dx}\frac{dx}{dz} = 2\frac{a}{a_z} a_x \frac{da}{dx} \tag{23a}$$

$$da_x/dz = (a/a_z)da/dx \tag{23b}$$

This is the second of Hamilton's equations. Differentiating further dx/dz in Eq.(20b) with respect to z, we
finally obtain the (exact) ray equation

$$\ddot{x} = \frac{d^2x}{dz^2} = (a/a_z^2)da/dx \tag{24}$$

It reduces in the paraxial approximation
$[a(x) \simeq a_z \simeq a(0)]$ to

$$\ddot{x} = -\frac{dU(x)}{dx} \; ; \; U(x) \equiv 1-\frac{n(x)}{n(0)} \equiv 1-\frac{a(x)}{a(0)} \tag{25}$$

or, more generally,

$$\ddot{x} = -\frac{\partial U(x,y)}{\partial x} \; ; \; \ddot{y} = -\frac{\partial U(x,y)}{\partial y} \tag{26a}$$

$$U(x,y) = 1-n(x,y)/n(0,0) \tag{26b}$$

U is analogous to a mechanical potential and Eq.(25) to Newton's law of motion.

Once a ray has been traced on the basis of Eq.(26) with given initial conditions: position x, y, and slope $\dot{x}$, $\dot{y}$, the time of flight along the ray is obtained integration

$$t(z) = \int_0^z u^{-1} ds \qquad (27)$$

where

$$ds = (dx^2+dy^2+dz^2)^{1/2} \simeq dz\left[1+\frac{1}{2}(\dot{x}^2+\dot{y}^2)\right] \qquad (28)$$

is the elementary arc length, and

$$u^{-1}(x,y) = \frac{\partial a(x,y,f)}{\partial f} \qquad (29)$$

is the reciprocal of the local group velocity. When the ray trajectory x = x(z), y = y(z) has been specified, $\dot{x}$, $\dot{y}$ and thus ds/dz and u^{-1} become known functions of z. We may proceed with the integration in Eq. (27) analytically or numerically.

Note that, even though a wave optics language has been used, all the results given in this section are strictly space-time ray-optics results. Diffraction effects are not accounted for. Alternatively, one may solve first the wave equation (perhaps the Helmholtz or the parabolic Fock equation to avoid the complications of the exact Maxwell's equations) for the modes, and apply the WKB approximation to the modal field. This brings us back to the ray equations. The relationship between ray manifolds and modes will be outlined in the next section.

5 - UNIFORM CIRCULARLY SYMMETRIC FIBERS

Most present-day optical fibers can be modeled as isotropic dielectric media uniform along the z axis and circularly symmetric. For the sake of clarity, losses are neglected. The variation of the refractive index n as a function of radius is kept arbitrary in the following analysis, except that it is a monotonically decreasing function of r from r = 0 ($n(0) \equiv n_o$) up to the core radius, r = r_c, where it reaches a constant

value, n_c. The cladding region is assumed to extend to
infinity. Finally, the whole theory is based on the
paraxial ray equations derived from plane-wave concepts
in the preceding section. Because the fiber is circu-
lary symmetric, it is natural to use an r, ϕ,z cylindrical coordinate system. The equations for the first and
second derivatives of r and ϕ with respect to z follow
from the ray equations, Eq.(26).

5.1. Ray Energy (E), Angular Momentum (ℓ) and Action (I)

Let us first define the "potential" function

$$U(r) = 1-n(r)/n_o, \quad n_o \equiv n(0). \tag{30a}$$

where

$$U(0) = 0 \; ; \; U(r_c) = \Delta \tag{30b}$$

There should be no confusion between this function $U(r)$
and the function of two variables $U(x,y)$ used earlier
in Eq.(26). We have

$$U(x,y) \equiv U(\sqrt{x^2 + y^2}) \tag{31}$$

The chain rule of differentiation gives

$$\frac{\partial U}{\partial x} = \frac{dU}{dr}\frac{x}{r} \; ; \; \frac{\partial U}{\partial y} = \frac{dU}{dr}\frac{y}{r} \tag{32}$$

The ray "energy" E is defined as

$$E = U + \frac{1}{2}(\dot{x}^2+\dot{y}^2) \; ; \; 0 < E < \Delta \tag{33}$$

and the ray "angular momentum" as

$$\ell = x\dot{y} - y\dot{x} \tag{34}$$

E and ℓ remain constant along any given ray. To prove
this assertion it suffices to calculate the total derivatives dE/dz and $d\ell/dz$, and use Eq.(26) **and** (32).

The mathematical identity

$$(x^2+y^2)(\dot{x}^2+\dot{y}^2) - (x\dot{x}+y\dot{y})^2=(x\dot{y}-y\dot{x})^2 \tag{35}$$

is equivalent to

$$2r^2(E-U) - (x\dot{x}+y\dot{y})^2 = \ell^2 \tag{36}$$

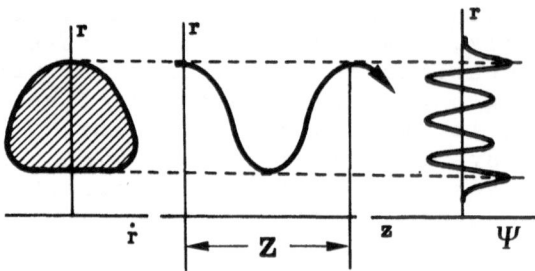

Figure 3 - On the left, ray trajectory in the "phase-
space" $\dot{r}$, r. In the middle, variation of r with z. On
the right, modal field associated with the ray shown
on the left.

if Eqs.(33) and (34) are used. Thus, the first derivative
of r with respect to z is

$$\dot{r} \equiv d(x^2+y^2)^{1/2}/dz = (x\dot{x}+y\dot{y})/r = \left[2(E-U)-\ell^2/r^2\right]^{1/2} \quad (37)$$

If we differentiate r a second time we obtain

$$\ddot{r} = -\frac{d}{dr}\left[U(r)+\frac{1}{2}\ell^2/r^2\right] \equiv -\frac{d}{dr}U'(r,\ell) \quad (38)$$

where we have introduced an effective potential $U'(r,\ell)$.
Rays can be trapped in the medium, for some value of ℓ,
only if the $U'(r)$ curve exhibits a minimum below Δ.

Next let us introduce the ray "action"

$$I(E,\ell) = \oint \dot{r}\,dr = \oint \{2\left[E-U(r)\right]-\ell^2/r^2\}^{1/2}dr \quad (39)$$

where the expression for $\dot{r}$ was taken from Eq. (37). The
circle on the integral refers to a full round trip in
the phase space shown in Figure 3. I is equal to twice
the integral of $\dot{r}(r)$ from r_1 to r_2, where r_1 and r_2
denote the lower and upper roots of the equation
$E = U'(r)$, respectively. They are called the "turning
points". From a straightforward differentiation under
the sum sign in Eq.(39), we obtain

$$\frac{\partial I}{\partial E} = \oint dr/\dot{r} = \oint dz = Z \quad (40)$$

where Z denotes the ray period.

Let us now consider the azimuthal angle

$$\phi = \arctan(y/x) \tag{41}$$

By differentiation with respect to z we obtain

$$\dot{\phi} \equiv d\phi/dz = \ell/r^2 \tag{42}$$

where the definition of ℓ in Eq.(34) has been used. Using the above expression of ϕ, we find that

$$\frac{\partial I}{\partial \ell} = -\oint \ell \, dr/r^2 \dot{r} = -\oint (\dot{\phi}/\dot{r}) dr = -\Phi \tag{43}$$

where Φ denotes the azimuthal ray period. In general, Φ is not an integral fraction of 2π and the projections of the ray trajectories on the transverse xy plane do not recycle after a round trip. Square-law media are exceptional in that, for all rays, $\Phi=\pi$, the projected rays being ellipses that may degenerate into straight lines or circles.

Finally, we shall need the relation

$$2Z(E-\overline{U}) = I + \ell \Phi \tag{44}$$

where Z, E, ℓ and Φ were defined earlier, and $\overline{U}$ denotes the axial average value of U(r), that is

$$\overline{U}(E,\ell) \equiv \frac{1}{Z} \int_0^Z U\left[r(z)\right] dz \tag{45}$$

For a square-law profile for example (harmonic oscillator in mechanics)

$$\overline{U} = \frac{1}{2} E \tag{46}$$

does not depend on ℓ.

For a given value of E, ℓ reaches its maximum value for helical rays, that is, for rays that remain at a constant distance from the z axis. Such rays project as circles on the transverse planes and as sinusoids on the meridional planes. Clearly, for such rays, $I = 0$. On the basis of the previous expressions one easily establishes that, for helical rays

$$E = \frac{d}{dR}(RU) \tag{47}$$

where $R \equiv r^2$ and U is considered a function of R.

5.2. Time of Flight

Let us go back to the basic Eqs(27) to (29).The latter is rewritten in term of n and of the free-space wavelength λ_o

$$u^{-1} = \frac{n}{c} - \frac{\lambda_o}{c} \frac{\partial n}{\partial \lambda_o} = u_o^{-1}(1-U-\frac{\lambda_o}{n_o} \frac{\partial n}{\partial \lambda_o}) \qquad (48a)$$

where

$$u_o^{-1} = n_o/c \qquad (48b)$$

denotes the time of flight per unit length on axis, if we neglect the dispersion on axis. We shall assume that $\partial n/\partial \lambda_o$ is a linear function of n as the dopant concentration varies. Note, however, that this assumption is open to question, particularly if two dopants are used. This linear relation is denoted

$$\frac{\lambda_o}{n_o} \frac{\partial n}{\partial \lambda_o} = 2(d-1) U \qquad (49)$$

The coefficient d is called the inhomogeneous dispersion parameter. Its departure from unity affects the optimum index profile, as we shall see. If we substitute Eqs(48a) and (49) into Eq.(27) we obtain the relative time of flight

$$\tau \equiv t/t_o = u_o t/z = \frac{1}{Z} \int_0^Z (1+E-U)\left[1-U-2(d-1)U\right]dz \qquad (50)$$

which simplifies to

$$\tau = 1+E-2\bar{U}d \qquad (51)$$

where $\bar{U}$ denotes the average value of U over a ray period, because E and U are smaller than $\Delta \ll 1$. It is very convenient (and sufficiently accurate) to restrict our attention to helical rays, because, along such rays, U is a constant. Using Eq.(47), we find that the profile is optimum ($\tau=1$) if

$$\frac{d}{dR}(RU) = \frac{2d}{R}RU \qquad (52)$$

or, by integration

$$U \propto R^{2d-1} \qquad (53)$$

The proportionality constant is easily found, and we finally obtain the optimum profile

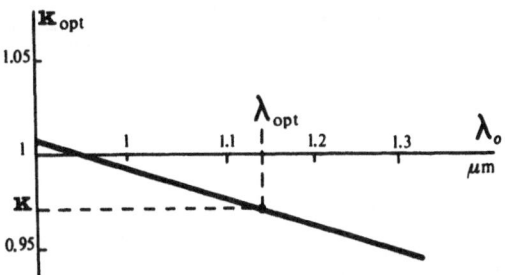

Figure 4 - Variation of the optimum value of the exponent κ with free-space wavelength for germania doping.

$$U(r) = \Delta(r/r_c)^{2\kappa_{opt}} \; ; \; \kappa_{opt} = 2d-1 \tag{54}$$

Under the assumptions made earlier [particularly in
Eq.(49)], the Δ of the fiber may vary with the optical
frequency, but the exponent κ that specifies the shape
of the profile is independent of frequency. However, it
turns out that the inhomogeneous dispersion parameter d,
and thus, the optimum exponent κ_{opt}, vary significantly with optical frequency. An optimum profile therefore
does not remain optimum at neighboring frequencies, as
Figure 4 shows. This effect is most conspicuous for
GeO_2 doping. It is smaller for P_2O_5 doping. The effect
may perhaps be reduced further if two dopants are used,
but the validity of our assumptions becomes in that case
questionable.

The pulse spreading characteristics of a fiber can
be summarized by a power-series expansion

$$\Delta t = a\Delta + b\Delta^2 + c\Delta f + d\Delta f^2 + e\Delta\Delta f \tag{55}$$

where Δ is the relative index change, as before, and Δf is
the linewidth of the source (for a LED for example,
$\Delta f \approx 9THz$). For a step-index fiber, the first term in
Eq.(55) is

$$\Delta t = 5000\Delta \text{ ns/km} \tag{56}$$

For example, $\Delta t = 50ns/km$ if $\Delta = 0.01$, a very fast spreading. This term disappears for the optimum profile in
Eq.(54). The second term $b\Delta^2$ can be accounted for only
by an exact ray theory. One finds

$$\Delta t \approx 300\Delta^2 \text{ ns/km} \tag{57}$$

At the present time, this term can be considered negligible. The term $c\Delta f$ is important mostly for LED's. It may be written

$$\Delta t = 5000 \ M_o \ \frac{\Delta f}{f} \ \text{ns/km} \tag{58}$$

where

$$M_o = \frac{\lambda_o}{n_o} \frac{d^2 n_o}{d\lambda_o} = \frac{f}{a_o} \frac{d^2 a_o}{df^2} \tag{59}$$

is the material dispersion parameter <u>on axis</u> (the value of M off-axis does not matter here). $\overline{M}_o$ vanishes at $\lambda_o = 1.383\mu m$ for 13.5 mole per cent GeO_2 and at $\lambda_o = 1.274\mu m$ for 9.1 mole per cent P_2O_5. At those wavelengths, we become interested in the quadratic term $d\Delta f^2$, first discussed in Ref.9. Measurements on bulk samples show that

$$\Delta t \simeq 0.0018\Delta f^2 \ \text{ns/km} \ (\Delta f \ \text{in terahertz}) \tag{60}$$

for almost all dopants. Finally, the term $e\Delta\Delta f$ is different from zero if optimum profiles do not remain optimum at neighboring optical frequencies. We have discussed this effect earlier in this chapter. This last term appears to be negligible for all practical purposes, particularly for P_2O_5 doping.

6 - RANDOMLY BENT FIBERS

Random distortions of the fiber axis can hardly be avoided in any practical fiber cable. These bends have the unfavorable effect of reducing the optical power available at the detector (microbending loss). But they also have the favorable effect of reducing the output impulse width. For the sake of clarity, we shall consider first dielectric slabs and assume that the curvature correlation is microscopic (white or uniform curvature spectrum).

6.1. Randomly Bent Dielectric Slabs

With respect to the curved slab axis z, the angle Θ that a ray makes with z evolves as a function of z according to the law

$$\Theta(z) = \Theta(0) + \int_0^z C(z') dz' \tag{61}$$

as long as θ does not exceed the critical angle $\theta_c = \sqrt{2\Delta}$.
This equation is well known, as it applies also to pro-
pagation along a curved earth, a problem that received
considerable attention early in the 1900's In
the present problem, the curvature C, defined as the reciprocal of the local radius of curvature, is a function
of z.

If $C(z)$ is gaussian distributed, so is $\theta(z)$ because
the equation, Eq.(61), which relates C to θ is linear. We
have, from Eq.(61), setting $\theta(0) = \theta_o$

$$< \left[\theta(z) - \theta_o\right]^2 > \, = \, < \int_0^z \int_0^z C(z')C(z'') dz' dz'' > \, = \, \gamma z \qquad (62)$$

for microscopic curvature correlations:

$$<C(z')C(z'')> \, = \, \Gamma(z'-z'') \, = \, \gamma\delta(z'-z'') \qquad (63)$$

The sign <> refers to an average over an ensemble of fibers. Note that the constant γ is also the power spectral density, defined over the $-\infty$ to $+\infty$ range of spatial frequencies. Since θ is gaussian and has the va-
riance in Eq.(63), the conditional probability density
of θ at z, given that $\theta(0) = \theta_o$, is

$$P(\theta, z \mid \theta_o) = (2\pi\gamma z)^{-1/2} \exp\left[-\frac{1}{2}(\theta-\theta_o)^2/\gamma z\right] \qquad (64a)$$

provided

$$|\theta| < \sqrt{2\Delta} \qquad (64b)$$

We notice next that $P(\theta, z \mid \theta_o)$ is the Green function of
the heat-diffusion-like equation

$$\frac{\partial P}{\partial z} = \frac{\gamma}{2} \frac{\partial^2 P}{\partial \theta^2} \qquad (65)$$

and account for the fact that rays are absorbed whenever
$|\theta| > \sqrt{2\Delta}$ somewhere along the fiber length, by setting the
boundary condition

$$P(\pm\sqrt{2\Delta}, z) = 0 \qquad (66)$$

To obtain the statistical modes, we look for solutions
of Eq.(65) of the form

$$P(\theta, z) = P_m(\theta)e^{-\lambda_m z} \; ; \; P_m(\pm\sqrt{2\Delta}) = 0 \qquad (67)$$

$$m = 0, 1, 2 \ldots$$

where the λ_m (not to be confused with wavelengths) correspond to the steady-state microbending losses. Numerically, we have

$$\alpha_m = \text{Microbending loss} = 4.34 \ \lambda_m \ \text{dB/unit length} \qquad (68)$$

The solution of Eqs. (65), (66) and (67) is straightforward

$$P_m(\theta) = \cos\left[(2m+1) \ \frac{\pi}{2} \ \frac{\theta}{\theta_c}\right] \qquad\qquad\qquad (69)$$

$$\lambda_m = (2m+1)^2 (\pi/2)^2 (\gamma/4\Delta) \qquad\qquad\qquad (70)$$

In particular, for the fundamental statistical mode

$$\alpha_0 = 2.65 \ \gamma/\Delta \ \text{dB/unit length} \qquad\qquad (71)$$

Because $\lambda_{m\neq0} > \lambda_0$, only the lowest statistical mode (m=0)
remains after a distance of the order of Δ/γ. The loss
given in Eq.(71) is applicable after a longer fiber
length. Equation (69) gives the steady-state radiation
pattern of the fiber (neglecting refraction at the fiber
tip). As we can see, the power radiated between θ and
$\theta + d\theta$, divided by $d\theta$, is simply a cosine function. But
this result rests on a number of assumptions that are
now relaxed.

6.2. Arbitrary Profiles and Arbitrary Curvatures

We now give the general result. The derivation,
which is based on paraxial ray optics, will be omitted.
An identical result can be obtained from modal theory
in the limit where the modes are so dense that they
form a continuum, if the WKB approximation is used.

Let $Q(I,\ell,z) \ dId\ell$ denote the probability that a ray
has action in the range $I, I + dI$ and angular momentum
in the range $\ell, \ell + d\ell$ at z. The Q function obeys the
partial differential equation

$$\frac{\partial Q}{\partial z} = \frac{\partial}{\partial \ell}\left(D_o\frac{\partial Q}{\partial \ell}\right) + \frac{\partial Q}{\partial \ell}\left(D_1\frac{\partial Q}{\partial I}\right) + \frac{\partial}{\partial I}\left(D_1\frac{\partial Q}{\partial \ell}\right) + \frac{\partial}{\partial I}\left(D_2\frac{\partial Q}{\partial I}\right) \qquad (72)$$

where

$$D_i(I,\ell) = \frac{1}{2}\sum_s (2\pi s)^i |f_s^2| \ G\left[(s+\phi/2\pi)/Z\right], i=0,1,2 \qquad (73)$$

the sum in Eq.(73) is extended over all integral values
of s(0,±1,±2...). Φ and Z denote as before the azimuthal
and axial ray periods. G(u) is the power spectral density of the C(z) process at spatial frequency u, that is

$$G(u) = \int_{-\infty}^{+\infty} <C(0)C(\xi)>e^{2\pi iu\zeta}d\zeta \qquad (74)$$

where C is assumed to have the same statistical properties in every meridional plane. The f_s coefficients are
defined from a ray trajectory x(z), y(z) in the unperturbed fiber

$$f_s=\frac{1}{Z}\int_0^Z \big[x(z)+iy(z)\big]e^{-2\pi i(s+\Phi/2\pi)z/Z}dz \qquad (75)$$

Closed-form expressions of the f_s are easily found for
square-law and for step-index profiles. For other profiles, they are found from simple numerical integrations.

   The statistical modes are obtained as in the previous subsection by setting

$$Q(I,\ell,z) = Q_m(I,\ell)e^{-\lambda_m z}, \quad m=0,1,2... \qquad (76)$$

and the boundary condition

$$Q_m(I(\Delta,\ell),\ell) = 0 \qquad (77)$$

where I is considered a function of E and ℓ. Because
Q(I,ℓ)has been defined as a probability, it is natural
to normalize the solutions by the condition

$$\int_{E<\Delta} Q_m(I,\ell)dId\ell = 1 \qquad (78)$$

Note that the Q_m are orthogonal

$$\int_{E<\Delta} Q_m(I,\ell)Q_n(I,\ell)dId\ell = 0, m\neq n \qquad (79)$$

If the correlation of C is microscopic, as we assumed in
section 6-1, G(u) has the constant value γ. If, on the
contrary, G(u) decreases very rapidly as a function of
u in the relevant range of spatial frequencies, it is
permissible to keep only the first term of the series
in Eq.(73) (adjacent-mode-neighbor approximation).

   For power-law profiles, (with exponent κ of r^2), a
set of statistical modes(which includes the fundamental) is described by Bessel functions of order $1/\kappa$. For
square-law profiles in particular (κ=1), we find

$$\alpha_0 = 7.9 \; \gamma/\Delta \; \text{dB/unit length} \qquad (80)$$

6-3. Impulse Response

Let $P(t)$ denote the output power when the source is modulated by a short pulse, with total energy equal to unity. We define

$$<t^n> \equiv \int_{-\infty}^{+\infty} t^n P(t) dt \tag{81}$$

and the root-mean-square (rms) pulse width

$$\sigma \equiv (<t^2> - <t>^2)^{1/2} \tag{82}$$

The transmission capacity of the fiber is of the order of $1/4\sigma$ bit/sec.

In order to obtain σ, one needs simply integrate time along the ray trajectories. For a particular fiber sample, we have

$$t(z) = u_o^{-1} \int_0^z \tau \left[I(z), \ell(z) \right] dz \tag{83}$$

where τ denotes the relative time of flight defined in section 5. This is a function of the ray action I and angular momentum ℓ, which are, in turn, functions of z for a particular ray **trajectory**. u_o is the group velocity on axis.

It follows from the central-limit theorem and Eq. (83) that the impulse response of a **distorted** fiber always tend to a gaussian for large lengths. The rms width σ defined in Eq.(82) is found to be given, for large lengths L, by

$$\sigma^2(L)/L = 2 \sum_{m=1}^{\infty} \mu_{Om}^2 \ a_o \ a_m \ (\lambda_m - \lambda_o)^{-1} \tag{84}$$

$$a_m^{-1} = \int Q_m^2 \ dId\ell, \quad \mu_{Om} = \int \tau \ Q_o Q_m \ dId\ell \tag{85}$$

For the step-index slab considered in section 6-1 for example, one finds

$$\sigma^2/L = 0.00431 \ \Delta^3/\gamma u_o^2 \ ; \ u_o \approx c/n_o \tag{86}$$

It is remarkable that the ratio $\mathcal{L} \equiv$ Microbending loss/ (transmission capacity improvement)^2 is a constant. For the slab, this ratio is

$$\mathcal{L} = 0.128 \ dB \tag{87}$$

This number means that if we are willing to tolerate a

microbending loss of 12 dB, the transmission capacity
of the system can be increased by a factor of 10, if the
curvature spectrum is uniform. A larger improvement
would be obtained if the curvature spectrum were band
limited.

For small fiber lengths L, σ is proportional to L
as if the fiber were undistorted. If, however, only
axial rays are excited (laser excitation), pulse broadening is initially negligible and increases according to
a $L^{3/2}$ law. Eventually the $L^{1/2}$ behavior is reached.

7 - EFFECTS NOT PREDICTED BY PARAXIAL RAY THEORY

As we have shown, the most important propagation
effects in multimode optical fibers are adequately explained by a simple paraxial ray theory. This theory,
incidentally, is formally identical to classical mechanics. A few secondary effects exist, however, which require more advanced methods. We have already mentioned
the $b\Delta^2$ term in pulse broadening, which is obtained only
from exact ray theory. Inhomogenous losses in the core
are easily incorporated in the paraxial ray theory, but
losses due to tunelling into the cladding, and cladding
or jacket losses, can be calculated only if a wave theory
or a generalized WKB approximation (which provides
an expression for the field beyond the turning points)
is used. Note also that steps in the index profile cause
a broadening of the pulses which is not properly accounted for by the ray theory. Scalar wave equations
predict that these steps should have a negligible effect
on pulse broadening if there are more than about one
hundred of them in the core. There are also in optical
fibers slight polarization effects, particularly when
strain-induced anisotropy is significant. Deviations of
the index profile from perfect circular symmetry is a
problem that has not been fully solved.

REFERENCES*

1 - K.C. Kao and T.W. Davies, J. Scient. Inst. 1 (ser.2)
    1063 (1968).
2 - F.P. Kapron, D.B. Keck and R.D. Maurer, Appl. Phys.
    Lett. 17 - 423 (1970)
3 - J.B. Mc. Chesney, P.B. O'Connor and H.M. Presby,
    Proc. IEEE 62, 1280 (1974)
4 - M. Horiguchi and H. Osanai, Electronics Letters, 12
    310 (1976)

5 - M.K. Barnoski and S.D. Personick, Proc IEEE, $\underline{66}$, 429
    (1978)
6 - J.W. Fleming, Electronics Letters, $\underline{14}$, n°11 (1978)
7 - J. Arnaud and R.M. Derosier, Bell Syst. Tech. J. $\underline{55}$,
    1489 (1976)
8 - C. Froehly, B. Colombeau and M. Vampouille - Proc.
    Int. Symp. Meas. Telecom. (Lannion, France, 3-7 oct.
    1977) p.436.
9 - J. Arnaud, Electronics Letters, $\underline{12}$, 654 (1976)
10 - J. Arnaud and M. Rousseau, Optics Letters, August
    1978 (to appear).

Text Books

D. Marcuse "Theory of Dielectric Optical Waveguides"
    Academic 1974
J. Arnaud "Beam and Fiber Optics", Academic 1976
M.K. Barnoski "Fundamental of Optical Fiber Communica-
    tion" - Academic 1976.

*Except for the first four historical papers we did not
attempt to cite original authors, the references being
given only for supplementary information. Reference to
earlier works can be found in those papers, particularly
in Ref.5.

FABRICATION OF OPTICAL FIBERS

A. de Panafieu

Thomson CSF
Laboratoire Central de Recherches
BP10, 91401 Orsay, France

INTRODUCTION

Optical fibers which are used for high bit rate transmission over long distances must meet two requirements: low transmission loss and low signal distortion.

To obtain such high quality optical fibers, new fabrication techniques have been developed which permit simultaneously:

1) the elaboration of extremely pure glasses, a prerequisite to low loss since even traces of certain impurities such as Fe and Cu increase the attenuation drastically;
2) the control of the refractive index profile in the fiber which is necessary for wide band transmission since it determines in large part the distortion of the signals transmitted along the fiber.

The first set of techniques which have been developed are referred to as the Chemical Vapor Deposition (or CVD) techniques. They use vapor phase reactions to produce ultra-pure glass rods, also called preforms, made of doped silica. To obtain the desired variation of refractive index, the concentration of dopant is varied radially. The preforms are then pulled into fibers at high temperature (typically 2000°C). The optical fibers made by this technique are referred to as "high silica fibers."

The classical glass making techniques, starting with powdered raw materials, have also been tremendously improved in the last decade in order to produce purer glasses. In these techniques, two glasses (core and cladding) of different chemical compositions are

melted separately and fed into two concentric crucibles with a hole
at the bottom through which a fiber is pulled. Either step index
or graded index fibers can be made by this double crucible method.
The fibers made by this process are referred to as "multicomponent
glass fibers."

Finally, other techniques have been developed which lead to
lower performance optical fibers which could still be used for their
lower cost in shorter transmission links and/or when wide band
transmission capacity is not needed. These are the plastic clad
fibers with cores of pure silica and cladding of a plastic resin,
and fibers made by phase separation for which cheap raw material can
be used, the glass being purified in the process.

It is the purpose of this paper to review the main optical
fiber fabrication techniques. The paper has the following sections:

    1. Attenuation and dispersion control
    2. High silica fiber fabrication techniques
        2.1. Dopants. Raw materials purification
        2.2. Flame hydrolisis
        2.3. Plasma torch CVD
        2.4. Thermal CVD
        2.5. Plasma activated CVD
        2.6. Fiber pulling
    3. Multicomponent glass fiber fabrication
    4. Plastic clad fibers
    5. Fibers made by phase separation

1. ATTENUATION AND DISPERSION CONTROL

1.1. Attenuation Controlling Factors

The transmission loss of an optical fiber is expressed in
dB/km and is defined by

$$\text{T.L. (dB/Km)} = (10/L)\log(I_0/I_L) \tag{1}$$

where I_0 and I_L are, respectively, the original light intensity and
the intensity after a length L (in km). The transmission loss which
is a function of wavelength comes from three main causes:

a) Rayleigh scattering caused by imperfections smaller than the
   wavelength. This is a decreasing function of the wavelength
   (λ^{-4}). This kind of scattering loss is always present in optical
   fibers but in the best cases it reduces to the intrinsic scattering which results from the disordered nature of glass and the
   associated fluctuations of the refractive index. Intrinsic scattering depends on the glass system which is used. In high silica
   fibers it may be as small as 4 dB/km at 0.63μ and 0.3 dB/km at

TABLE 1. Absorption Loss of Transition Metal Ions at 0.8μ
(db/km ppm)

	Soda-lime silicate (3)	Sodium-borosilicate (4)	Fused silica (5)
V	–	39	1050
Cr	20	50	610
Mn	80	11	20
Fe	100	10	50
Co	–	9	20
Ni	150	130	30
Cu	620	500	10

1.2μ (1). When increasing the dopant concentration, the intrinsic scattering is also increased (2).

b) Absorption of light by impurities such as transition metal ions to produce electronic transitions in these ions. Typical absorption values caused by 1 ppm of these impurites are indicated in Table I for three different glass systems. It is clear from this table that impurity levels well below 0.1 ppm must be obtained for most of these ions for the absorption loss to become negligible.

c) Absorption of light by hydroxyl radicals. This is associated with overtones of the fundamental OH vibration at about 2.8μ. The second overtone occurs around 0.95μ with a peak intensity, varying with glass composition, on the order of 0.5 to 1 dB per ppm OH (6).

Other loss mechanisms can be encountered in optical fibers but they are usually negligible in the best fibers. These are:

a) intrinsic absorption which comes from electronic transitions for the short wavelengths and from molecular absorption to induce vibrations at longer wavelengths – this loss mechanism is negligible in the spectral region 0.8-1.4μ;
b) stimulated scattering which does not occur in multimode fibers when conventional light sources are used;
c) scattering by imperfections large with respect to the wavelength (e.g. bubbles and platinum precipitates).

1.2. Dispersion Control

To use optical fibers for high bit rate transmission the distortion of the transmitted signals must be minimized. The two main factors controlling distortion are modal dispersion and material dispersion. Modal dispersion can be eliminated when the group

velocities of all trapped modes are made equal. This result is
achieved by carefully controlling the refractive index profile in
the fiber. The optimum profile is still a matter of controversy
but is closed to a power law

$$n^2(r) = n^2(o)[1 - 2\Delta(r/a)^\alpha] \quad \text{for} \quad r \leq a$$

$$n^2(r) = n^2(o)[1 - 2\Delta] \qquad\quad \text{for} \quad r > a$$

(2)

where n(r) is the refractive index at a distance r from the axis,
Δ is the relative refractive index difference between the axis and
the cladding and a is the core radius. The exponent α depends on
the glass system which is used and the wavelength of the light
signal but is closed to 2 (7). Material dispersion comes from the
fact that the refractive index of glass varies with wavelength.
Since available light sources (LED's and lasers) have a finite
spectral width, different spectral components of the light signal
travel with different velocities and shift with respect to each
other along the fiber. This effect is negligible at the so-called
"zero dispersion wavelength" (where $d^2 n/d \lambda^2 = 0$) which is around
1.3 μm (8-9).

2. HIGH SILICA FIBER FABRICATION TECHNIQUES

2.1. Dopants. Raw materials purification

High silica fibers are typically made of 90% silica and 10% of
at least one other compound whose incorporation in an amount varying radially permits the control of the refractive index profile in
the fiber. Several dopants can be incorporated in the silica network. The most common ones include:

- cation replacers, Ge, P, and B, which are incorporated in the
  glass as GeO_2, P_2O_5, and B_2O_3;
- anion replacer, F, which is incorporated in the glass as SiF_4.

As can be seen in Fig. 1, Ge and P increase and B and F decrease
the refractive index.

High silica preforms are made by vapor phase reactions, e.g.
high temperature oxidation, in which vaporized compounds of silicon
and the doping elements are converted to their respective oxides.
These are collected as fine particulates on a suitable substrate,
e.g. a silica rod. Typical reactions can be written

$$SiCl_4 + O_2 \rightarrow SiO_2 + 2Cl_2$$

$$2POCl_3 + (3/2)O_2 \rightarrow P_2O_5 + 3Cl_2$$

(3)

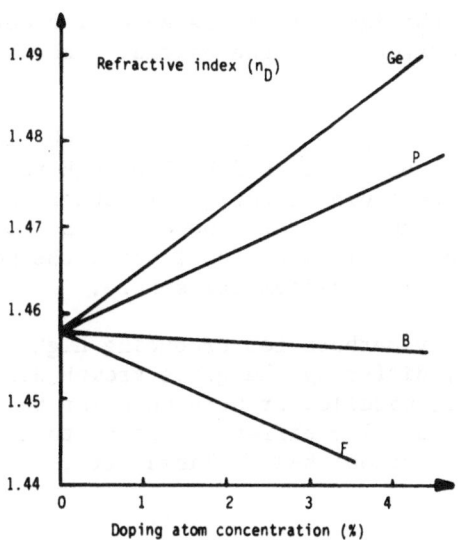

Fig. 1. Variation of the Refractive Index of Doped Silica as a
Function of the Doping Atom Concentration

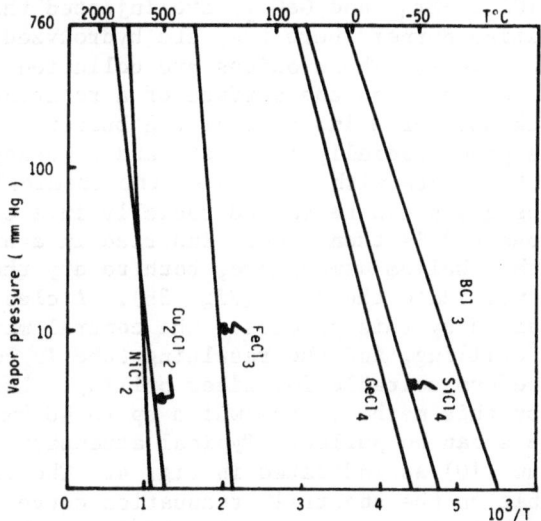

Fig. 2. Partial Pressure of Various Halides as a Function of
Temperature

The raw materials ($SiCl_4$, BBr_3, ...) are chosen to be liquids
at room temperature. Carrier gases such as N_2, O_2, or He are
bubbled through them and entrain vapors to the reaction zone. This
mechanism purifies the raw materials by distillation in a simple
manner which is very effective since, as can be seen in Fig. 2, the

partial pressures of the impurities are several orders of magnitude smaller at room temperature than the partial pressures of the raw materials.

In some cases, especially when Al or F is the dopant, the substrate temperature can be maintained high enough for a vitreous bubble-free glass to be grown directly. In other cases, to avoid the volatilization of the dopants, a porous glass must be grown in a first step and then consolidated (collapsing the pores) in a second step by heating in a helium atmosphere.

There are now many methods for producing high silica graded index preforms. They differ by the glass growth direction, either axial or radial, i.e., parallel or perpendicular to the preform axis respectively. They also differ according to the mechanism which activates the reaction: heat, plasma, etc.

2.2. Flame Hydrolysis

Flame hydrolysis (also called Corning Soot Process or external CVD) was historically the first method to yield graded-index low-loss high-silica fibers. In this process, the vaporized reaction products (e.g. $SiCl_4$, BBr_3, and $GeCl_4$) are injected through the flame of a combustion burner where they are hydrolyzed to produce the corresponding oxides. These oxides are collected as fine particulates, called soot, on the surface of a rotating mandrel which travels back and forth in front of the burner (Fig. 3a). A porous glass is grown radially in layers and by changing the respective amounts of dopants with each pass, the chemical composition of the porous glass can be varied radially in a controlled manner. The porous rod is then slowly inserted in a high temperature furnace with a helium atmosphere, both to dry the porous preform and to consolidate the soot (Fig. 3b). A clear, bubble-free glass can be obtained by this process. The central mandrel is eliminated, e.g. by drilling, and the resulting tube is collapsed to become a solid preform suitable for fiber pulling. Large preforms can be obtained by this method, from which up to 40 km of low loss graded index fibers can be pulled. Typical attenuation values are 4 dB/km at 0.85 μm (10) as indicated in Fig. 4. The three attenuation peaks visible on the spectral attenuation curve come from absorption by OH radicals not completely eliminated by the consolidation step.

Several modifications of the Corning soot process have also been developed (11,12). One is the axial soot process, or Verneuil method, developed by NTT in Japan. In this process, the soot is collected on a rotating silica plate which moves upwards from the combustion burner (Fig. 5) To obtain the desired variation in chemical composition, two burners are used: a central one through which $SiCl_4$ and $GeCl_4$ are sent and a lateral one with $SiCl_4$ and BCl_3. Radial variation of chemical composition is obtained by

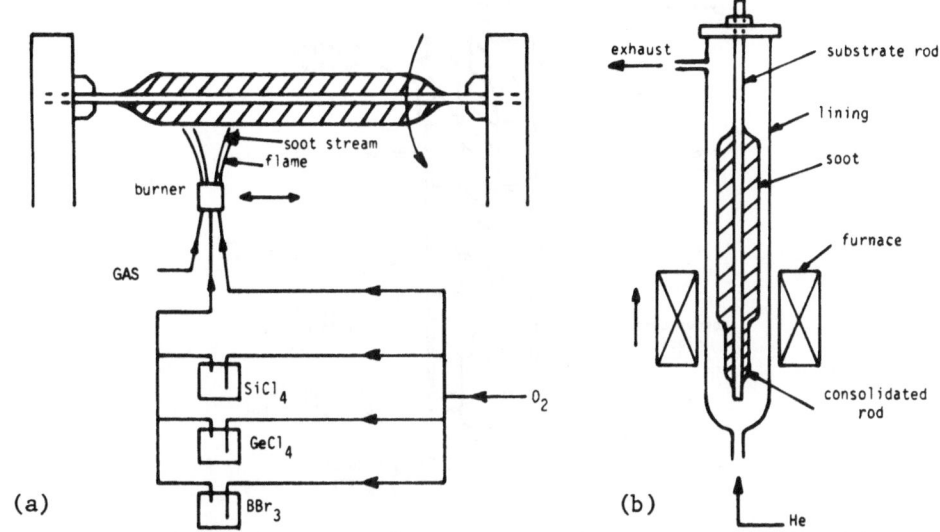

Fig. 3. Schematics of the Corning Soot Process. (a) Deposit;
         (b) Consolidation.

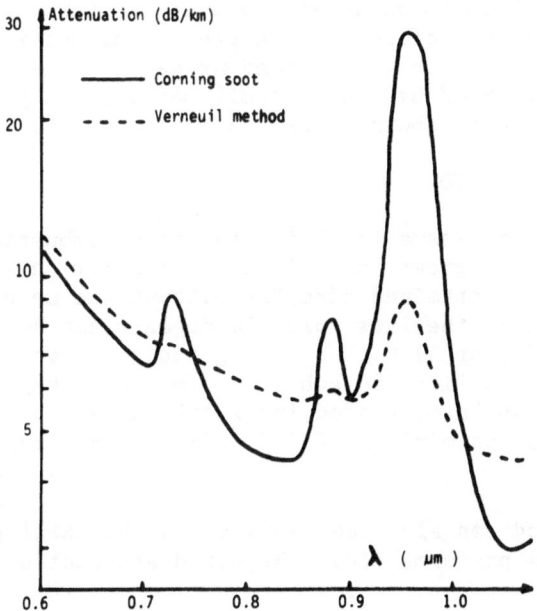

Fig. 4. Spectral Attenuation of Optical Fibers Made by Flame
         Hydrolysis.

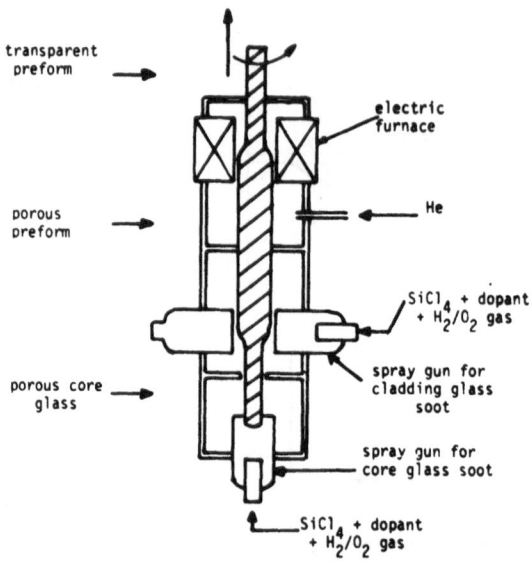

Fig. 5.   Schematic of the Axial Flame Hydrolysis Process

diffusion in the flame. The preform is also consolidated under
helium but the growth and consolidation steps can be conducted
continuously, which is an advantage over the Corning soot process.
However, the index profile and the preform diameter may be more
difficult to control in this second process. Fiber lengths up to
40 km can be obtained by the Verneuil method. A typical spectral
attenuation curve is shown in Fig. 4.

2.3. Plasma Torch CVD

When using a plasma torch instead of a combustion burner, water
free oxides can be grown and collected on a hot substrate so that
a glassy layer is obtained directly without the need for a consolidation step. Only the less volatile dopants can be used in this
method, e.g. SiF_4 or Al_2O_3. By using radial growth or F-doped
silica on a pure silica rod substrate (which is kept in the final
product) up to 40 km of graded index fibers have been obtained (13)
with an attenuation minimum of 2.5 dB/km and an impulse broadening
of 2 nsec/km.

This method has also been applied to the axial growth of F- or
Al-doped silica preforms (14). Reported attenuation values reach
5 dB/km.

2.4. Thermal CVD

In another set of techniques called thermal CVD or internal
CVD, the reaction zone is confined to the inside of a silica tube.

The vaporized raw materials (typically halides of Si, B, Ge, and P) together with oxygen and eventually another carrier gas such as He are sent through a rotating silica tube which is heated locally by a burner (or furnace) moving back and forth along the tube (Fig. 6). In the moving hot zone, at a temperature around 1300°C, the reactants are oxidized either homogeneously in the gas phase or heterogeneously on the inside wall of the tube and are deposited downstream as a soot which is immediately vitrified. The composition of the gas stream is varied at each travel to create the desired variation of refractive index. After the deposit steps the tube is collapsed to a solid preform.

It is by this method that optical fibers with the lowest reported attenuation have been made, reaching 0.5 dB/km at 1.2 μm (1). Typical spectral attenuation curves are shown in Fig. 7. In this method, many combinations of dopants can be used (15-19). It is also possible to produce low loss single mode fibers (20).

There are, however, two main limitations to this method:

1) the glass growth rate is small, on the order of 0.2 g/min;
2) the production of large preforms is difficult without a deterioration of the fiber band pass or a loss of circularity. Typical fiber lengths are presently around 2 km per preform.

2.5. Plasma Activated CVD

Two main methods using a plasma to activate the reaction have been reported.

In one method, a microwave cavity maintains a low pressure (20 torrs) non-isothermal plasma inside a silica tube which is

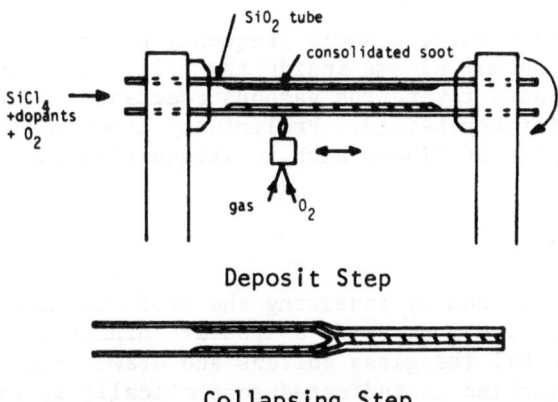

Deposit Step

Collapsing Step

Fig. 6.   Schematic of the Thermal CVD Process

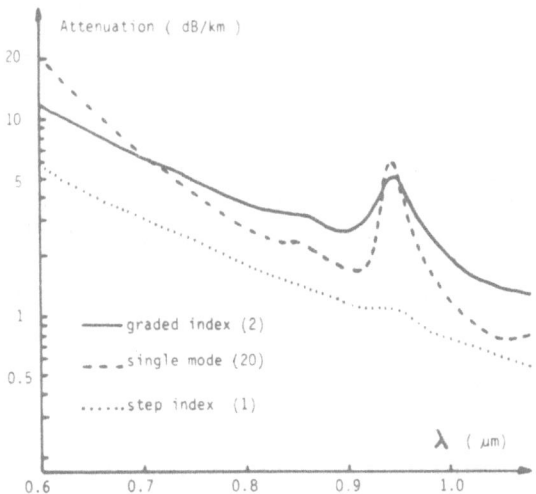

Fig. 7. Spectral Attenuation of Optical Fibers Made by the Thermal
        (or internal) CVD Process

heated at a temperature around 1100°C. The plasma enhances the
reaction such that 100% reaction efficiency can be obtained, which
eases the control of the refractive index profile since the incoming
gas streams can be adjusted precisely. The plasma acts as a local
reaction zone, about 1 cm long, and is moved back and forth at high
speed such that many layers typically 1 μm thick are deposited.
This method has been used with B and Ge as dopants and has yielded
low loss fibers with an attenuation minimum of 1.4 dB/km at 1.05 μ
(21). More recently, this method has been used with F as a dopant
(22) with an attenuation minimum of 5 dB/km. In both cases, very
low impulse broadening, 0.8 and 2 nsec/km, respectively, have been
obtained, which shows the efficiency of this method for controlling
the refractive index profile.

    In the second method, a radio frequency generator (3.4 MHz)
generates an isothermal plasma inside the silica tube such that
heat is provided directly to the gaseous reactants. This increased
the reaction efficiency 5-fold. Preliminary experiments (23) have
yielded long lengths of fibers with an attenuation minimum of 6.2
dB/km at 1.05 μm.

2.6. Fiber Pulling

    Fibers are produced by inserting the preforms made by any of
the methods reported above inside a tubular furnace which is heated
to about 2000°C until the glass softens and draws into a fiber. A
typical pulling machine is indicated schematically in Fig. 8. The
preform is lowered at a small speed v while the fiber is pulled at
a much larger speed V either with a capstan or with a take up drum.

After a few minutes, an equilibrium is reached in temperature and position and the fiber diameter becomes stable, its value being given by

$$d = D(v/V)^{1/2} \tag{4}$$

where D is the preform diameter. The fiber is a faithful reproduction in miniature of the preform. In particular the ratio ϕ_{core}/ϕ_{ext} and the refractive index profile are maintained.

The heat source is usually a graphite furnace (either resistance or induction furnace). A neutral atmosphere such as argon is maintained inside the furance both to protect the heating element and to avoid any contamination of the fiber surface. Alternatively a CO_2 laser can be used with adequate optics to heat the preform (24).

A fiber diameter monitor is often used just below the furnace with a feedback loop which acts on the capstan speed and helps to keep the fiber diameter constant.

The fiber is coated on line with a plastic resin to keep its high pristine strength (ca. 1.4×10^4 N/mm^2), which would otherwise decrease with time and handling mostly due to surface reaction between moisture and silica. The plastic resin is either a thick thermally- or UV-cured resin coating, about 40 μm thick, or a thin coating in epoxy or kynar about 3 μm thick. With these coatings the typical strength of a fiber 150 μm in diameter is on the order of 2.5 kg.

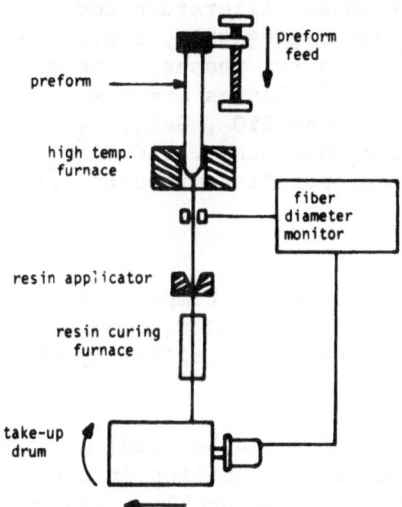

Fig. 8. Schematic of an Optical Fiber Pulling Machine

3. MULTICOMPONENT GLASS FIBERS

3.1. Glass Systems. Raw Material Purification

Conventional glass making techniques starting with powdered raw materials are also used to produce optical fibers. In these cases, core and cladding glasses of different chemical compositions and refractive indices are melted separately and combined at the end of the process in a specially designed double crucible which permits pulling either step index or graded index fibers. For low loss fibers to be obtained, ultrapure raw materials (especially purified for that purpose) have to be used and sophisticated melting methods have to be developed to maintain this high purity during the glass elaboration.

The desirable properties which govern the choice of a pair of glasses (core and cladding) are: 1) similar coefficients of thermal expansion and pulling temperatures; 2) good chemical and thermal stabilities. These requisites are not very stringent so that many matchable glasses can be found, even with very different refractive indices, which will yield high numerical aperture fibers. Many glass systems have been investigated, the main ones being sodium borosilicates (25,26), soda-lime silicates (27,28) and alkali germanosilicates (29).

In the production of multicomponent glass fibers the first problem to be solved is the preparation of ultrapure raw materials with impurity concentrations well below 0.1 ppm (see 2.1). The techniques used to purify the raw materials are numerous and their description would be beyond the scope of this course. The reader is referred to the relevant literature for a detailed description (30). Most methods use segregation, i.e., the unequal distribution of the impurities between two phases. The different compounds necessary to make typical glasses are now commercially available; they include oxides such as SiO_2, GeO_2, B_2O_3, and As_2O_3, carbonates and nitrates such as K_2CO_3, $NaNO_3$, and $TlNO_3$. Typical umpurity concentrations of these purified products are indicated in Table II.

3.2. Melting

Most glasses are melted at temperatures between 1200 and 1450°C. At these high temperatures, special care must be taken to avoid contamination of the glass either by the crucible itself or the refractories.

Two types of crucibles can be used: platinum crucibles (with a pure Pt grade containing impurities in the ppm range) and silica crucibles. With platinum, two problems are encountered: the corrosion of platinum by the glass, which may cause unacceptably high loss increase (32) and the leaching of impurities from the crucible.

TABLE II. Transition Metal Concentrations (ppb) in Purified Raw
          Materials (from ref. 31).

MATERIAL	Cr	Fe	Co	Ni	Cu
Acid washed milled quartz	0.9	34	0.06	<8	<100
Synthetic silica	<0.5	4.6	0.04	<30	6
B_2O_3 (M)	<50	<50	<10	<10	<10
Na_2Co_3 (B)	4	20	<5	10	10

(M) E. Merck Co., FO optipur grade
(B) J.T. Baker Chemical Co., Ultrex low attenuation grade

Even though these problems can be overcome, silica crucibles have
been preferred in most of the recent works on low-loss glass fabrication. Partial dissolution of silica also occurs; it does not
introduce many impurities, but rather silica-rich striae in the
glass. These striae can be maintained at an acceptable level by
choosing glasses with low melting temperatures.

An alternative method which completely prevents contamination
consists of heating the glass directly by R.F. induction while at
the same time cooling the crucible (33) with an air or water stream.
For sufficient R.F. power to be coupled in the glass, a graphite
susceptor is first introduced in the field to preheat the glass at
a temperature of 1000°C.

To prevent other sources of contamination, special furnaces are
built with only very pure oxides such as silica and recrystallized
alumina (34,35) surrounding the glass and protecting it from the
other refractories and the ambient atmosphere.

We saw above in Section 1.1. that the absorption losses of a
glass are determined by the impurity concentrations. It is also
determined by a second set of parameters, the valency states of
those impurities such as Fe and Cu which are multivalent. For instance, Fe^{3+} and Cu^+ have absorption bands outside the spectral domain concerned with optical fibers while Fe^{2+} and Cu^{2+} have absorption
bands centered around 1.1 and 0.8 μ, respectively. By changing the
redox conditions of the furance atmosphere and adding a redox buf-
fering agent such as As_2O_3 in the glass, it is possible to modify
the distribution of the impurities according to their different
valency states and in so doing to lower their corresponding absorp-

tions. This possibility has been studied in detail for the sodium
borosilicate glass system (26,34).

A careful control of the furnace atmosphere during melting also
permits the lowering of the hydroxyl content in the glass. It was
found that OH content is proportional to the square root of the
water partial pressure in the furnace (6). Dry gases are thus sent
through the furnace; they are even often bubbled through the glass
to speed the drying and homogenize the glass. By this method, the
absorption of the OH radicals at the 0.95 μ peak has been reduced
to 15 dB/km in borosilicates (26) and 6.5 dB/km in a soda-lime
silicate glass (36) which was prepared by wet mixing.

Several methods can be used to remove the glass from the crucible (30). The most common one is to draw a rod from the glass surface. The rods must be stored in a clean environment until used for
fiber pulling to avoid surface contamination.

3.3. Fiber Pulling

To convert the core and cladding glasses to step index or
graded index optical fibers, a specially designed double crucible
is used (Fig. 9a). This crucible can be platinum since fiber
pulling temperatures are lower than melting temperatures and the
aforementioned platinum corrosion and leaching of impurities are
not encountered at these lower temperatures (800-1100°C). The glass
rods are fed into the core and cladding crucibles (the center and

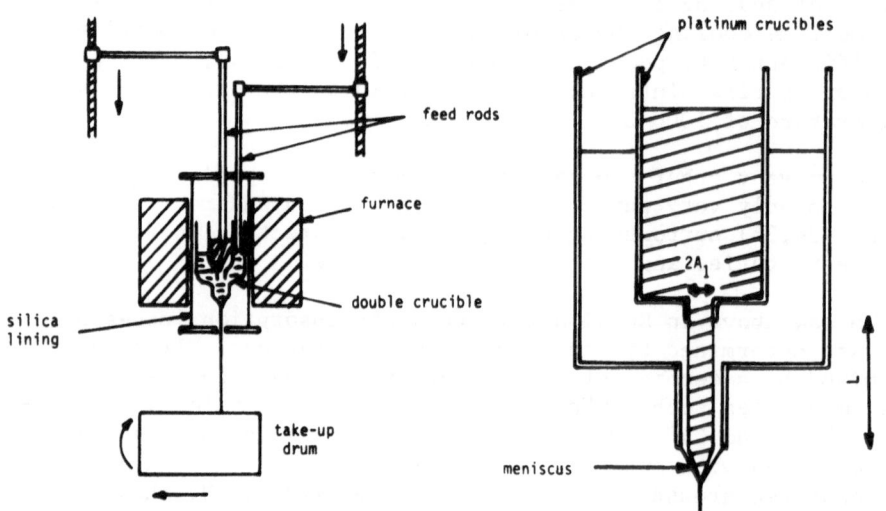

Fig. 9a. Schematic of Multicom-   Fig. 9b.   Detail of the Double
         ponent Glass Fiber   Crucible
         Pulling

outside crucibles respectively) and a fiber is pulled through the holes at the bottom. To obtain graded index fibers, interdiffusion between the core and cladding glasses is allowed by increasing the time t the two glasses are in contact at high temperatures (25). The best fit to the desired parabolic index profile is obtained when the parameter $\emptyset = Dt/A_1^2$ is equal to 0.081 (37) where D is the diffusion coefficient and A_1 is the core nozzle radius. The parameter $\emptyset$ can also be related to the crucible dimensions, the fiber core radius a, and the pulling speed V:

$$\emptyset = \frac{DL}{Va^2}$$ (5)

Typical attenuation results for multicomponent graded index glass fibers are indicated in Fig. 10. The attenuation minimum around 0.85 µ lies in the range 5-7 dB/km (38-40). An even lower attenuation of 4.23 dB/km has been obtained for a soda-lime silica step index fiber (36). The techniques of glass melting and fiber pulling have been improved to a point where the impurity level in the fiber is practically the same as that of the starting materials.

The pulse dispersion results in multicomponent glass fibers have also reached very low values, on the order of 1 nsec/km (38, 40) which shows that a good control of the refractive index profile has been reached.

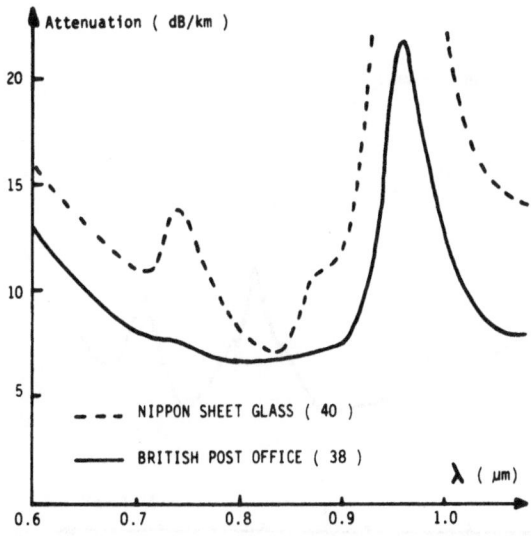

Fig. 10. Spectral Attenuation of Graded Index Multicomponent Glass Fibers

4. PLASTIC CLAD FIBERS

Very simple step index fibers can be obtained by pulling pure silica rods into fibers and coating them on line with an optical resin whose refractive index is smaller than that of silica. The silica starting rod is usually made by a plasma torch technique.(14) as explained in 2.3, which permits us to produce ultrapure silica with an impurity content typically on the order of ppb.

The resin is a silicone resin which has a fairly large attenu- ation with a minimum of 900 dB/km at 0.77 μ and a maximum of 70,000 dB/km at 0.91 μ. Its refractive index is close to 1.4 (41). Des- pite the strong attenuation of the resin, low loss optical fibers can be obtained by this method since only a small part of the light energy propagates in the cladding. A typical spectral attenuation curve is shown in Fig. 11.

The higher-order guided modes in the fiber are more strongly attenuated than the lower-order modes because of the lossy cladding. As a result:

a) the steady-state numerical aperture of the fiber (after 1 km) is
   about 0.2, smaller than the calculated value of 0.37;
b) the attenuation increases when the fiber diameter is lowered;
c) the band-pass of the optical fiber is typically 36 MHz for 1 km
   of 200 μm core fibers, larger than the calculated value of 8 MHz.

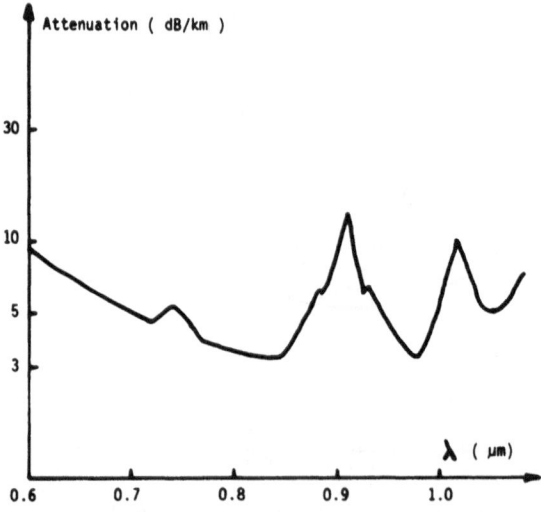

Fig. 11. Spectral Attenuation of Plastic-Clad Fibers

5. FIBERS MADE BY PHASE SEPARATION

In these methods, a base glass is made by standard techniques using cheap raw materials which may contain as much as 10 ppm impurities, the glass being purified afterwards by chemical processes.

The base glass, a sodium borosilicate, is chosen for its ability to separate during thermal annealing into two interconnected phases: one rich in silica and one containing the impurities. The phase-separated glass is then treated by an acid solution which extracts the phase containing the impurities.

It is thus possible to use this method to obtain porous purified rods which may be consolidated to produce low loss preforms. A dopant can be introduced in the process, either in the base glass (42) or in the porous rods (43), to create the desired variation of refractive index. Step index or graded index optical fibers have been made by this process with a typical attenuation minimum of 15 dB/km (43).

ACKNOWLEDGEMENT

The author wishes to thank Dr. M. Faure for valuable discussions during the preparation of this review.

REFERENCES

1. Horiguchi, M., Electr. Lett. 12, 310-312 (1976).
2. O'Connor, P.B., et al., 2nd Europ. Conf. on Opt. Fiber Comm., Paris, 1976, pp. 55-58.
3. Newns, G.R., et al., Opto Electr. 5, 289-296 (1973).
4. Ikeda, Y., et al., Proc. Xth Int. Congress on Glass, Kyoto, 1974, 6, pp. 82-89.
5. Schultz, P.C., J. Amer. Ceram Soc. 57, 309-313 (1974).
6. Beales, K.J., 1st European Conf. on Opti. Fiber Comm., London, 1975, pp. 30-32.
7. Presby, H.M., and Kaminow, I.P., Appl. Optics 15, 3029 (1976).
8. Payne, D.N., and Gambling, W.A., Electronics Letters, 11, 176-178 (1975).
9. Kobayashi, S., et al., Int. Conf. on Integrated Optics and Optical Fiber Comm., Tokyo, 1977, pp. 309-312.
10. Montierth, M.R., J. Electr. Materials 6, 349-372 (1977).
11. Dabby, F.W., et al., Appl. Phys. Letters 25, 714-715, (1974).
12. Izawa, T., et al., Int. Conf. on Integrated Optics and Optical Fiber Comm., Tokyo, 1977, pp. 375-378.
13. Muhlich, A., et al., 3rd Europ. Conf. on Optical Communication, Munich, 1977, pp. 10-11.

14. Achener, M., and Habert, M., L'Onde Elect. 56, 603-605 (1976).
15. Payne, D.N., and Gambling, W.A., Electronics Letters, 10, 289-290 (1974).
16. MacChesney, J.B., et al., Proc. I.E.E.E. 62, 1280-1281 (1974)
17. French, W.G., et al., Appl. Optics 15, 1803-1807 (1976).
18. Sommer, R.G., et al., Electronics Letters 12, 408-409 (1976).
19. Abe, K., 2nd European Conf. on Opt. Fiber Comm., Paris, 1976, pp. 59-61.
20. Kawana, A., et al., Elect. Letters 13, 188-189 (1977) and 13, 442-443 (1977).
21. Geittner, P., et al., Appl. Phys. Letters 28, 645-646 (1976).
22. Kuppers, D., et al., 3rd European Conf. on Opt. Comm. Munich, 1977, pp. 12-14.
23. Fujiwara, K., et al., ibid, pp. 15-17.
24. Jaeger, R.E., Ceram. Bull. 55(3) (1976).
25. Koizumi, K., et al., Appl. Opt. 13, 255 (1974).
26. Newns, G.R., 2nd European Conference on Opt. Fiber Comm., Paris, 1976, pp. 21-26.
27. Shibata, S., and Takahashi, S., J. Noncryst. Solids 23, 111 (1977).
28. Pinnow, D.A., et al., Appl. Phys. Letters 22, 527 (1973).
29. Van Ass, H.M., et al., Electron. Letters 12, 369 (1976).
30. Gossink, R.G., J. Noncryst. Solids 26, 114-157 (1977).
31. Gossink, R.G., et al., Mat. Res. Bull. 10, 35-40 (1975).
32. Shibata, S., and Takahashi, S., J. Noncryst. Solids 23, 111-122 (1977).
33. Scott, B., and Rawson, H., Opto. Electron 5, 285-288 (1973).
34. Beales, K.J., et al., Proc. I.E.E. 123, 591-596 (1976).
35. Scott, B., and Rawson, H., Glass Tech. 14, 115-124 (1973).
36. Takahashi, S., et al., Electron. Letters 14, 151-152 (1978).
37. Dyott, R.B., and Brain, M.C., Electron. Letters 10, 8 (1974).
38. Newns, G.R., et al., International Conf. on Integrated Optics and Optical Fiber Communication, Tokyo Osaka, 1977, pp. 609-612.
39. Imagawa, H., and Ogino, N., ibid, pp. 613-615.
40. Yamazaki, T., and Yoshiyagawa, M., ibid, pp. 617-620.
41. Inada, K., et al., 1st European Conf. on Optical Communication, London, 1975, pp. 57-59.
42. de Panafieu, A., and Faure, M., 3rd European Conference on Optical Communication, Munich, 1977, pp. 21-23.
43. Macedo, P.B., et al., 2nd European Conference on Optical Fiber Communication, Paris, 1976, pp. 37-39.

EXPERIMENTAL EVALUATION OF OPTICAL FIBERS: A REVIEW

Carlo G. Someda

Istituto di Elettronica

Università di Bologna, Italy

1. INTRODUCTION

A survey paper on optical fiber evaluation must contain at least three elements:

a) indications of fiber parameters whose measurements are of practical interest, and why they are important;

b) a description of suitable techniques for these measurements with specific emphasis on those which are peculiar to this technology;

c) state-of-the-art results.

The parameters of practical relevance may be listed in various ways, according to different criteria. Rather arbitrarily, to follow the same order as the lecture sequence in this A.S.I., we will deal first with those measurements whose results are usually fed back to the technology stage. Then we will move on to those measurements whose results are fed forward to the system design. Of course, there is no sharp boundary between these two categories, but this should not cause us too much trouble. Description of techniques and set-ups will be restricted to the most innovative ones. This will result in a considerable inhomogeneity, in length and importance, of the paper sections. But this too seems to be a reasonable trouble, in order to cover as much up-to-date material as possible within a given space.

2. GEOMETRICAL MEASUREMENTS

As known [1], there is a fundamental dimensionless quantity in the theory of ideal optical fibers, defined as

$$V = (2\pi a/\lambda)\surd(n_o^2 - n_2^2) \simeq (2\pi a n_o/\lambda)\surd(2\Delta) \qquad (2.1)$$

where a = fiber core radius, λ = light wavelength, n_o = refractive index on the fiber axis, n_2 = refractive index in the cladding, Δ = $(n_o - n_2)/n_o$. The core radius a is the only geometrical parameter in this formula. But other geometrical quantities matter in practical fibers. The cladding radius, b, affects mechanical properties and microbending loss. In order to estimate fiber splicing characteristics, core-cladding concentricity, ellipticity, and fluctuations of these parameters along the fiber length are important. As to the so-called single-mode fibers, they are not really single-mode if their cores become elliptical. Many other examples could be provided.

For multimode step-index fibers, geometrical parameters can be measured on discrete cross sections by simple techniques, typically with a microscope. Complications can arise with graded-core fibers and with single-mode fibers. For example, a sharply defined core radius is hard to measure in a smoothly graded profile. On the other hand, it could be of limited interest, compared with an "effective" radius, defined by its relationship to the optical power distribution in the fiber cross section. The question then becomes closely related to that of measuring the index profile, a subject to be dealt with in the next two sections. For single-mode fibers, again, the use of microscopy can be unsatisfactory because of the very small geometrical dimensions (core radii of a few μm) and because of the high accuracy that some measurements require. For instance, core ellipticities a_{min}/a_{max} = $1-\delta$, with δ of only a few percent, can be important. There is a suitable technique for such a measurement [2], and it shows how a test can be conceived on the grounds of propagation theory. The starting point is the fact that a single-mode fiber behaves as a birefringent medium. If circularly polarized light is sent into it, then the polarization along the fiber changes, gradually and periodically, to linear, and then back to circular. The azimuthal pattern of the scattered radiation which leaves the fiber through its wall depends on polarization of the guided light. A detector, moved around the fiber in the azimuthal direction and also, but much more slowly, along the fiber axis, gives an amplitude-modulated signal, from whose period one can deduce the fiber ellipticity. Fig. 2.1 shows a schematic diagram of the set-up. It has been stated [2] that this technique (with its additional advantage of being nondestructive) can detect diameter differences as small as a few percent, which would cause pulse spreadings in the order of tens of picoseconds per kilometer.

3. MEASUREMENT OF REFRACTIVE INDEX PROFILE

The lectures on theory of propagation in optical fiber [1] have

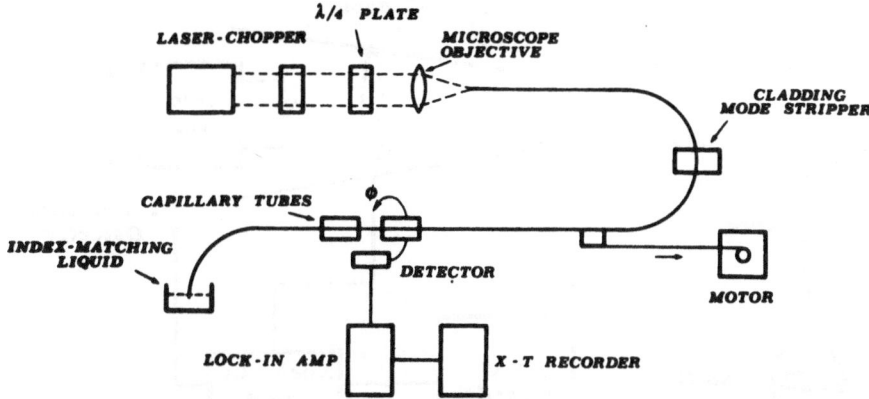

Fig. 2.1. Schematic diagram of single-mode fiber ellipticity measurement (from ref. [2]).

stressed how important and critical the refractive index profile is and, to a large extent, have covered the problem of its measurement. Despite the risk of overlapping with ref. [1], it is important to review at least the most important techniques for performing these measurements.

3.1. The Near-Field Scanning Technique

Let us start with a technique which is very popular, because of its simplicity both in principle and instrumentation, in spite of some uncertainties and limitations that will be pointed out later. It is based on the following theoretical statement. If all the guided modes of a large-V-value fiber are equally excited by a narrow-band, temporally incoherent source, then the optical power density in the fiber cross section is one-to-one replica of the index profile. This result was first proved [3] with a WKB approach, for the so-called α-type fibers, i.e. for index profiles of the type

$$n^2(r) = n_o^2\{1-2\Delta(r/a)^\alpha\} \qquad r < a$$
$$n^2(r) = n_o^2\{1-2\Delta\} \qquad\qquad r > a \tag{3.1}$$

Ray-optics arguments generalize it to all "weakly-guiding" (Δ<<1) profiles. It is virtually impossible to list all the published implementations of this technique. Our references [4-8] are just some of those which deserve specific credit.

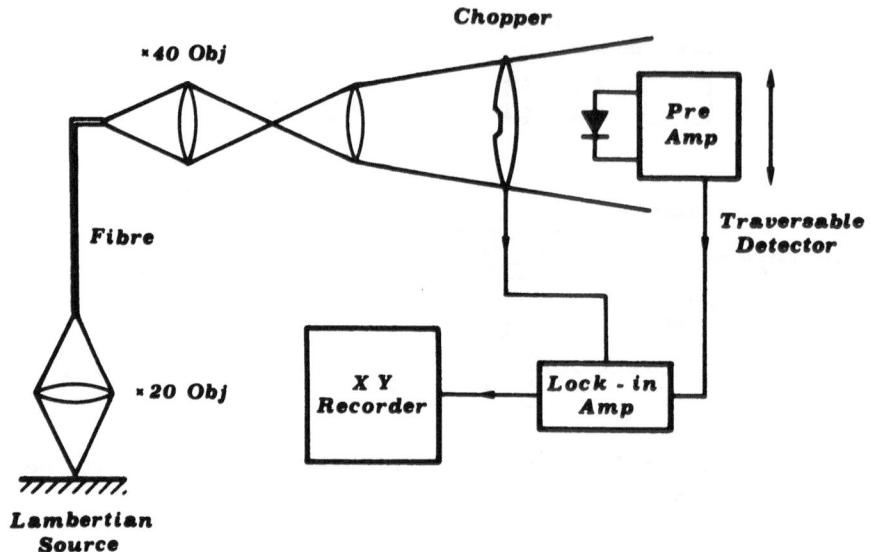

Fig. 3.1. Set-up for the near-field-scanning (n.f.s.) measurement of index profiles.

In principle, the set-up (see Fig. 3.1) consists just of: a source, incoherent both in space and time, but filtered to a narrow spectral width (for reasons to be discussed below); a short fiber sample; a system which scans the near-field intensity distribution at the fiber output, and is comprised of a magnifying lens and a small-area traversable photodetector (e.g. on a step-motor driven positioner).

The sample length, L, is the result of a trade-off between contradictory requirements. It should be short enough to prevent unequal modal powers as an effect of differential attenuation and/ or mode coupling. It should be long enough to loose the power fraction that the source launches onto the fiber radiation field, which modifies the intensity distribution inside the fiber. The influence of this radiation field (which, as known, can be described very well by means of the so-called leaky modes or leaky rays) is still, to some extent, a controversial matter. Starting from the viewpoint that, for all practical values of L, a finite radiation field would be inevitable, theoretical length-dependent correction factors, to be applied to the n.f.s. measurements, have been cal-culated [9,5]. Now, while there is a general consensus on the use of quasi-step-index fibers (see Fig. 3.2), for near-parabolic fibers ($\alpha \simeq 2$ in eq. (3.1)), some measurements [6] agree with the results of other techniques when the correction factors are used, but other results [5,8] behave in the opposite way, as shown in Fig. 3.3.

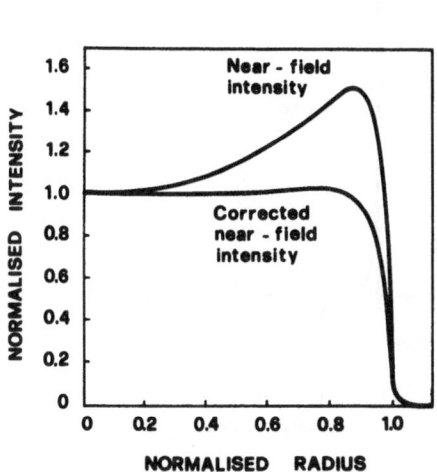
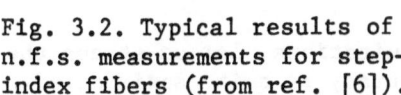
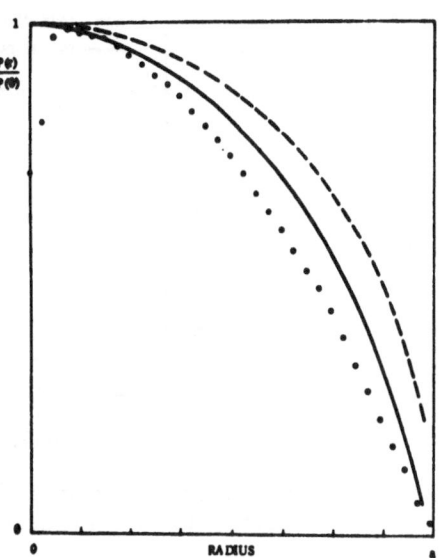

Fig. 3.2. Typical results of
n.f.s. measurements for step-
index fibers (from ref. [6]).

Fig. 3.3. n.f.s. results in a
parabolic profile: comparison of
measured values (dots) and theoretical predictions including
correction factors (from ref.
[5]).

Possible explanations of this discrepancy take into account actual
fiber defects: bends [8] or core ellipticity [10] cause much higher
attenuation rates for leaky rays than in perfect fibers. However,
some uncertainty remains about correction factors and their use.

    Another limit to the n.f.s. technique results from the fact
that the guided modes form a discrete set. The power density equals
the index profile under the approximation of a continuum of guided
modes. The summation of power densities over the actual discrete
set, at a given wavelength λ [11], yields patterns that deviate
from the index profiles because of "ripples," whose spatial frequency has an upper limit given by

$$K = (2\pi n_o/\lambda)\{\Delta\alpha/(\alpha+2)\}^{1/2} \qquad\qquad (3.2)$$

An example is shown in Fig. 3.4. In practice, this effect is observed [7] with a narrow-band light source, but is blurred when the
source spectral width increases [12]. It sets the resolution limit
of the n.f.s. technique: in fibers produced with CVD technology
[13], when the thickness of each layer becomes smaller than the
inverse of (3.2), fine details of the profile are ambiguous if

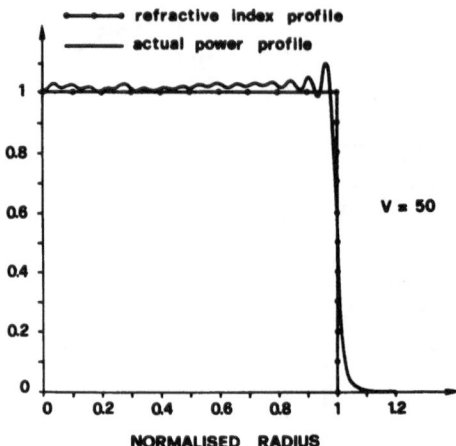

Fig. 3.4. Comparison between index profile and theoretical power
          distribution over guided modes in step-index fibers.

measured in this way. Against these limitations, let us stress the
simplicity of this technique, whose accuracy is quite satisfactory
for many purposes.

3.3. Reflection and Transmission Methods

     Another optical method for measuring the index profile in a
fiber, suggested and implemented by several authors [14,15,5], is
shown in Fig. 3.5. A laser beam is focused to a small spot (diameter on the order of 1 μm) onto the fiber input face. The reflected light intensity, R, is measured as a function of the radial
coordinate of the spot position. In this way n(r) is determined
by means of Fresnel formula for normal incidence:

$$ R = \left[\frac{n(r) - n_m}{n(r) + n_m} \right]^2 \qquad (3.3) $$

where n_m is the index of the surrounding medium. An important indi-
cation of how to implement this technique is given by the logarithmic differential of (3.3)

$$ \frac{\Delta R}{R} = \frac{4n_m}{n^2 - n_m^2} \, \Delta n \qquad (3.4) $$

The uncertainty $\Delta R/R$ is instrumental (electrical noise, surface
imperfections, defocusing, etc.) and independent of n_m; typically,
it is on the order of a few percent. Hence, if we wish to measure

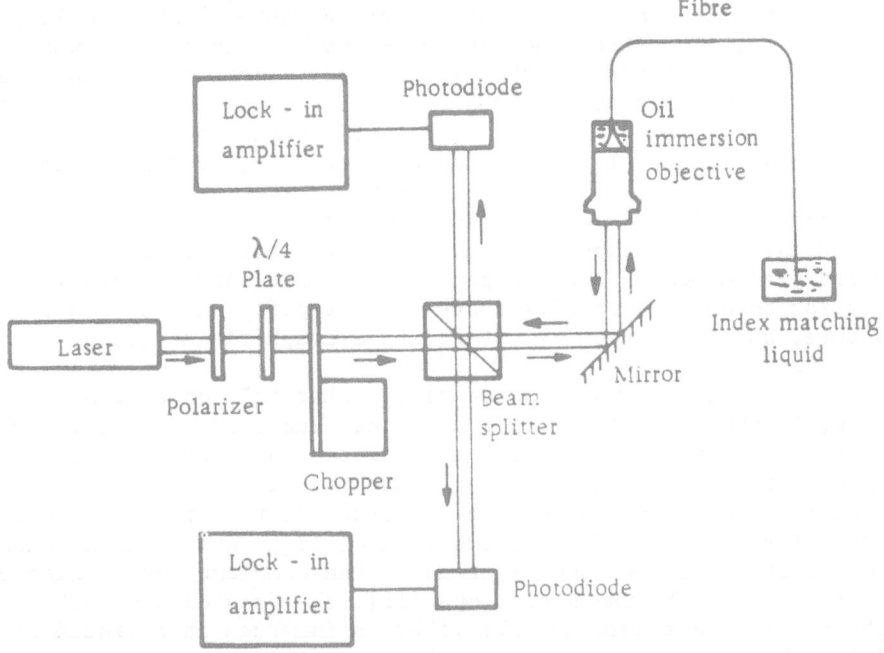

Fig. 3.5. Schematic diagram of a reflection set-up for index profile
measurements (from [5]).

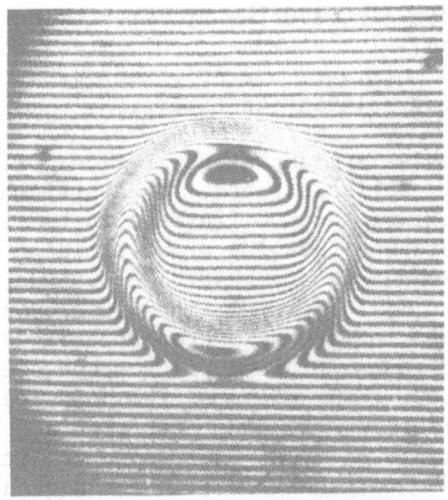

Fig. 3.6. An example of optical fiber microinterferogram (from [20]).

Δn with sufficient accuracy, the factor $4n_m/(n^2-n_m^2)$ must be large, i.e., $n_m \simeq n$. The fiber end has to be immersed in a fluid whose index matches well that of the fiber itself. This is the purpose of the oil-immersion objective shown in Fig. 3.5. In addition, it has been pointed out [16] that inadequate preparation of the fiber input surface can yield unacceptable inaccuracy. Polished samples give worthless results because polishing creates a thin surface layer sith slightly altered dopant concentration and this, in turn, causes substantial perturbation in the reflection coefficient. Fractured samples behave much better in general, but for some dopants (typically for boron) the results change very rapidly with time after fracture, as the surface is altered by atmospheric moisture.

Let us mention two more techniques that rely on similar principles, in the sense that fiber coupling, under small-spot launching conditions, is tested while the spot scans the fiber end. Unfortunately, there will not be enough space to discuss these techniques in detail, and the reader will be referred to the original papers. In the first method [17], the fraction of light which is launched into the fiber is measured, instead of the complementary reflected fraction (3.3). In the second one [18], the focused beam has a high numerical aperture and the fiber is immersed in a liquid of slightly higher refractive index than the cladding. A hollow cone of light emerges from the fiber cylindrical wall, and a disc shields the detector from its innermost part, consisting of tunnelling leaky rays. The remaining part (refracting rays) can be measured as a function of spot position; from this, one gets the index profile.

3.4. The Interference Microscopy Method

Viewing an index distribution by interference microscopy is classical in optics [19]. However, application to fibers gave rise to new problems and motivated new solutions [20]. We will concentrate on what is most relevant to fibers.

A delicate problem is sample preparation. Fiber slices must have flat, parallel, polished faces. Their thickness, t, depends primarily on the maximum index difference in the fiber, Δ; it must be thin enough to prevent focusing or waveguiding effects. As a result, typical thicknesses are less than 0.05 mm, and sometimes as small as 0.01 mm [21]. However, it should be remembered that the final uncertainty is inversely proportional to t. When a good sample is placed in an interference microscope, results look typically as in Fig. 3.6. Index changes from point to point are related to fringe displacements in the interferogram. This presentation gives an immediate way to understand qualitatively what the profile is. In order to make it useful for a quantitative evaluation, data processing is needed. There is a fairly simple approach to this, with the following steps. Circles are drawn on micro-

interferograms (see Fig. 3.6) with their center on the core axis.
Next, a straight reference line is drawn through the midpoints in
the cladding of that fringe which passes through the core center.
Then the displacement of the same fringe from the reference line
at its intersection with each circle is measured under a low-power
microscope. The ratio between this distance and the distance of
two uniformly spaced fringes in the cladding, q, gives the index
change with respect to the cladding,

$$\Delta n = q\lambda/t \qquad\qquad\qquad\qquad (3.5)$$

The measurement accuracy has been reported to be about 2 parts
in 10^4. Small improvements are possible with rather classical
optical techniques, but they are not satisfactory for all purposes.
Indeed, we know from theory [1] how critical the profile is, and
furthermore [22] how critically the "optimum profile" depends on
the profile dispersion, i.e., on the wavelength-dependence of the
profile dispersion, the accuracy of index measurements at each
individual wavelength has to be better than about 5 parts in 10^5.
Presby and Kaminow [23] have developed a very sophisticated set-up,
shown in Fig. 3.7, whose performance satisfies this requirement.
The eyepiece of the interference microscope is replaced by a silicon
camera tube so that an electrical signal is extracted from the magni-
fied interference pattern. Individual scan lines can be displayed
(for example, on an oscilloscope) and processed by suitable electronics. The increased resolution in reading the half-tones yields the
required improvement in accuracy. Drawbacks of this method are the
high cost of instrumentation and the long time involved, first of
all, by sample preparation.

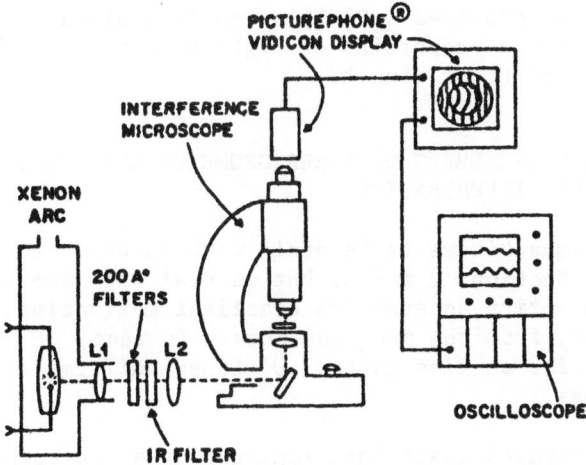

Fig. 3.7. An interference-microscopy set-up with electronic data
processing (from [23]).

3.5. Further Possibilities

Especially in the early days of low-loss fiber technology, other well-known instruments have been successfully applied to the measurement of index profiles. We can cite scanning electron microscopy and chemical microprobing, among others. However, at least as concerns the latter method, a straightforward relationship between the refractive index and the dopant concentration has not yet been fully proved and therefore this technique is less fashionable than those we have discussed so far.

Other possibilities [24,25,26] have been suggested by the cut-off behavior of individual fiber modes. In those experiments, light that leaks either from the fiber outer wall or from a prism-taper coupler is studied as a function either of wavelength [24] or of the output angle [25,26]. However, recent work on these methods is scarce, and this could be due to the increasing favor of the refracted-power method described in an earlier section [18].

Conversely, the power launched by an incoherent source onto the whole set of guided modes is proportional to $n_0^2 - n_2^2 \simeq 2\Delta n^2$. Therefore, its measurement as a function of wavelength [27] yields information on profile dispersion. Fig. 3.8 shows schematically how this test is performed. Small-spot launching conditions (Fig. 3.8a) avoid the excitation of leaky modes. Fig. 3.8b shows how a reference level is taken. We devote a very limited space to this technique only because its full capabilities - as compared with Section 3.4 - will be reported in the near future [28].

Let us add that there seems to be some space for further innovative ideas in the field of index profile measurements. A suitable example might be provided [29] by an optical-signal-processing technique, based on the properties of light partial coherence at the output of a short fiber length.

4. CONTINUOUS MEASUREMENT OF FIBER GEOMETRY AND PROFILE BY TRANSVERSE ILLUMINATION

Physical quantities to be dealt with in this section coincide with those of Sections 2 and 3, but we deal with another technique as a separate entity because its practical motivations differ, to a large extent, from the previous ones. In fact, it provides a suitable tool for on-line control of fiber uniformity during the pulling process.

If a collimated laser beam impinges on a fiber, perpendicular to its axis, then the scattered light pattern contains information on fiber geometry and index profile. Implementations of this principle differ substantially from one another because of the amount

of information extracted from this test. In principle, at least
for weakly-guiding fibers (Δ<<1), one can get the entire profile,
n(r), 0<r<a, but complication and inaccuracy grow along with the
amount of information looked for.

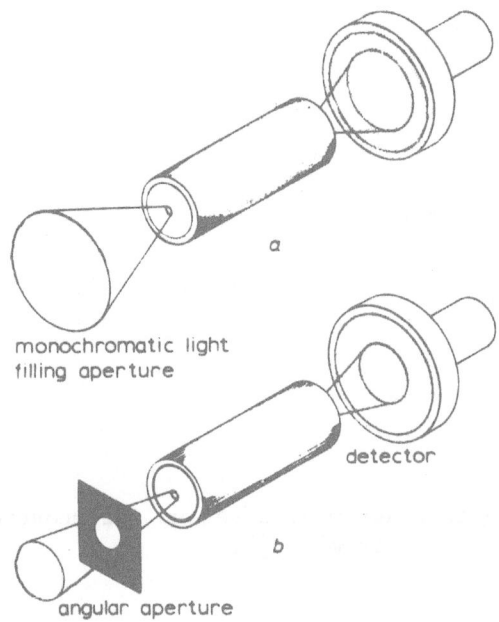

Fig. 3.8. Schematic representation of the measurement of profile
dispersion (from [27]).

    Early work was limited to index measurements in unclad fibers
[30] or to core ellipticity measurements in step-index fibers [31].
Fig. 4.1 shows a simple set-up that performs ellipticity measurements
by recording the fringe displacement when the fiber is rotated
around its axis, as shown in Fig. 4.2. This set-up can detect
ellipticities as small as 1-0.997. Cumbersome calculations arose
when it was suggested that core and cladding diameters and indices
in step-index fibers could be measured in this way with just one
experiment [32]. Finally, the evaluation of the whole n(r) in a
graded-index fiber, again with a single experiment, was shown [33,
34] to be possible, but this implied a complicated integral over
the fiber diffraction pattern. The experimental procedure and the
computational effort became very heavy, and very little has been
reported on the accuracy of the final results. In conclusion, this
method is not recommended for a single measurement of n(r), but it
is attractive for checking the uniformity of n(r) along the fiber
length. In fact, fluctuations in the scattering pattern, as the

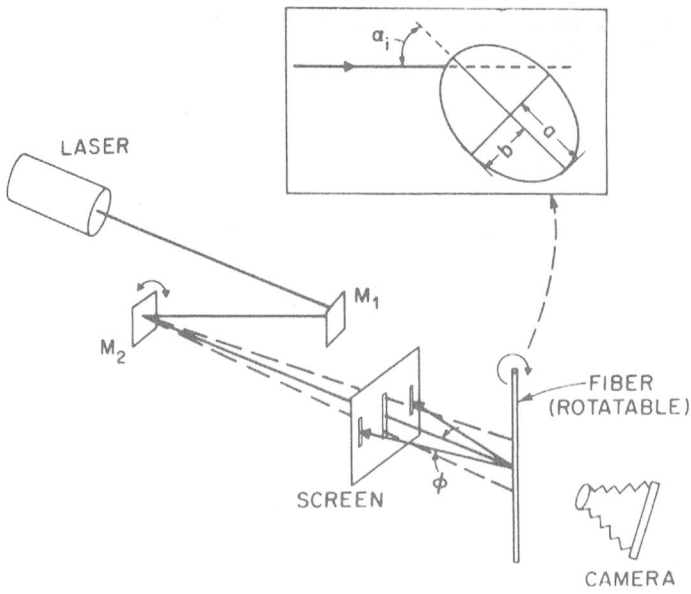

Fig. 4.1. Schematic diagram of a set-up for measurements of fiber
          ellipticity (from [31]).

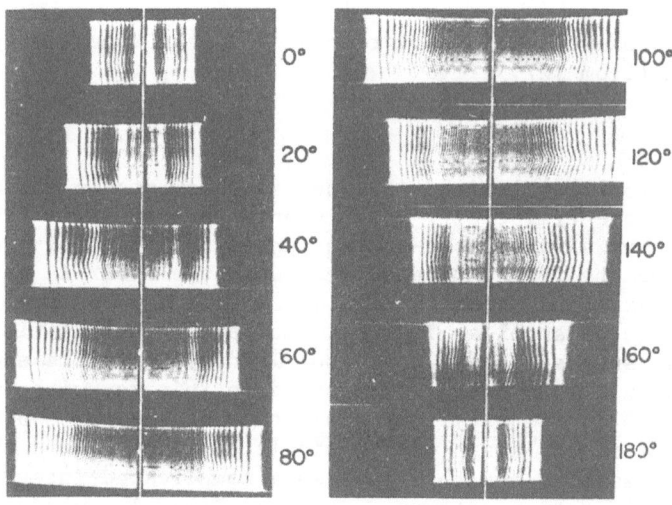

Fig. 4.2. An example of fringe pattern in the backscattered field
          of the set-up fig. 4.1 (from [31]).

fiber is pulled through the laser beam, are much easier to detect and to process than the whole pattern at a fixed value of the axial coordinate. Such signals can be used for feedback control of fiber pulling, and also processed (e.g., Fourier-analyzed) in order to extract statistical information on fiber imperfections.

5. ABSORPTION AND SCATTERING LOSS MEASUREMENTS

Attenuation of light along an optical fiber can be tested in many different ways and for several purposes. According to our general framework, from tests for technology towards tests for system design, we will deal first with measurements of separate physical components of intrinsic fiber loss, which are absorption and scattering. Later, we will discuss tests of fiber behavior with respect to additional loss due to microbending. Measurements of total attenuation, relevant to system design, will conclude the discussion of loss in optical fibers.

5.1. Absorption Loss

In this section, we will rely on the assumption that both absorption and scattering incremental power losses are proportional to the local value of the guided power, and that the two loss contributions add linearly. This means an exponential decay of the guided light power along the fiber length, z:

$$P(z) = P_o \exp\{-(\alpha_a+\alpha_s)z\} \tag{5.1}$$

with α_a and α_s the absorption and scattering power loss constants, respectively.

Intrinsic absorption in pure silica glass is negligibly small over the whole wavelength range exploited or studied for optical communication (600–1300 nm). Hence, absorption loss in fibers is a consequence only of glass impurities, the most dangerous of which are ions of transition metals (in particular Cu^{2+} and Fe^{2+}) and OH-residuals.

The micro-calorimetric method [35] is a rather peculiar way to measure absorption in bulk glass samples and in fibers, and is a little outmoded by the rapid progress in the fabrication of very-low-loss fibers. A light beam which travels through a short sample of length L (i.e., $\alpha_a L \ll 1$) has an essentially constant power P_o and the sample thermal balance at steady state is expressed by the following equation:

$$\alpha_a LP_6 = hS\Delta T \tag{5.2}$$

where S is the sample surface, ΔT is the difference in temperature between sample and ambient, and h is the surface heat transmission coefficient. ΔT is measured with a thermopile and P_o with a calibrated photometer (remembering, at any rate, that P_o is defined inside the sample and is different from outside because of reflections). In practice, though, h is not known with sufficient accuracy and must be eliminated from (5.2) with another measurement. Normally, the thermal time constant of the sample is measured, switching off the beam. We then have

$$\Delta T = \Delta T_o \, \exp\{-(hS/cV)t\} \tag{5.3}$$

where c is the specific heat and V is the volume of the sample; h is eliminated from (5.2) since V and c are known with much better accuracy.

Ingenious solutions increase the measurement sensitivity. For a bulk test at a laser wavelength (typically, at 1.06 μm), placing the sample inside the laser cavity increases P_o by about two orders of magnitude, compared with the laser beam. However, in this case extreme care is necessary to avoid sample surface contamination: dust particles can cause errors in excess of 100%. For a fiber test [36], heat transfer to the ambient is strongly reduced and a smaller power ($\sim$100 mW) is sufficient. Another substantial decrease in P_o, close to the possibility of spectral measurements with a lamp source, is obtained by placing the fiber in a vacuum chuck. A light power of 1 mW with a high-responsivity thermopile should then give a sensitivity limit of 1 dB/km. Finally, another improvement consists [37] of using two distinct heat propagation models in the sample or in the fiber; one holds for infinite length and the other one includes two heat sinks at the ends. The corresponding time constants can be observed during two distinct portions of the heating transient, when light is shed on the sample. The change occurs when the heat generated by light absorption reaches the heat sinks at the sample ends. The corresponding result is a measurement uncertainty of ± 0.2 dB/km over 3 dB/km, with a laser power $P_o \simeq$ 450 mW.

5.2. Scattering Loss

Several physical mechanisms can scatter part of the guided light out of a fiber. In low-loss fibers, Rayleigh scattering [38] is by far the most important one, and actually sets the ultimate theoretical limit to fiber attenuation. Testing α_s requires a measurement of the scattered power over a short fiber length, $\alpha_s LP_o$, and the knowledge of the guided power, P_o.

Total scattered powers, integrated over all directions, are often measured in optics. A classical instrument that has been

proved suitable for fiber scattering tests is the so-called Ulbricht
integrating sphere [39]. But there are also new developments.
Tynes [40] built an integrating detector with six silicon solar
cells, forming a cube, with a suitable resistor network compensating
for differences in responsivity among individual cells. Later on,
it was shown that the scattering pattern of low-loss fibers allowed
the number of cells to be reduced to two [41], if they were arranged
to make a narrow channel around the tested fiber. Later (Section
7.3), we will show the application of these devices to measurements
of total fiber attenuation and discuss further improvements of the
detector itself. As for the scattering measurements, the linear
dynamic range of the detector sets an important problem, since in
good fibers the scattered powers are smaller than the guided powers
by several (typically 5) orders of magnitude. Linear dynamic range
and sensitivity of the detector both depend on its electrical load,
but in opposite ways, and an optimum load exists for a given sensitivity or for a given dynamic range. Its value can be determined by
numerical analysis of a nonlinear circuit [42].

6. TEST OF MICROBENDING LOSS

As is known, microbending means a random sequence of small
curvatures in the axis of an optical fiber, induced by a lateral
pressure on the fiber itself (due to its weight or to externally
applied stresses) against a randomly rough surface. Theory [43]
indicates that microbending-induced mode-coupling leads to an
extra loss added to those of a loose fiber. We will not discuss
the measurement of this loss, because it is a test that does not
require a peculiar procedure, but it can be done with some of the
techniques to be described in Section 7. Instead, we will outline
how the parameters that affect this microbending loss can be controlled in the laboratory in order to predict the fiber sensitivity to practical cabling conditions.

When a fiber is wound on a drum whose surface roughness is
characterized by an rms value σ and by a power spectral density P_{ν},
microbending depends essentially on the lateral force that presses
the fiber against the drum. This lateral force is constant if the
fiber is wound under a constant axial stress. This can be obtained
[44] with a suitable electromagnetic brake on the pay-out drum
shaft. The same result can be obtained [45] with the set-up shown
in Figure 6.1, which operates at constant angular velocity and
constant braking torque after a short initial transient. Fig. 6.2
shows results of microbending loss tests vs. axial tension for
silica-core, silicone-resin-clad fibers. The shapes of these
curves are typical, while the values of the loss are very high
because of the peculiarly lossy cladding material.

Further information on microbending can be collected by

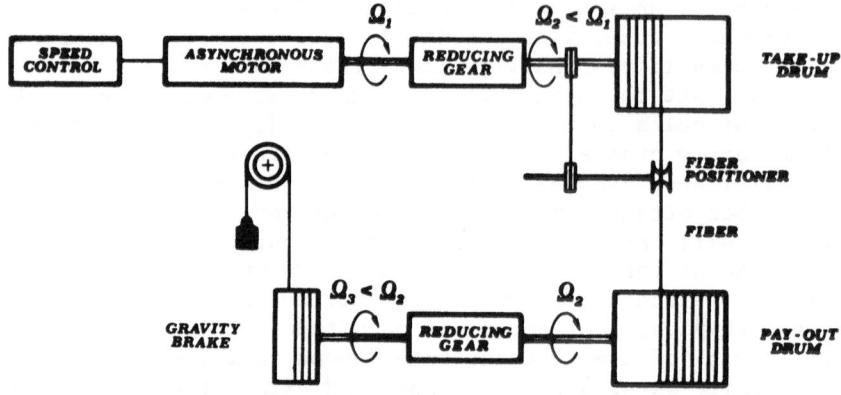

Fig. 6.1. Layout of a mechanical set-up for microbending tests
        (from [45]).

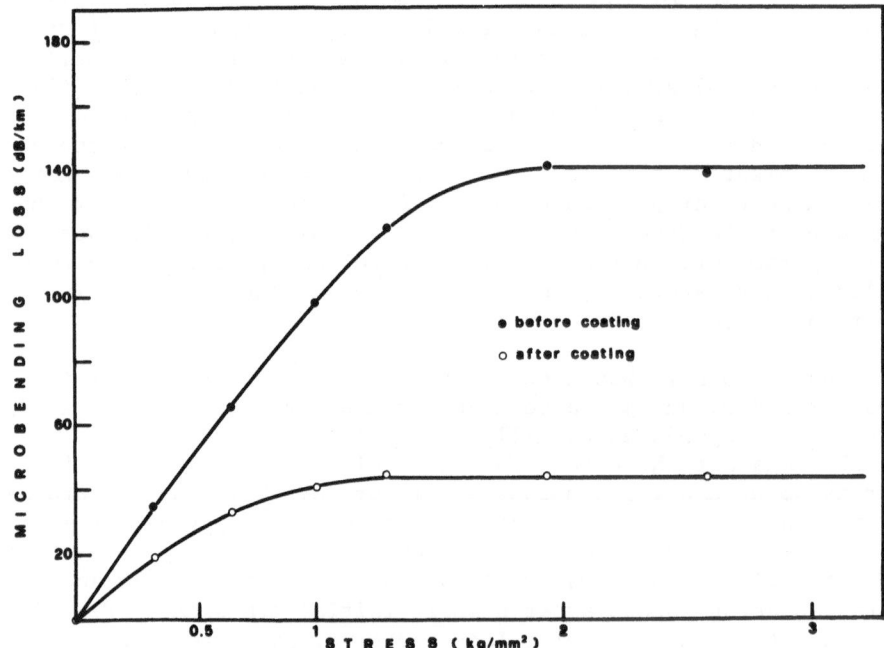

Fig. 6.2. Microbending loss vs. tensile load for silica-core,
        silicone-clad fibers wound on a drum (from [45]).

monitoring the far-field pattern at the fiber output as a function
of axial stress. This indicates whether mode mixing in the fiber
has reached a steady state. Loss and far-field tests must be
correlated with the drum roughness, which is monitored by a mechanical or an optical instrument whose output is usually Fourier-
analyzed on a numerical computer.

7. ATTENUATION MEASUREMENTS

It is well known that the signal power loss along an ordinary
transmission line is well described as an exponential decay, $P(z) = P_o \exp(-\alpha z)$. While the same behavior can be assumed without any
doubt for a single-mode fiber, for a multimode fiber this simpli-
fying assumption is not immediate, for theoretical reasons that are
known from previous lectures (e.g. ref. 1). Let us summarize them
and show their influence on analysis of measurements of multimode
fiber attenuation.

If absorption and scattering are not uniform throughout the
fiber cross section, then individual modes have different attenuation constants. As long as mode coupling is negligible, the total
guided power (i.e., the sum of all modal powers) is not an exponential function of fiber length. In actual fibers, mode coupling
has a major influence over long lengths. Regardless of the particular assumptions (typically, the "nearest-neighbour-coupling"
model [46]) that originated them, the following results are now
accepted as general, although in a statistical sense. There is a
set of coupled-mode-powers equations [47], with a set of discrete
solutions, corresponding to positive real attenuation constants.
The smallest among these eigenvalues, α_s, corresponds to the
asymptotic loss and characterizes the asymptotic power distribution
among the fiber modes. For long fiber lengths, this will be the
output distribution, regardless of the input distribution. Thus,
α_s is crucial for system design. At any rate, further information
on the initial spatial transient, where power decay is not expressed
by $\exp(-\alpha_s z)$, is extremely important as well.

7.1. The Cut-Back Technique

In the simplest - and yet the most popular - way to measure
its attenuation, a fiber is excited with a monochromatic (or
filtered) source, and the power output P_o is measured at its far
end. Then, a length, L, measured from the far end, is cut away.
Without changing the input conditions, the new output, P_i, is
measured. Then, within the theoretical limitations that we just
pointed out,

$$\alpha L = \ln(P_i/P_o) \tag{7.1}$$

A very important laboratory tool in these cw optical measurements is the light-chopper lock-in-amplifier combination [48] (see
Fig. 7.1) where the chopper transforms the photodetected signal
from dc to low-frequency ac (typically, 10-1000 Hertz). In the
lock-in amplifier, this ac signal undergoes a narrow-band homodyne
detection, with a reference signal generated by the chopper as well.
The overall result is a very consistent reduction in noise (the
equivalent bandwith becomes a fraction of Hz).

Simplicity, reliability and low cost of instrumentation are
the main advantages of the cut-back method. There is no major
difficulty in implementing a spectral measurement over a broad
wavelength range (e.g., 0.7-1.3 μm) with a white source (e.g., an
arc lamp) followed by a monochromator. Output power from the
initial fiber length is recorded first over the whole wavelength
range, and then, after cutting the fiber, the range is swept again.
This measurement would be quite sensitive to long-term drifts, both
in the source and in the optical set-up, unless a part of the input
beam, split by a partially reflecting mirror, were used as a reference.

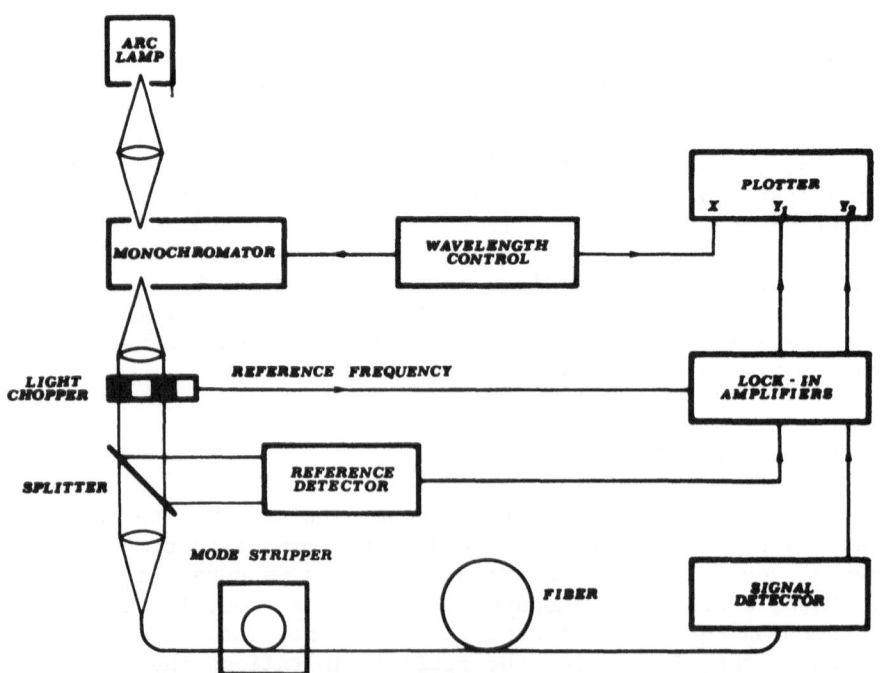

Fig. 7.1. Scheme of the set-up for cut-back spectral loss measure-
          ments.

The main drawbacks of this technique are easy to grasp. If power outputs are measured for several lengths of the fiber (for a better accuracy and for gaining confidence in an exponential decay), then the test destroys the whole fiber. This is particularly bad for cabled fibers, and totally impossible in the field. On the other hand, with just one cut in the vicinity of a fiber énd, the accuracy decreases, especially if the cut is near the output end. If it is close to the input end, it is hard to know whether the measurement gives the steady-state loss α_s and no information is collected on the initial transient. Furthermore, this test is still unpleasant in the field, since it requires operation at both ends of the link. The sensitivity of this test to the input power distribution is easy to demonstrate: α changes when source-fiber coupling conditions (e.g., numerical aperture or the angle between the fiber axis and the input beam axis) change. To minimize this effect, some precautions have become part of common laboratory practice. First of all, the light guided by the cladding outer surface is stripped off [49] by immersion of a bent section of fiber into a fluid with a higher index of refraction. In addition to that, the influence of launching conditions and the length of the initial transient can be reduced in either one of the following ways. One is a so-called "mode scrambler," that is to say a short fiber where tight coupling involves all the modes. A typical realization [50] consists of a sandwiched fiber, between random rough surfaces (e.g., abrasive paper), under a light pressure. A mode scrambler is good if near-field and far-field distributions at its output are similar to those at the output of a long fiber length (typically, 1 km). From this, one can infer the other way to shorten the input transient. The launching optics is built so that the input beam half-power spot radius and half-power angular width are equal to those at the output of long fibers [51]. At any rate, it is still hard to say whether this is sufficient to reproduce the actual steady-state distribution, which has other implications (like complete phase uncorrelation among the guided modes).

Finally, sophisticated versions of this method measure the so-called differential mode attenuation, i.e., the attenuation constants of individual modes or mode groups as a function of mode order. In a step-index fiber, this is not too complicated: the angle of incidence of a collimated laser beam is varied and the power output is measured (on a long fiber and then again on a short one). For graded-index fibers, the measurement is much more complicated. It involves the test of the radiation pattern of each portion of the output fiber face, up to the angle x surface diffraction limit [52]. The scheme of the corresponding apparatus is shown in Fig. 7.2. This measurement is meaningful only for fibers where mode-coupling characteristic lengths are much larger than the inverse of the differential attenuation. Otherwise, the "history" of each ray pencil at the fiber output cannot be associated uniquely with a fiber mode.

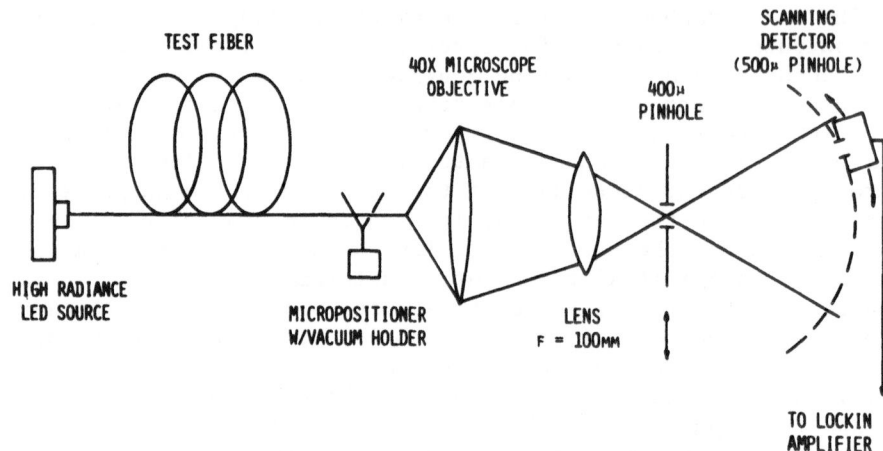

Fig. 7.2. Schematic diagram of modal attenuation measurements in
graded-index fibers (from [52]).

7.2. The Backscattering Technique

An impressive list of successful implementations [53-57] indicates that the technique to be described now is a very tough compe-
titor with the classical method we dealt with in the previous
section. This new technique is based on the following principle.
When a short pulse (typical half-maximum full widths from a few ns
to some tens of ns) of laser light travels along a fiber, a fraction of the pulse power is scattered in the backward direction and
guided to the fiber input. There it can be detected by a fast
photodetector and recorded as a function of time. This signal
(amplified, and processed by a boxcar integrator to improve the
signal-to-noise ratio) yields information on the traveling power
(in the forward direction) vs. fiber length. In a semi-log plot,
the fiber attenuation constant equals 0.5 times the curve slope,
where the factor 0.5 results from the fact that both the forward
pulse and the backscattered signal undergo the fiber attenuation.

The main trouble in implementing this test is the need to
detect a signal well below (50 to 60 dB) the input signal, at the
same point where the input signal is launched into the fiber. With
a direct laser-to-fiber coupling, Fresnel reflection from the input
face ($\sim 10^{-2}$ × input power) saturates the receiver and then makes
it blind for a long time after the initial pulse. Hence, ingenious
coupling schemes are absolutely needed.

The first successful idea [53] consisted of coupling the laser
light into the fiber through a suitably tapered section of its outer
wall (Fig. 7.3). In this way, one avoids delicate index-matching at
the input face, and the photoreceiver does not need a gate circuit.

The price one pays for this is launching conditions that are quite
far from the steady-state power distribution among the fiber modes
because only high-order modes are strongly excited at the launch
point and power is coupled to low-order modes over long lengths.
The steady-state attenuation is then given only by a rather low
tail in the recorded backscattering signal. This can be an
important drawback because of the limited dynamic range of the
receiver.

Another parallel implementation [54] by-passed the coupling
problem with the use of a gated photomultiplier front-end. The
inconvenience of a high-voltage pulser and of costly electronics
should not be underestimated, especially for field tests on
installed cables. It is balanced by the advantages of a much better
control on the launch conditions and of displaying the return signal
over selected time intervals.

Getting rid of the Fresnel reflection echo by means of very
accurate impedance matching at the fiber input was another solution,
presented in [56]. As shown in Fig. 7.4, a 3 dB beam splitter is
embedded in a cell full of an index-matching liquid and terminated
by spherical lenses that match the curvature of the incoming wave-
fronts. This arrangement seems to be more suitable than the others
for a systematic study of the influence of launching conditions.

A simpler device, described in [57], consisted of just a beam
splitter, glued directly onto two 45-degree cuts in the middle

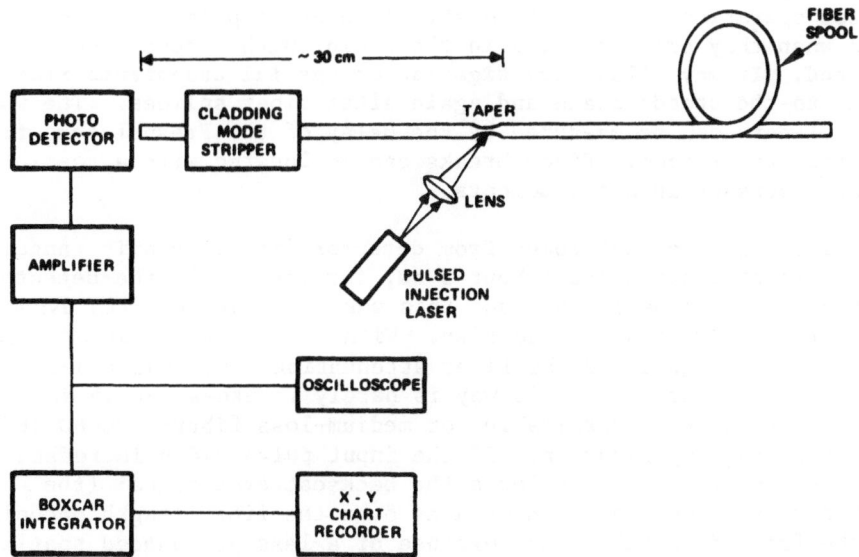

Fig. 7.3. Side-wall taper launching implementation of the backscattering technique (from [53]).

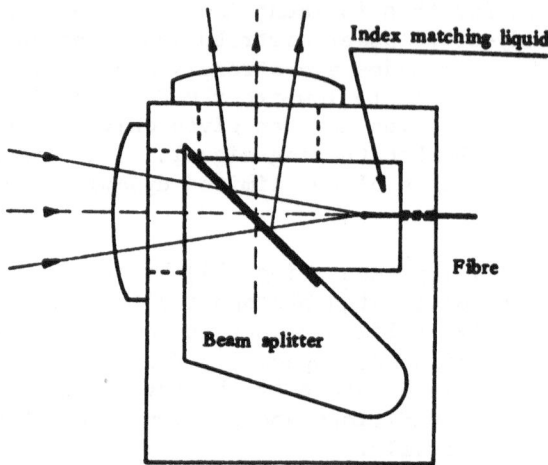

Fig. 7.4. Impedance-matching cell, embedded with a 3 dB splitter,
          for the backscattering test (from [56]).

of a "pig tail" fiber, i.e., a short section of the same fiber as
the one to be measured, located between the latter and the set-up.
An interesting combination of index-matching and gating, which
relaxed the requirements on either one, was presented in [55].

     The following features of the backscattering technique make it
quite appealing. The test can be performed at the wavelength of
system operation. It is non-destructive and applicable to cables,
even when they are installed in the field, with access from just
one end. It may allow investigation of spatial transients from the
input to the steady state and again after fiber splices. The whole
fiber length can be measured by the delay of the Fresnel echo from
the far output face. Fiber breaks can be located with a resolution
to be discussed in a few moments.

     The main drawback comes from detector linear dynamic range.
With a short input pulse (about 5 ns, for instance), the detector
must be fast and sensitive, and this makes it hard for its dynamic
range to be more than ∿ 3 decades. With the known factor of 2 between
the detected signal and the fiber attenuation rate, the fiber loss
that can be explored in this way is hardly in excess of 15 dB. In
particular, this is unsuitable for medium-loss fibers (20–60 dB/km)
for short-haul application. If the input pulse width increases,
a proportional increase affects the backscattered signal (the
backscattering mechanism integrates over the fiber length occupied
by the forward pulse) at the expense of a less pronounced spatial
resolution. The receiver bandwidth can be reduced with a subsequent
improvement in the dynamic range. However, pulse widths in excess
of ∿50 ns have not been reported, because the corresponding reso-

lution ($\sim$10 meters) has been considered as a necessary minimum.

Another limitation of the backscattering technique comes, at least for the time being, from the following argument. Rayleigh scattering in a quasi-homogeneous medium is known to be isotropic [38]. If this statement is applicable to fibers, then, even when the forward signal has reached the steady-state distribution, the backscattered fraction does not have the same distribution. Hence, in the backward signal path there is a spatial transient which can not be detected by the receiver, so that its length and its actual loss remain unknown. However, actual discrepancies with results given by other techniques are always small enough to limit the practical weight of these theoretical doubts.

Finally, backscattering-based tests have been performed, very recently, in the so-called "second spectral window," at 1.06 μm [58]. The use of a Q-switched Nd-YAG laser (see Fig. 7.5) has negative and positive implications. The minimum pulse width is $\sim$300 ns, with a bad loss in resolution. On the other hand, the available peak power (1 kW) is such that the backscattered signal remains large enough, if the input pulse is spatially filtered in a suitable way, to study attenuation vs. launching conditions. Also, a new interesting approach to the input-output coupling problem consists of a polarizer-analyzer combination which minimizes Fresnel echo from the input face.

7.3. The Lateral Scattering Technique

In Section 5, we saw that a suitable instrument for measuring the scattered light leaving a fiber through its outer wall consists of two flat, parallel photosensitive surfaces, with a fiber in between [41]. An index-matching oil insures that all the power, which is scattered out of the fiber core over the detector length D, reaches the detector itself, without being partially trapped by

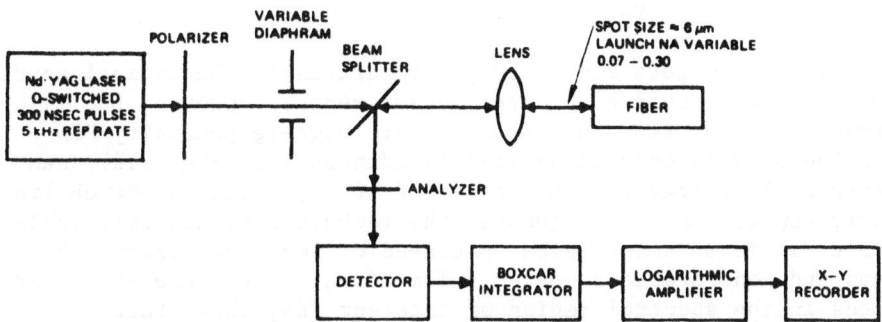

Fig. 7.5. Schematic diagram of a backscattering-based set-up for loss tests at 1.06 μm (from [58]).

the cladding outer wall. The fiber attenuation measurement to be
described now relies on an assumption that we introduced in Section
5: namely, that the scattered power, leaving the fiber over the
length D, is supposed to be proportional to the guided power that
travels along the fiber. This is true, on the average, with respect
to the axial coordinate, z. Often, though, it is not correct on a
small scale, because of fluctuations in the scattering constant, α_s
("scattering centers" are present, especially in bad fibers).
However, let us first go on as if the assumption were correct.
Later we will show how to handle the statistics of α_s in order to
reduce the consequent error.

Early implementations of a side-scattering technique [59,60]
required one scattering detector, or one photomultiplier [61], which
was located, sequentially, at different distances from the fiber
input. A light-chopper and a lock-in amplifier were used, for the
same reasons as in Section 7.1. The detector sensitivity was not
good enough to perform spectral measurements: tests at the He-Ne
laser wavelength on a high-loss fiber were reported. Results were
sensitive to drifts in launched power, but this inconvenience could
be fixed by monitoring the fiber end output. In addition, the test
could be in error, because of the presence of scattering centers,
if the measurement sections were chosen on a purely geometrical
basis (i.e., at a given distance from one to the next). Last,
moving the detector over the whole fiber length would be impossible
if the fiber were part of a cable.

Most of these inconveniences have been by-passed by a recent
implementation [62] which is illustrated by Fig. 7.6. Two detectors,
equal to each other, give two signals, V_1 and V_2, proportional to
the scattered powers in two fiber sections separated by a length,
L. Their difference is measured first. Then, one of the detectors
is short-circuited and the remaining signal is measured. This yields

$$\alpha = \frac{1}{L} \ln\left(\frac{V_1 - V_2}{V_2} + 1\right) \qquad (7.2)$$

This set-up gets rid of long-term drifts in the optical source,
even for time-consuming spectral measurements. However, the main
improvement comes from the sensitivity: high-responsivity PIN
photodiodes, with very large load impedances (see Fig. 7.6), can
measure optical powers on the order of 10^{-13} W, with a $\sim$60 dB linear
dynamic range. As a consequence, the technique is now applicable
to spectral measurements even on coated or jacketed fibers [63] with
a lamp and monochromator source. The advent of tunable dye laser
sources in the spectral region of interest [64] shows further
possibilities for advancement.

The main surviving error consists, at this stage, of fluctu-

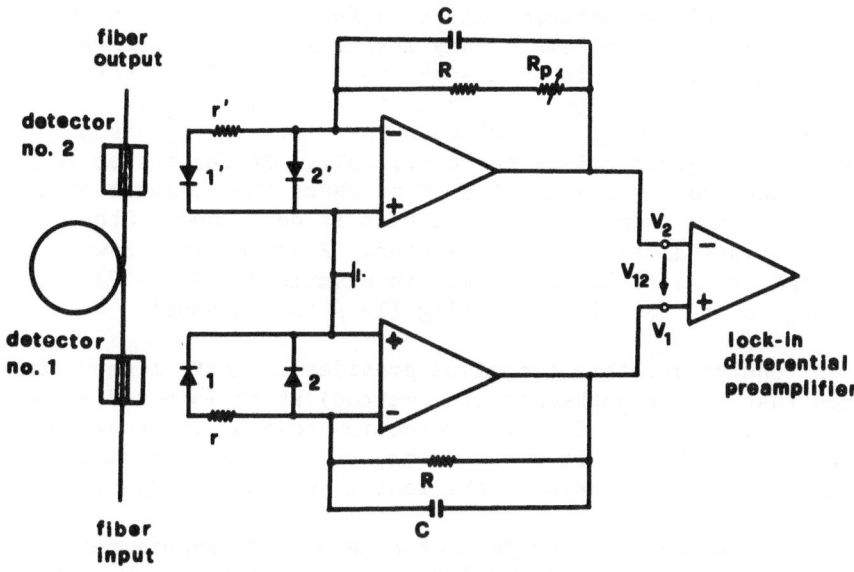

Fig. 7.6. An improved set-up for attenuation measurements based on
the side-scattering technique (from [62]).

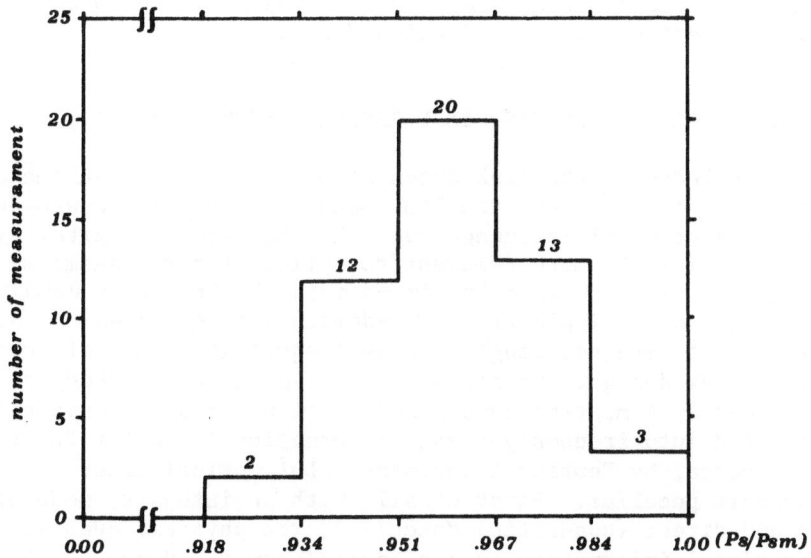

Fig. 7.7. Histogram of scattered light power values over 1 meter of
uncoated fiber (from [63]).

ations in the scattering coefficient α_s. However, this can be
reduced to within acceptable limits (a few percent), by a statistical treatment, as follows. Many measurements of scattered power
are performed at a single wavelength on a short fiber section around
a selected measurement point (e.g., 50 measurements over 1 meter of
fiber). A χ^2 test will show whether their average is a significant
value within a given tolerance (e.g., ±5%). If the answer is yes,
then the detector is placed at a point where the measurement equals
the average, and then the spectral test is performed. Otherwise, a
larger statistical sample is collected. Even for fibers with a
broad statistical fluctuation (see an example in Fig. 7.7) the whole
procedure is not too time-consuming (less than ½ hour).

In conclusion, this technique provides a spectral measurement
(in contrast to the backscattering method) which is non-destructive
for the fiber and even for its protective coating (in contrast to
the cut-back method). It allows, very simply, an investigation of
the spatial transient between the source and the steady-state regime.

Before we leave the subject of attenuation measurements, let
us mention two more techniques [57]. In one of them, the fiber
insertion loss is measured directly with a calibrated detector and
a reference short fiber (or a light beam). In the other one [65],
a light pulse travels twice through the fiber with a mirror placed
at its far end. The return pulse is compared with the original
one. So far, neither one of these methods has received too much
attention. In particular, the second one seems to be ruled out by
the backscattering method of Section 7.2.

8. PULSE-DISTORTION AND FREQUENCY-RESPONSE TESTS

If knowledge of the link attenuation is essential in the design
of a communication system, the link bandwidth and its length-depen-
dence have a critical influence too. In the case of digital systems,
which are by far the more relevant to optical fiber transmission,
the designer makes use of both time-domain and frequency domain link
characterizations. Typically, time-domain data speed up calculation
of maximum bit rate vs. length, while frequency-domain data can
help equalizer design. Of course, it is known to everybody that,
for a linear system, time-domain information (impulse response) can
be converted into frequency-domain information (transfer function),
and vice-versa, by Fourier transforms. But optical-fiber systems
are somewhat peculiar. First of all, with an intensity-modulated
source and direct (quadratic) detection, the physical quantity that
carries signal information is the optical power. Hence, the
designer is interested first in establishing whether a fiber link
behaves linearly with respect to optical power and second, in
knowing the fiber behavior with respect to the envelope of a modulated optical carrier (i.e., the so-called baseband frequency response), without any interest in the optical carrier itself. A

discussion on the theoretical proof of fiber linearity with respect to power [66,67] is beyond the scope of this paper. What must be said here is that, so far, even if linearity with respect to power seems well grounded, at least for incoherent sources and/or for long fibers (kilometer lengths) over bandwidths from dc to ∿1 GHz, nevertheless the practical equivalence between time-domain and frequency-domain measurements is not fully assessed. For instance, very recently [68], doubts have been raised about fiber behavior as a minimum-phase system [69], which had been said before to be verified experimentally. All this induces us to present time-domain and frequency-domain measurements as if they were independent of each other.

8.1. Time-Domain Measurements

These tests require a short light pulse, generated by a suitable laser source (a pulse-driven or self-pulsing Ga-As laser or a mode-locked Nd-YAG or He-Ne laser). While a part of this pulse is detected as a reference, most of its energy is injected into the tested fiber. The fiber output pulse is detected and its change with respect to the input pulse is measured. The basic scheme is shown in Fig. 8.1.

If the input pulse is short enough to approximate a Dirac δ-function (i.e., if its width is negligible with respect to that of the output pulse), then the output can be taken directly as the fiber impulse response. But in practical fibers, output pulse widths are so short (always less than a few tens of ns, and sometimes less than 1 ns, after kilometer lengths) that a δ-function would mean an initial pulse width in the picosecond range. This requires tremendous complication [70] in pulse detection, and test wavelengths that are not in the spectral range of interest for optical communications. Then, usual time-domain operation makes

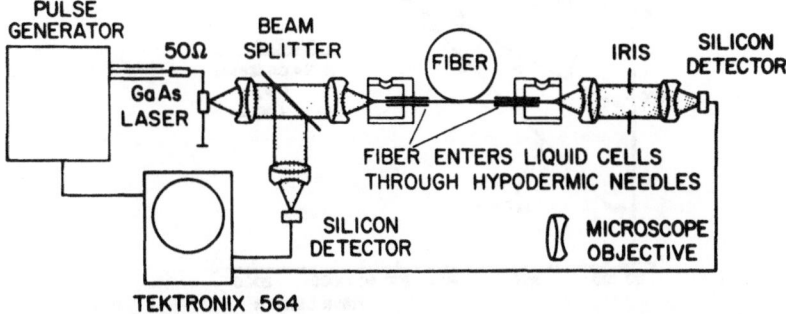

Fig. 8.1. Schematic diagram of a set-up for time-domain pulse
          distortion tests in optical fibers (from [61]).

use of input pulses that should be deconvolved from the output
pulses. Nevertheless, the output pulse evolution as a function of
fiber length is a good information source per se, even without such
a deconvolution. For example, it permits us to see whether pulse
broadening is proportional to the fiber length L (as predicted by
a simple model with uncoupled modes) or to $L^{\frac{1}{2}}$ (as predicted) [66]
for lengths L beyond a critical value that corresponds to complete
mixing among guided modes). Fig. 8.2 is a very suitable example,
with measurements performed at three different laser wavelengths.
It should be remarked that, for short fiber lengths, the results
are blurred by the nonnegligible initial pulse widths.

Early tests of this kind, including [71], were destructive:
the fibers were cut back, similar to the attenuation test. Pulse
width vs. length was tested without breaking the fiber by means of
a shuttle-pulse set-up [72]. Fig. 8.3 illustrates the principle:
two partially transparent mirrors are placed at the fiber ends,
with very careful alignment. The pulse shape is detected at the
far end after one, three, ... trips of the propagating pulse.
Successive shapes (see an example in Fig. 8.4) are compared and the
relationship between their widths and fiber length is obtained.

As an alternative to the shuttle-pulse set-up, a pulse circulation set-up was developed [73] as shown in Fig. 8.5. In the
shuttle-pulse method, when the i-th pulse (i = 1,3,...) is displayed
on the scope, there have been i-1 reflections, accompanied by use-
less energy loss through the mirrors. If, on the other hand, the
two fiber ends are separated by just a short beam deflector D, then

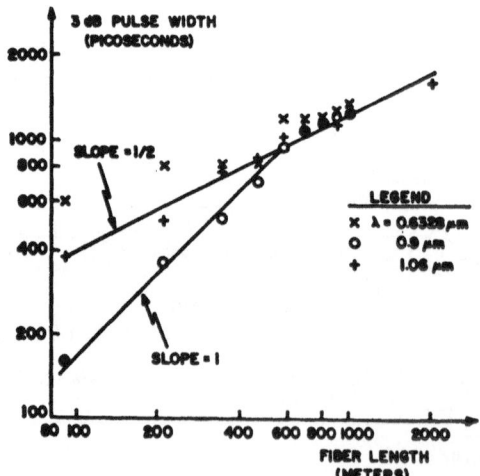

Fig. 8.2. Examples of measured pulse widths vs. fiber length, at
three different laser wavelengths (from [71]).

the laser pulse is coupled into the fiber by D, circulates i-1 times
through the fiber + D loop, and finally is extracted by D at its i-th
arrival. In [73] an acousto-optic deflector was used. It was driven
by 150 MHz rf double pulses and had a deflection efficiency of 30%.
This allows, on average, the detection of pulses after a total distance which exceeds by two times the fiber length allowed by the
shuttle-pulse method before getting down to the noise level.

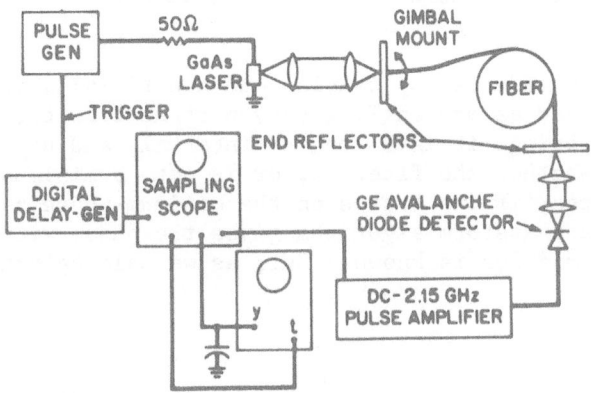

Fig. 8.3. Schematic illustration of shuttle-pulse dispersion test
          (from [72]).

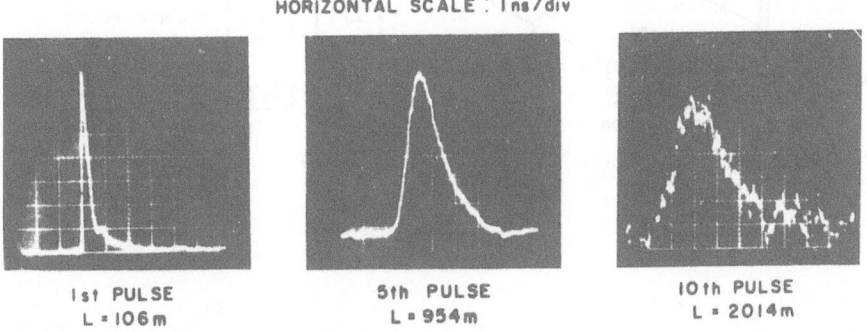

Fig. 8.4. An example of detected pulses in a shuttle-pulse dispersion
          test (from [72]).

The measurements of [73] were extensive enough to separate contributions of intermodal dispersion and of material dispersion.

     For a rigorous interpretation of pulse distortion, one must
deconvolve the initial pulse shape from the final one.   The current
procedure is to get the Fourier transforms of both pulses by processing time-domain data and then to calculate the fiber transfer
function as their ratio.   This is rather straightforward as far as
the modulus is concerned, but the phase gives troubles that will be
outlined in a short while.   Even the modulus, though, is sensitive
to pulse "tails", where the signal-to-noise ratio can be quite poor,
especially with low-cost laser diodes driven by low-cost pulsers
[74] that give a considerable jitter.   Substantial improvements [75]
come from filtering the signal with a low-frequency light-chopper
and lock-in-amplifier combination, in cascade with the higher frequency pulse repetition system.

     The phase is affected by trivial and nontrivial ambiguities.
If it is calculated as arctan(Imaginary part/Real part), then the
computer always brings it into the 0-π interval, and no information
is gathered on whether the fiber is, or is not, a minimum-phase
system.   Of course, if one starts on the minimum-phase assumption,
then the Hilbert transform algorithm gives the phase without ambi-
guity, once the modulus is known.   But, as we said before, this

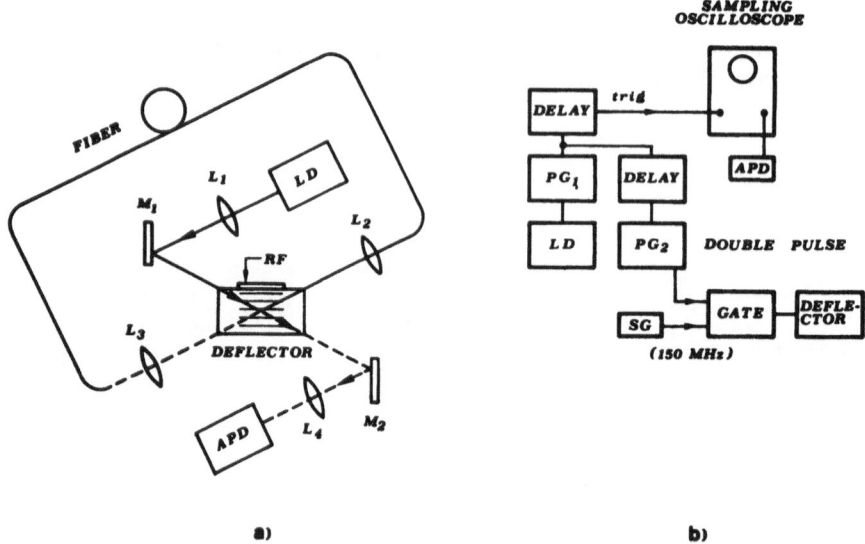

Fig. 8.5. Layout of the pulse circulation setup: a) scheme of the
          optics; b) block diagram of the electronics.   (From [72]).

assumption, looked at with some favor over the last few years, has
been re-discussed in a recent paper [68] by the same authors who
used it for the first time. We will touch on this problem again in
Section 8.2.

Unfortunately, it is still doubtful how data collected in these
ways have to be handled in system design. This statement is based,
in particular, on some interesting anomalies that were observed at
the British Post Office [76,77]: data from individually measured
stretches led to an erroneous prediction for the bandwidth of a
long link built as the cascade connection of the individual cables.
Surprisingly, the overall bandwidth was considerably larger than
the expected one (see Fig. 8.6). A detailed theoretical interpretation of this phenomenon (which is, broadly speaking, a partial
compensation between index profiles of successive fibers) is beyond
the scope of this paper. We just wish to point out some disappoint-
ing limits of the state-of-the-art of fiber evaluation. In this
respect, let us recall also that shifts in the emission wavelengths
of Ga-As lasers, which can occur during individual pulses, if
combined with material dispersion, may lead to pulse compression
along a fiber [78], and so to erroneous interpretation of time-
domain measurements.

Despite its limitations and drawbacks, the time-domain pulse
distortion test is very attractive. It has a relatively low cost,
compared to the frequency-domain techniques of the next Section.
It can test an entire installed link at the wavelength of system

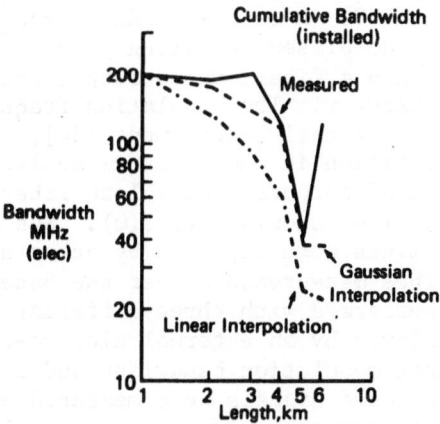

Fig. 8.6. Comparison between measured bandwidth of a long fiber
       link and predictions based on tests of individual cables
       (from [76]).

operation, and for immediate use of the results if the system is digital. It is a powerful laboratory tool for studying in detail the effects of different index profiles [79,80]. Finally, another laboratory application of very high interest [81] is the investigation of the "zero-material-dispersion" spectral region (1.12-1.55 µm), where suitable input pulses ($\sim$ 0.2 ns width) can be generated by multiple-order stimulated Raman scattering in single-mode silica fibers pumped by a Q-switched mode-locked Nd-YAG laser.

8.2. Frequency-Domain Tests

The measurement of fiber transfer function in the frequency domain has received much attention, but implementations are somewhat less correlated to one another than those in the time domain.

Some basic ideas on frequency-domain tests were outlined in an early paper by Gloge et al. [82]. It suggested the beat spectrum of a free-running multimode laser as baseband signal for testing the fiber transfer function $G(\omega)$. In principle, one should be able to measure both modulus and phase of the transfer function, and to explore a very broad frequency range (e.g., from 100 MHz to several GHz) by suitable frequency mixing. In practice, the result reported in [82] was just the amplitude response of a short fiber (30 m in length) over the 0-2.5 GHz band. It is interesting to point out (Fig. 8.7) that the tested fiber is not a low-pass system in a strict sense, in contrast to the statistical theory [83], which is then suitable only for longer fibers where the guided modes undergo a wider variety of coupling and scrambling events.

This test was not performed at the wavelength where a system operates. To satisfy this demand, another technique was developed [84] as shown in Fig. 8.8. A Ga-Al-As LED source was modulated sinusoidally around a dc-biased operating point. The resulting signal was injected into a fiber, whose output was detected and sent to a spectrum analyzer. The modulation frequency could sweep a frequency range (in the early experiment [84], $\sim$0-200 MHz) and the corresponding variation in the spectrum analyzer signal was recorded. Comparison of outputs from a long fiber (typically, 1 km) and from a short one (1 m) should give $G(\omega)$. Substantial developments along similar lines were reported by other authors. In particular, [85] describes measurements over the baseband frequency range 0.1-110 MHz, performed with three different optical sources: a cw He-Ne laser followed by an external electro-optical modulator [86], a LED with large modulation bandwidth and a cw heterostructure laser diode. The detected signals were measured by means of the vector voltmeter of a network analyzer. A typical result on the modulus of the transfer function is shown in Fig. 8.9. No results were reported on the troublesome measurement of the phase.

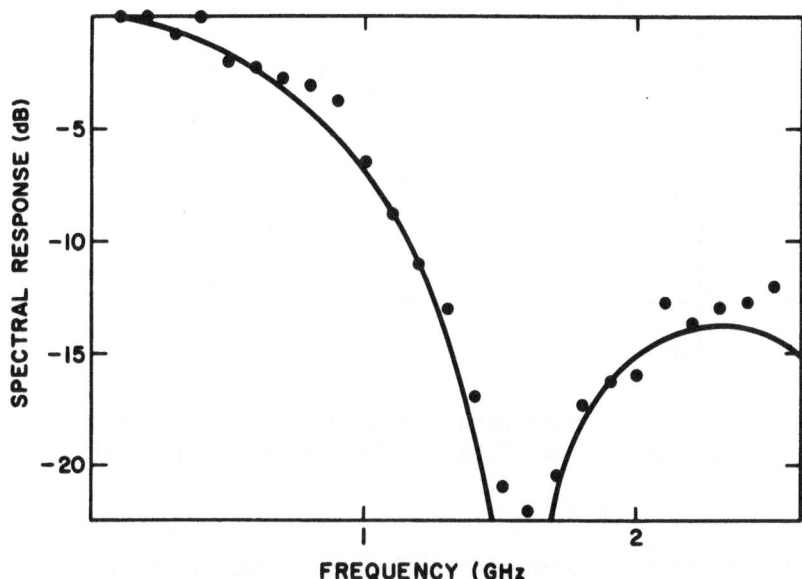

Fig. 8.7. Modulus of the transfer function of a fiber measured with a free-running Krypton laser (from [82]).

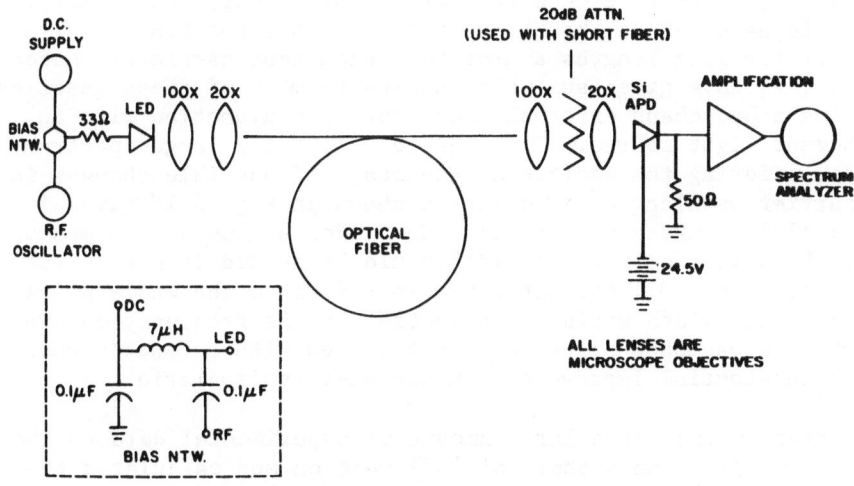

Fig. 8.8. Set-up for frequency-domain baseband response tests at the Ga-Al-As emission wavelength (from [84]).

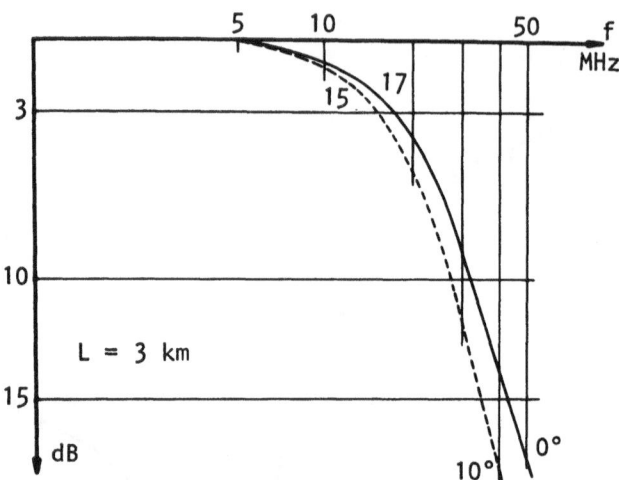

Fig. 8.9. Typical low-pass behavior of the baseband frequency response for a long fiber (3 km in length; from [85]).

This sinusoidal modulation technique requires large modulation bandwidth, because if the source bandwidth is comparable to the fiber transmission bandwidth the accuracy in measuring the latter is poor. Availability and cost of cw laser diodes with modulation bandwidths in excess of 1 GHz are improving rapidly, so the weight of this limitation is decreasing. Some suspicion, however, remains, especially on the reference test of a short fiber, when a coherent source is used, since the theory tells us that the fiber is linear in power for long lengths and/or for incoherent carriers. Cohen et al. [87] have given suitable answers to most of these questions by using a broadband external modulator in conjunction with an incoherent light source. This can be a filtered large-spectrum source, offering the additional advantage of possible changes in the carrier wavelength. The set-up shown in Fig. 8.10 gives $|G(\omega)|$ with a ±10% accuracy from dc to 1 GHz. For a time-domain measurement, the same receiver bandwidth would be needed if pulse rise times were about 0.1 ns, but while in this case the whole noise from this bandwidth would be collected, in the frequency-domain set-up, the baseband signal can be filtered with a tunable receiver with a substantial improvement in signal-to-noise ratio.

After collecting a large amount of experimental data on the modulus $|G(\omega)|$, the authors of [87] went on and calculated the phase $\underline{/G}(\omega)$ as a Hilbert transform on the ground of the minimum-phase assumption [69]. However, as we said before, they did return to this point [68], so that doubts remain on the validity of this procedure. It is clear, then, that a breakthrough in the phase measurement is highly desirable. Promising results are expected in

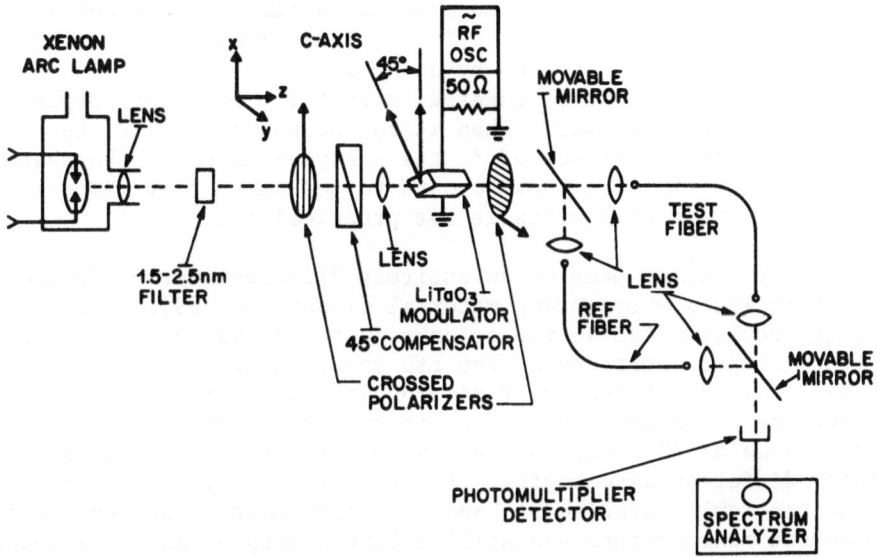

Fig. 8.10. A set-up for broadband measurements of the fiber transfer function (from [88]).

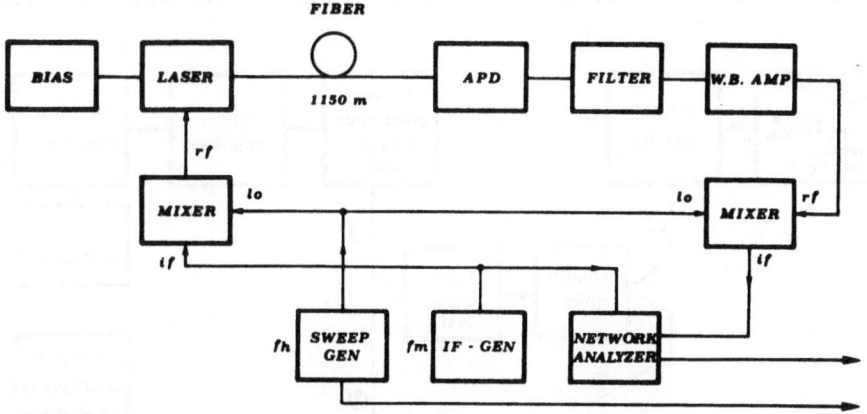

Fig. 8.11. A set-up for differential phase measurements in the frequency domain (from [88]).

the very near future [88] from the following technique (see Fig.
8.11). Two sinusoidal signals (from a sweep generator, f_h, and from
a tunable generator, f_m) beat in a balanced mixer, whose outputs
($f_h \pm f_m$) modulate an LED source. The optical signal travels through
a tested fiber and is detected by an APD. The APD output is sent
to a second balanced mixer, together with f_h. The mixer output is
sent to a network analyzer. Then it can be shown that in the neigh-
bourhood of discrete values of f_h the analyzer measures the phase
as a function of f_m. We look forward to seeing whether this appealing principle is really suitable for practical use.

Finally, let us mention an entirely different use of frequency-
domain techniques - measuring material dispersion [89]. A sinusoidally modulated LED source is coupled to the tested fiber; at
the fiber output, portions of the LED spectrum are filtered, and
the phase shifts between their ac modulations are measured. This
technique seems to be very simple, if compared with early work in
the time domain [90] based on pulses from two lasers operating at
slightly different wavelengths. But even if this is the conclusion
for the particular problem at hand, it seems fair to say that today
time-domain measurements are still a little simpler and more popular
than the frequency-domain ones.

9. CONCLUSIONS

This tutorial presentation is intended to be an introduction
to the subject of fiber evaluation rather than a review paper for
specialists. This relaxes the requirement of completeness with the
unavoidable consequence of covering the author's personal interests.

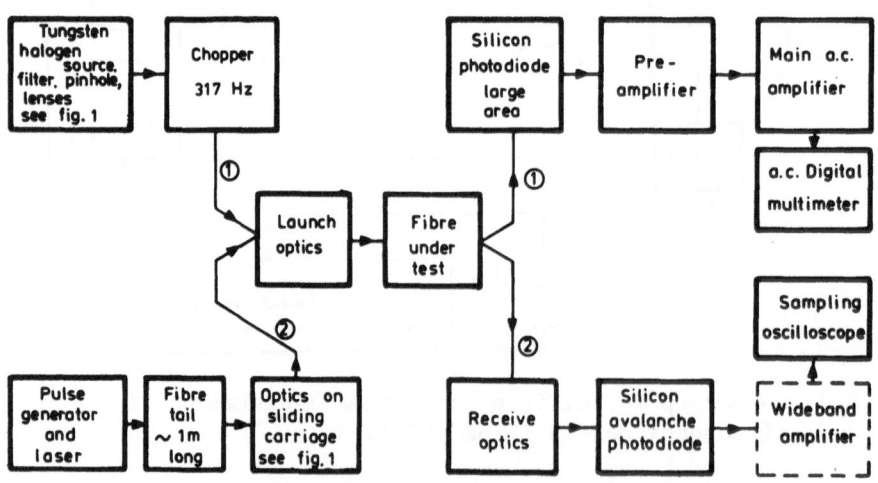

Fig. 9.1. Schematic diagram of a multi-purpose set-up for field
tests on optical cables (from [92]).

Because of this an important area had to be left out of the paper,
namely, that of stimulating theoretical ideas which can draw the
attention of experimentalists. As an example of this, let us
mention a new proposed test [91] of the attenuation constants and
of the coupling coefficients of individual guided modes. Another
area in which our attention was marginal is that of industrialized
set-ups. To make up in part for this omission, let us show at
least an example [92] of a rugged, versatile, highly automatic
set-up that permits one to perform several tests on installed fiber
cables in the field (see Fig. 9.1).

    It is hoped, however, that these notes will help the newcomer
to get started without a "random walk" through this field. No
paper, however, will ever be an acceptable replacement for direct
laboratory experience.

    A very helpful critical review of the manuscript by P. Bassi,
E. Bianciardi, and G. Cancellieri is gratefully acknowledged.

REFERENCES

1. J. Arnaud, this volume.
2. W. Eickhoff, O. Krumpholz, Electron. Lett., 12, 405 (1976).
3. D. Gloge, E.A.J. Marcatili, B.S.T.J., 52, 1563 (1973).
4. D.N. Payne et al., Proc. 1st ECOC (IEE Publ. 132), 43, London
    (1975).
5. B. Costa, B. Sordo, Proc. 2nd ECOC, 81, Paris (1976).
6. F.M.E. Sladen et al., Appl. Phys. Lett., 28, 255 (1976).
7. D.N. Payne et al., Proc. 3rd ECOC, 25, Munich (1977).
8. K. Behm, ibidem, 28.
9. M.J. Adams et al., Opt. Comm., 17, 2 (1976); Electron. Lett.
    12, 281 (1976).
10. K. Petermann, AEO, 31, 201 (1977).
11. E. Bianciardi et al., Electron. Lett., 13, 25 (1977);
    E. Bianciardi, V. Rizzoli, Opt. Quant. Electr., 9, 121 (1977).
12. B. Costa, private communication.
13. A. de Panafieu, this volume.
14. W. Eickhoff, E. Weidel, Opt. Quant. Electr., 7, 109 (1975).
15. M. Ikeda et al., Appl. Opt., 14, 814 (1975).
16. J. Stone, H.E. Earl, Opt. Quant. Electr., 8, 459 (1976).
17. G.T. Summer, Opt. Quant. Electr., 9, 79 (1977).
18. W.J. Stewart, Proc. IOOC Conference, p. C2.2, 395, Tokyo
    (1977); K.I. White, to be presented at the 4th ECOC, Geneva
    (1978).
19. M. Born, E. Wolf, "Principles of Optics", Pergamon, Oxford
    (1964), pp. 311-312.
20. C.A. Barrus, R.D. Standley, Appl. Optics, 13, 2365 (1974).
21. H.M. Presby et al., Rev. Sci. Instrum., 47, 348 (1976).
22. K. Petermann, C.G. Someda, NTZ, 31, 119 (1978).

23. H.M. Presby, I.P. Kaminow, Appl. Opt., 15, 3029 (1976).
24. L.A. Jackson et al., Tech. Digest Top. Meeting, TuC1,
     Williamsburg (1975).
25. W.J. Stewart, ibid., TuD8.
26. W.J. Stewart, Proc. 1st ECOC (IEE Publ. 132), 21, London
     (1975).
27. F.M.E. Sladen et al., Electron. Lett., 13, 212 (1977).
28. F.M.E. Sladen et al., to be presented at the 4th ECOC, Geneva
     (1978).
29. B. Daino et al., to be published.
30. H.M. Presby, J.O.S.A., 64, 280 (1974).
31. H.M. Presby, Appl. Opt., 15, 492 (1976).
32. P.L. Chu, Electron. Lett., 12, 155, (1976).
33. T. Okoshi, K. Hotate, Opt. Quant. Electron., 8, 78 (1976);
     Appl. Opt., 15, 2756 (1976).
34. E. Brinkmeyer, Appl. Opt., 16, 2802 (1977).
35. D.A. Pinnow, T. C. Rich, Appl. Phys. Lett., 20, 264 (1972).
36. D.A. Pinnow, T.C. Rich, Appl. Opt., 14, 1258 (1975).
37. K.I. White, J.E. Midwinter, Optoelectron., 5, 323 (1973).
38. See ref. 19, pp. 652 ff.
39. I. Broquet, Ann. Télécomm., 29, 195 (1974).
40. A.R. Tynes, Appl. Opt., 9, 2706 (1970)
41. J.P. Dakin, Opt. Comm., 12, 83 (1974).
42. G. Cancellieri, M. Zoboli, unpublished material.
43. D. Gloge, B.S.T.J., 54, 245 (1975).
44. W.B. Gardner, ibid., 457.
45. M. Papa, C.G. Someda, to be presented at the 4th ECOC, Geneva
     (1978).
46. D. Gloge, B.S.T.J., 51, 1767 (1972).
47. D. Marcuse, "Theory of Dielectric Optical Waveguides,"
     Academic, New York (1974).
48. S. de Vito, Proc. XXIII Congr. Intern., 395, Rome (1976).
49. D.B. Keck, A.R. Tynes, Appl. Opt., 11, 1502 (1972).
50. M. Eve et al., Proc. 2nd ECOC, 143, Paris (1976).
51. E.A.J. Marcatili, Tech. Digest Top. Meeting, TuE1,
     Williamsburg (1977).
52. R. Olshansky et al., ibid., TuE5.
53. M.K. Barnoski, S.M. Jensen, Appl. Opt., 15, 2112 (1976);
     M.K. Barnoski et al., Proc. 2nd ECOC, 75, Paris (1976).
54. S.D. Personick, B.S.T.J., 56, 355 (1977).
55. B. Daino, D. Sette, Proc. EUROCON 77, Venice (1977).
56. B. Costa, B. Sordo, Proc. 3rd ECOC, 69, Munich (1977).
57. O. Krumpholz, ibid., 38.
58. M.K. Barnoski et al., to be presented at the 4th ECOC, Geneva
     (1978).
59. A.R. Tynes et al., J.O.S.A., 61, 143 (1971).
60. S. de Vito, B. Sordo, LXXV Riun. AEI, B.5, Rome (1974).
61. D. Gloge et al., Electron. Lett., 8, 526 (1972).
62. G. Cancellieri, M. Zoboli, Alta Freq., 47, 424 (1978).
63. E. Bianciardi et al., to be presented at the 4th ECOC, Geneva
     (1978).

64. Spectra-Physics Review, 4 (April, 1977).
65. B. Hillerich, Electron. Lett., 12, 92 (1976).
66. S.D. Personick, B.S.T.J., 52, 1175 (1973).
67. C. Vassallo, IEEE Trans., MTT-25, 572 (1977).
68. I.W. Sandberg et al., B.S.T.J., 57, 99 (1978).
69. E.A. Guillemin, "Theory of Linear Physical Systems," Wiley, New York (1963).
70. D. Gloge et al., IEEE J.Q.El., QE-8, 217 (1972).
71. E.L. Chinnock et al., Proc. IEEE, 61, 1499 (1973).
72. L.G. Cohen, Tech. Digest Top. Meeting, TuD1, Williamsburg (1975); Appl. Opt., 14, 1351 (1975).
73. T. Tanifuji, M. Ikeda, Appl. Opt., 16, 2175 (1977).
74. J.R. Andrews, Rev. Sci. Instr., 45, 22 (1974).
75. P. Bassi et al., to be published; P. Jeppesen et al., unpublished material of Tech. Univ. Denmark.
76. M. Eve et al., Proc. 3rd ECOC, 53, Munich (1977).
77. M. Eve, Electron. Lett., 13, 315 (1977).
78. J.V. Wright, B.P. Nelson, ibid., 361.
79. L.G. Cohen, Appl. Opt., 15, 1808 (1977).
80. R. Olshansky, S.M. Oaks, to be presented at the 4th ECOC, Geneva (1978).
81. L.G. Cohen, C. Lin, Appl. Opt., 16, 3136 (1977).
82. D. Gloge et al., Appl. Opt., 11, 1534 (1972).
83. K.H. Steiner et al., Appl. Opt., 12, 2732 (1973).
84. S.D. Personick et al., Appl. Opt., 13, 266 (1974).
85. L. Jeunhomme et al., Prov. 2nd ECOC, 123, Paris (1976).
86. I.P. Kaminow, "An Introduction to Electrooptic Devices," Academic, New York (1974).
87. L.G. Cohen et al., B.S.T.J., 55, 1509 (1976); Appl. Phys. Lett., 30, 17 (1977); Tech. Digest Top. Meeting, TuE3, Williamsburg (1977).
88. E. Nicolaisen, J.J. Ramskov Hansen, to be presented at the 4th ECOC, Geneva (1978).
89. T. Ozeki, A. Watanabe, Appl. Phys. Lett., 28, 382 (1976).
90. D. Gloge et al., Appl. Opt., 13, 261 (1973).
91. S. Kawakami, Electron. Lett., 13, 706 (1977).
92. K.C. Byron et al., Proc. 3rd ECOC, 34, Munich (1977).

CABLE FABRICATION

M. Carratt

Compagnie Lyonnaise de Transmissions Optiques
170, quai de Clichy
92111 Clichy, France

A system of transmission by optical fibers is comprised mainly of three parts:

- the emission system that ensures the conversion of electrical signals into modulated light signals;

- the transmission support formed by the cable and the connection systems;

- the reception system that ensures conversion into electrical signals.

Thus, the functions that have to be considered as characteristic for optical fiber cables are transmission, which should create the least possible disturbance, and reliability with respect to aging and mechanical strength.

Requirements for Optical Fiber Cable

In manufacturing the cable, the main requirement is to maintain the advantages of the optical fiber and avoid the disadvantages, which are mostly mechanical. The specification should take into account the elements composing the cable to ensure that the manufacturing process is possible. The specification varies according to the desired purpose and the conditions of use. The parameters to be observed are the following:

• choice of the type of fibers, determined by the loss of the transfer function and the coupling efficiency;

• the mechanical characteristics and protection with respect to the environment.

Transmission Properties

For long distance systems like those used for P.T.T. services, the distances between repeating stations must be as long as possible, from 5 to 10 kilometers and more. The attenuation will be the fundamental criterion and will determine the choice of the fiber - doped silica fiber. As attenuation is very sensitive to pressure phenomena and microbends, it will be necessary to provide for high performance conditioning in order to obtain cables with attenuation near 6-8 decibels per kilometer at a wavelength of 0.8 μm. For short distance systems, the length does not exceed 500 meters and the attenuation could reach several tens of decibels. The fiber will be a glass or plastic type.

The transfer function of the fiber is determined by the maximum capacity with respect to the cable length. Step-index fibers with bandwidths of approximately 50 MHz will be used for short distance systems. More important links will use graded-index fibers whose bandwidths can reach 1 GHz x km and higher. It should be noted that cabling processes increase the mode coupling, so the bandwidth can only increase during the manufacturing process.

The fiber diameter and the refractive index of the materials determine the coupling efficiency. Losses are about 20 dB for the coupling of a L.E.D. and a fiber of the same dimensions. They are greatest for short distance systems and require the use of glass fibers with numerical apertures between 0.3 and 0.5. Long distance systems, using material dispersion to connect the bandwidth with the numerical aperture, require an aperture of approximately 0.2 (half angle of about 10 degrees).

Finally, the ratio of the coating diameter to the core diameter has an influence on the microbending loss. In fact, the coating protects the core against mechanical stress from outside which can modify the geometry of the core and the conditions of propagation.

Mechanical Properties

We will now examine the mechanical properties of bare and precoated fiber. Resistance to tension and flexion is the main consideration in the manufacture of fibers. Thermal problems and aging must also be considered.

Glass and silica fibers have an extremely high resistance to tension. Unfortunately, the fibers obtained immediately after drawing have a large number of microcracks at the circumference. The fiber breaks under stress by the propagation of these microcracks through the cross section. This results in important variations in the tensile strength. A fiber of 125 μm diameter and average tensile strength of 0.7 daN will break with an extension of 0.6-0.8%. This disadvantage is prevented by application of a precoating carried out in situ with the drawing process. This precoating makes a mechanical bridge for the microcracks. Thus, we obtain tensile strengths of about 2-10 daN, depending on the thickness of the precoating applied. The deposition of these materials must be very regular and polymerization extremely fast. Furthermore, in normal atmospheric conditions, fibers are very sensitive to dust and humidity which can damage the silica. Precoating can also prevent this effect.

Figure 1 presents the cumulated histograms of tensile strength for bare and precoated fibers. Aging tests have been carried out for the fiber and the polymer deposit in a stove at high humidity (80%). The evolution of the resistance to tension is very encouraging.

Changes in the attenuation of bare fibers are negligible in the temperature range of -60°C to +90°C. However, differential dilatations result in considerable variations after deposition of the first coating. The thickness of the application and the nature of the deposit have an important influence. A silica fiber with KYNAR R of 4 μm thickness has a variation of less than 0.01 dB/°C/km. A silica-plastic fiber varies by a factor of 0.07 dB/°C/km.

Multifiber Bundle

In short distance systems, especially for data processing and aircraft, the reliability of the system is extremely important. Signals are transmitted on several fibers placed in parallel in order to increase the redundancy of the system (1). It is well known that abrasion between uncoated fibers dangerously limits the lifetime of cables. Therefore, each fiber is protected individually by plastic coating. The manufacturing of optical multifiber bundles is similar to that of optical cables for telecommunications discussed in the following paragraph. Seven fibers are generally used, which allows hexagonal piling up. This arrangement allows considerable reduction of losses with an L.E.D. source (the emitting diameter is currently 200 μm) and simplifies the connecting process.

Telecommunication Cables

The most generally used technique in manufacturing optical cables is the individual protection of each fiber. The main purpose

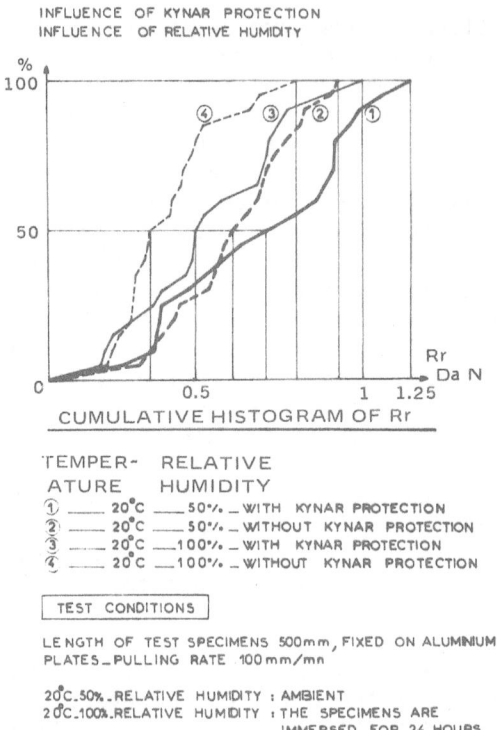

INFLUENCE OF KYNAR PROTECTION
INFLUENCE OF RELATIVE HUMIDITY

CUMULATIVE HISTOGRAM OF Rr

TEMPER-   RELATIVE
ATURE   HUMIDITY

① ____ 20°C ____50%/. _WITH KYNAR PROTECTION
② ____ 20°C ____50%/. _WITHOUT KYNAR PROTECTION
③ ____ 20°C ____100%/. _WITH KYNAR PROTECTION
④ ____ 20°C ____100%/. _WITHOUT KYNAR PROTECTION

TEST CONDITIONS

LENGTH OF TEST SPECIMENS 500mm, FIXED ON ALUMINIUM
PLATES_PULLING RATE 100mm/mn

20°C.50%.RELATIVE HUMIDITY : AMBIENT
20°C.100%.RELATIVE HUMIDITY :THE SPECIMENS ARE
                             IMMERSED FOR 24 HOURS

Fig. 1.   Tensile strength on bare fiber and precoated fiber.

of this procedure is to increase the tensile strength in order to
obtain compatibility with the cabling techniques.

Coating Fiber

Tensile strength is improved by repartition of the stress between the fiber and the plastic coating. The thickness of the
coating is determined by the thermal dilatation factor of the material
used. The latter must also have a high Young modulus.

The materials generally used for the coating are polyesters,
nylons, and polypropylenes. They are applied by an extrusion process
along the traditional insulation line of the cable. The coating can
be either tight or loose on the fiber (2).

The manufacturing described here consists of a polyamide 6-10
coating in tight deposit on the fiber. The inner and outer diameters
of the coating are 0.125-0.85 mm. Fig. 2 shows a cumulated histogram
of the tensile strength of fibers after coating. The attenuation

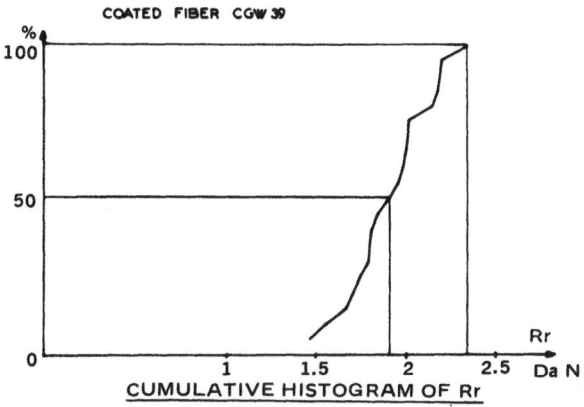

Fig. 2.   Tensile strength on coated fiber.

variations of this structure related to temperature are about 0.01 dB/°C/km.

Monofiber Cable

An optical fiber coated with thermoplastic material can easily be used as a monofiber cable. The extrusion of a loose thermoplastic tube on the coated fiber makes it independent of radial pressures. It is easy to include one or more strength members in order to obtain the correct tensile strength. These tensile members should have a high Young modulus. They are generally used as bundle types to retain sufficient flexibility in the cable. The main types of materials used are steel wires, textile fibers, and glass fibers.

In the example shown in Fig. 3, the monofiber cable is made with nylon coated fiber of 0.85 mm diameter. It is protected against transverse stress by a polyethylene tube of 2.5 mm diameter. Three strength members of glass fiber are added and kept in place with a slight overcoating. The cable has a 3.5 mm diameter for a breaking strength of 25 daN.

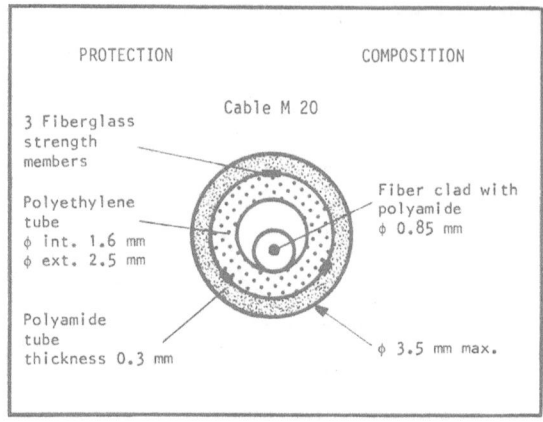

Fig. 3.   Monofiber cable (C.L.T.O.).

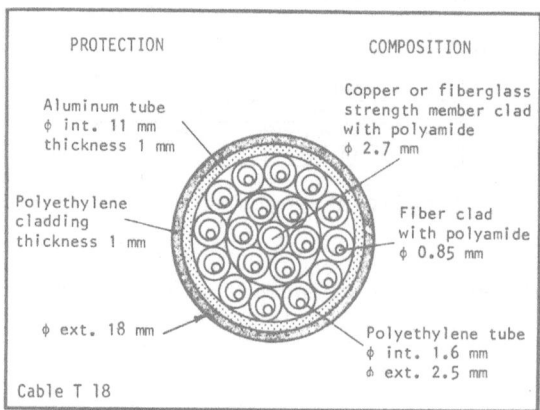

Fig. 4.   Cross section of an 18 fiber optic cable (C.L.T.O.).

Multifiber Cable

If a fiber is individually protected against tensile stress and a second coating is provided as protection against radial pressure, it is possible to assemble these elements to obtain an optical cable (3). Different examples are shown in Figs. 4-7. The individual protection of the fibers in the cable is very effective. It is formed by two sheaths, one sheath and one cushion, or one sheath and one spacer. The attenuation performances are generally very interesting. For the example in Fig. 4, the additional average loss is less than 1 dB compared to the initial values of the fibers. The thermal

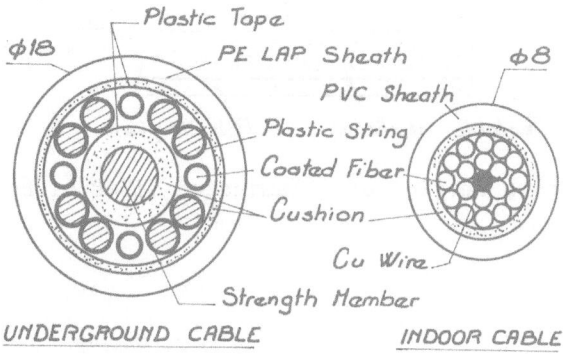

Fig. 5. Cable construction (Sumitomo).

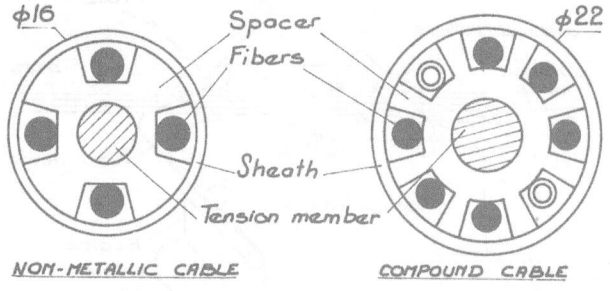

Fig. 6. (Hitachi).

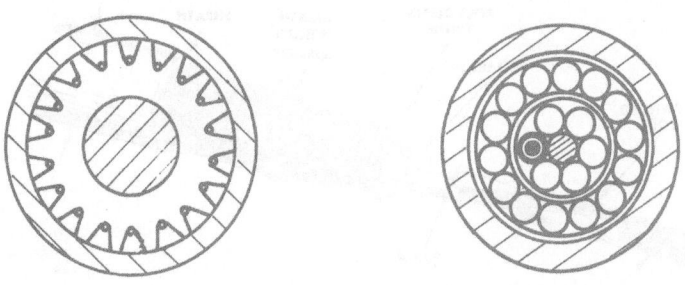

Fig. 7. Cable structure 19 x 18 fibers, ϕ 47 mm (C.N.E.T.).

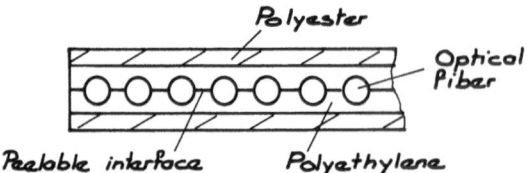

Fig. 8. Cross section of a laminated fiber ribbon (Bell).

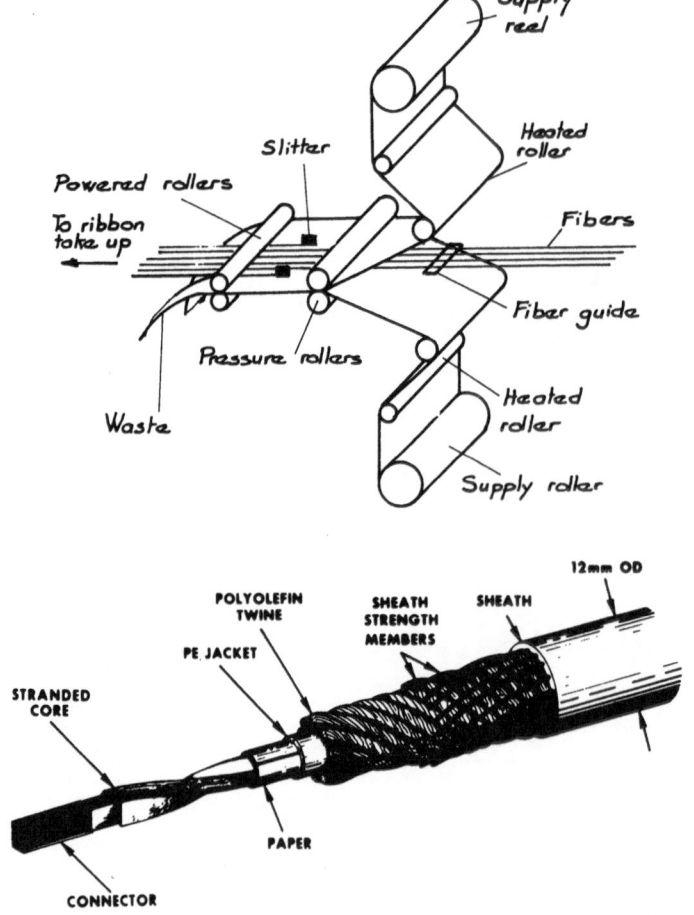

Fig. 9. Fabrication and structure of the ribbon cable (Bell).

problems are generally well resolved, which ensures stability of performance after the cable is layed. The external coating can be made of thermoplastic coating or metal coating, so there is a wide range of applications for this type of cable.

The use of optical fibers as telephone cables in urban systems shows the problem of dimensions. In order to take advantage of the small dimensions of optical fibers, cables have been developed with improved fiber density. The structure shown in Fig. 7 is made of a cylinder-shaped grooved core that ensures the protection of uncoated fibers (4). They are placed, without tension, in the helicoidal grooves (0.4 to 1 mm deep) of a cylinder shaped core. The helicoidal shape is obtained by extrusion. Correct geometrical regularity is important and is obtained by means of a central cylindrical strength member. The dimensions, the helical thread and the surface structure of the grooves are designed to maintain the transmission qualities of the fiber. The result is a cable element with 18 fibers in a diameter of 7 millimeters.

Ribbon Cable

Cables of very high density can be obtained when a ribbon of optical fibers is used (Figs. 8 and 9). The first structure of this type has been developed by Bell Laboratories (5). The fibers, protected by their primary coating, are unrolled in parallel between two plastic sheets that adhere to the fibers due to action of heat. The ribbons are assembled by piling up in order to obtain a rectangular array of 12 x 12. The matrix is protected by thermoplastic reinforced coating. The cable can contain 144 fibers in a diameter of 12 mm.

The advantage of these structures is their very reduced dimensions. The positioning of the fibers inside the ribbon permits the use of a global solution for connection. Furthermore, cost of manufacturing for this type of cable is reduced because of the protection of several fibers in only one process.

References

(1) T.A. Hawkes and J.C. Reynaud, "Optical communication systems for aircraft," 2nd European Conference on Optical Fiber Communication, September, 1976, pp. 399-409.
(2) H. Murata, "Broadband optical fiber cable and connecting," 2nd European Conference on Optical Fiber Communication, September, 1976, pp. 167-175.
(3) R. Jocteur, "Optical fiber cable for digital transmission systems," 2nd European Conference on Optical Fiber Communication, September, 1976, pp. 193-201.

(4) G. le Noane, "Optical fiber cable and splicing techniques,"
     2nd European Conference on Optical Fiber Communication, Septem-
     ber, 1976, pp. 247-253.
(5) M.I. Schwarz, R.A. Kempf, and W.B. Gardner, "Design and charac-
     terization of an exploratory fiber optic cable," 2nd European
     Conference on Optical Fiber Communication, September, 1976,
     pp. 311-315.

FIBER OPTIC COMMUNICATIONS SYSTEMS:

SUBJECTIVE NATURE OF HUMAN COMMUNICATIONS

A. Cozannet (CNET/Lannion) and
P. Cochrane (UKPORC/Ipswich)

Centre National d'Etudes des Télécommunications
CPM/PMT, BP 40, 22300 Lannion, France

INTRODUCTION

Electronic communication between machines is subject to both mathematically defineable and electrically measurable parameters. Shannon provided a complete theory in his classic paper of 1946, where he stated that

$$I = BT[1+k(S/N)]$$

where:

    I = the information
    B = the bandwidth
    T = the transmission time
    k = a system constant
  S/N = the signal-to-noise ratio

It is thus possible to define the necessary bandwidth, transmission time and signal-to-noise ratio for a given information rate.

No such analytic relationship exists (or seems likely to exist) for electronic (or other) communications between humans. We rely entirely on subjective assessments to define the required channel parameters for human communications, then in most cases, Shannon's relationship may be applied directly.

For $S/N \gg 1$, we can write

$$\overline{I} = kBT\frac{S}{N}\Big|_{dB} \text{ bits}$$

In real time human communications, T is large and fixed, so we can change only S/N and the bandwidth: for analogue channels, S/N≈40 dB and for digital, S/N ≃ 10 dB. Between machines, trade-off between all three parameters is usually employed.

WHY DIGITAL COMMUNICATIONS?

Digital communications have several advantages over an analogue operation:

- transmission standards are more easily maintained;
- they have greater utility and flexibility for a given size communication network;
- they are compatible with computer controlled switching systems;
- hardware realizations are more economic;
- there is a higher degree of standardization;
- translation for operation over cables, radio, optical systems, etc., is easier;
- noise limitations are less severe.

However, there are also disadvantages:

- they require a far greater bandwidth for a given analogue signal;
- the circuitry required is far more complex (this objection has been somewhat negated by the introduction of MSI or LSI).

HOW TO GO FROM ANALOGUE TO DIGITAL SIGNAL

We will use the phone signal as an example.

The subjective spectrum required for one channel is 300-3400 Hz. In order to simplify the description, we shall assume a bandwidth of 4 kHz. The most common process for generating the digital signal requires 3 operations.

Sampling. We take a sample of the analogue signal at a clock rate whose inverse is at least two times the required bandwidth (Shannon's rule), i.e., 4 x 2 = 8 kHz. So we must take 8000 samples each second, or one sample each 125 μs.

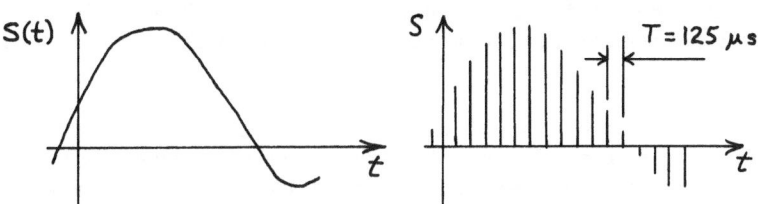

Quantification. The value of the sample is then measured and,
to simplify the transmission, taken equal to the nearest standard-
value among a set of discrete values. For example, the set is the
entire range of values 0, 1, 2, 3, 4 ... The measured value of
5.3 volts is then rounded off to 5.

Coding. The etalon value of the sample is written in a binary
system. In the chosen example, 5 is written 00000101. With 8 bits,
we can separate 2^8 = 256 different values, 128 positive and 128
negative. To transmit the information, we must then transmit 8 bits
per sample, or 8 x 8000 = 64 kbit/s. Suppose we use a faster clock
rate. In the remaining time before the next sample, we can transmit
sequences of 8 bits from other channels (we are multiplexing the
signal in time). 32 channels require 32 x 64 = 2048 kbit/s; these
new signals can then be multiplexed: 4 times, we obtain 8 Mbit/s
transmitting 128 channels, 34 Mbit/s or 512 channels, 140 Mbit/s or
more than two thousand channels.

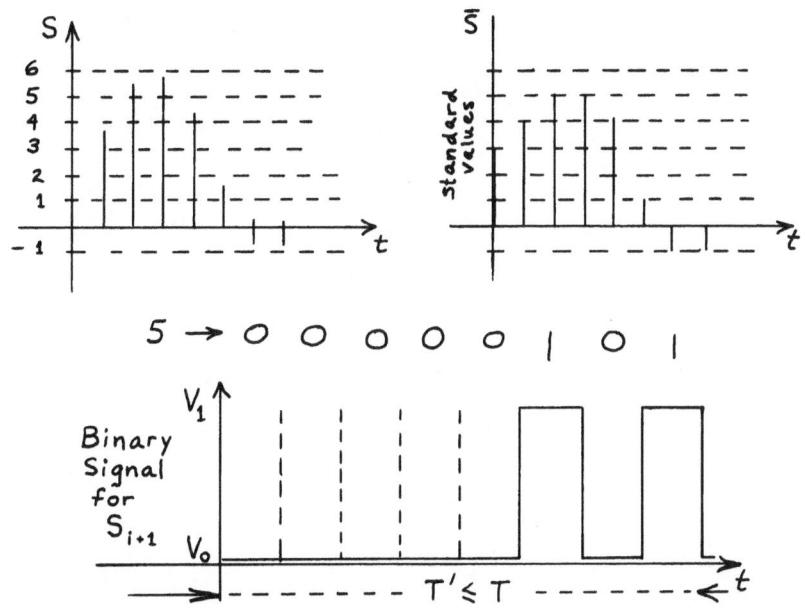

At the other end of the fiber, one must be able to reconstruct the clock and the beginning of each sequence of 8 bits in order to recover the information: we assume that each pulse is properly detected.

The next lecture will be devoted essentially to pulse coded information and transmission. The messages will then be continuous streams of two-level pulses (light and dark) and any stream will be a long series in time of random sequences with a small amount of redundancy.

FIBER OPTIC COMMUNICATION SYSTEMS

C. Boisrobert

Centre National d'Etudes des Télécommunications
CPM/PMT, BP 40
22300 Lannion, France

I. INTRODUCTION

I.1. General Considerations on Fiber Optic Links

Industrial manufacturing of low loss optical fiber cables and their specific splicing toolings, together with recent developments of optoelectronic components will lead to real systems installations in the 1980's. The requirements for these new systems are related to some of the fiber characteristics:

- protection against electromagnetic perturbations
- radiation leakage
- small size
- low weight
- electrical isolation
- high bandwidth
- low attenuation
- low cost expectation

Understanding fiber optic transmission systems requires a knowledge of the characteristics of optoelectronic transmitters and receivers, which do not differ much from traditional copper pairs (choice of analog to digital modulation, total attenuation, etc.). Therefore, throughout this course we will limit our effort to descriptions of specific elements, essentially those whose parameters are directly connected to system performance.

The next section is devoted to digital communications: pulse regeneration and spectrum width, transmission code, and basic func-

97

tions. The receiver and transmitter will be described and investigated later, just before some specific applications.

I.2. Digital Transmission

        Before using code and modulation theories, one usually begins
with pure binary signals. Other more sophisticated pulse shapes
and spectra become useful when binary signal transmission weakens for
functions like clock and sampling, error detection, and alarm trig-
gering.

        I.2.1. Baseband spectrum (Figs. 1, 2, and 3). The fundamental
condition for baseband spectra is that message and channel bands
must be compatible. Other considerations are: 1) low bandwidth leads
to receiver bandwidth and noise reduction and 2) "0-low frequency
components" means "no DC level restitution" (clamping technique)
and no gain automatic control (trigger level change on word).

        We will now turn to clock timing and decision circuits. Before
we get to the definitions of these, let us look at what every transmission expert calls "eye-diagram" and what we mean when we say
"the message and channel bands must be compatible."

        Consider a simple, pure, binary NRZ message without any tricky
sequence. If the pulses are assumed to be square, their frequency
spectrum contains lobes in the high frequency region, but most of
their energy is contained in the high first lobe (T = pulse duration;
1/T = frequency limit of the first lobe).

        The transmission channel generally acts as a linear low pass
filter and cuts down part of the frequency spectrum of the pulses.
If the channel characteristics have been well established and properly measured, this filtering can be accounted for and compensated
for by means of an "equalizer." The "channel + equalizer" transfer
function modulus can then be represented by an almost rectangular
function VS frequency along a logarithmic "decibel" scale. In our
simple binary case the frequency drop of this total transfer function modulus must theoretically be around 0.7 times clock frequency.
The output pulse is given by

$$f(t) = \frac{\sin\left(\frac{\pi t}{T}\right) \cdot \cos\left(\frac{\pi \beta t}{T}\right)}{\frac{\pi t}{T}\left(1 - \left(\frac{2\beta t}{T}\right)^2\right)}$$

This is a raised cosine function. The "eye diagram" is the superposition of such functions corresponding to all the possible configurations of a binary pulse series synchronized on the same clock.
It can be observed on an oscilloscope at the output of the equalizer.
The wider the eye is opened, the easier a decision will be made (Fig. 3

CODE	SIGNAL SHAPE	90% ENERGY BAND
N R Z	 1 0	$\dfrac{.86}{T}$
R Z	 1 0	$\dfrac{1.72}{T}$
BI ∅ . L	 1 0	$\dfrac{2.96}{T}$
EP 1	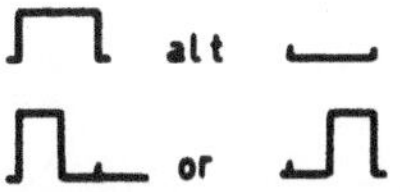 alt 1 / or 0	$\dfrac{2.96}{T}$
BI ∅ S	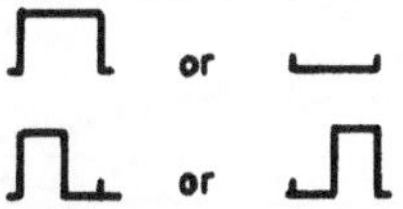 or 1 / or or 0	$\dfrac{1.7}{T}$
C M I	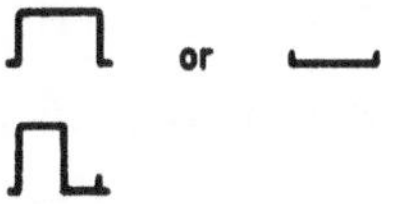 or 1 / 0	$\dfrac{1.52}{T}$
E P 2	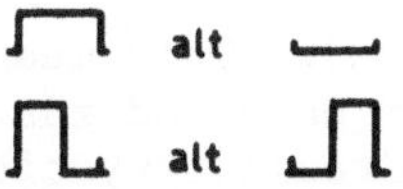 alt 1 / alt 0	$\dfrac{1.70}{T}$

Fig. 1. Definition of some usual codes.

This eye shaped curve allows for easy observation of transmission quality, excess noise, or total bandwidth.

This particular example makes it easy to understand the equalization purpose, but it must not be forgotten that the pulse amplitude is only one of two items of information – we have not yet mentioned the phase and great care should (or might) be taken on jitter and sampling time to signal clock synchronization. In some specific cases a phase equalizer (or group delay distortion compensator) might be required.

I.2.2. Clock timing. Decisions are taken on the tops (or the valleys) of the equalized "eye diagram" raised cosine pulses. The sampling pulse is derived from a clock signal which must be locked

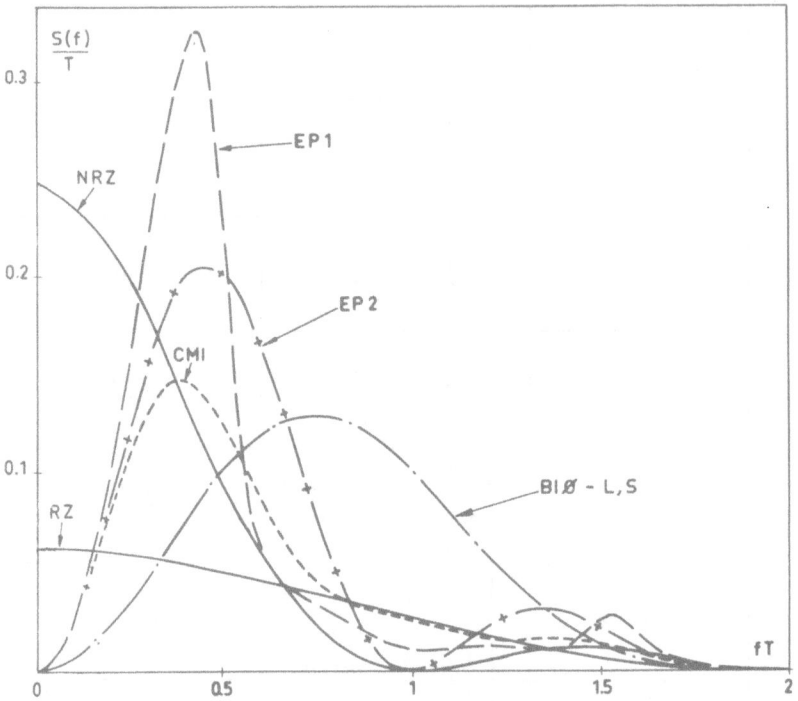

Fig. 2.   Spectral densities of different codes.

to the preceding slice of word. This means that the number of signal
transitions must be high for easy and fast synchronization. Clock
timing and retriggering a circuit very often result in a high degree
of complexity (and also cost) in repeater design and freezes the bit
rate. If jitter and signal-to-noise ratio are not drastic, a
straightforward comparison of the amplified signal and a threshold
may be used in most cases. It has been observed that under identical
conditions the repeater spacings are related as follows:

$$L_{3R3R} = 1.05 L_{3R2R} = 1.25 L_{2R2R}$$

in which 2R means Reshaping Regenerating and 3R means Reshaping
Regenerating Retiming.

     I.2.3. <u>Binary message configurations</u>. The code must not exclude
any binary sequence.

     I.2.4. <u>Error detection</u> (Fig. 4). The transmission quality must
be evaluated permanently when the whole system is under operation.
This implies redundancy and specific sequences in the code scheme -
any violation of parity rules or specific sequences will be thrown

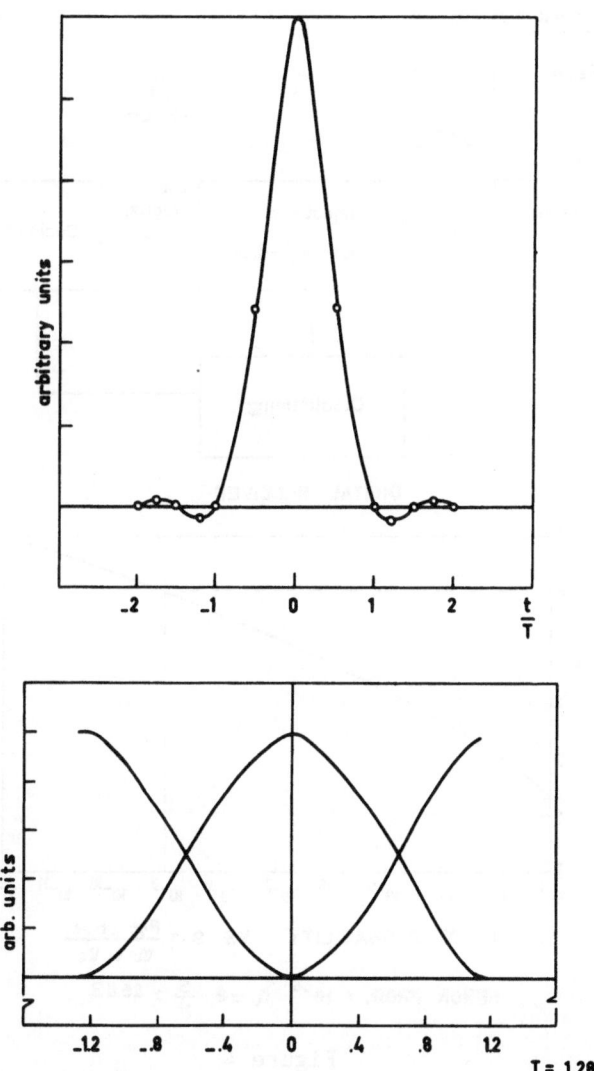

Fig. 3. Ideal pulse shape and eye diagram (from Personick).

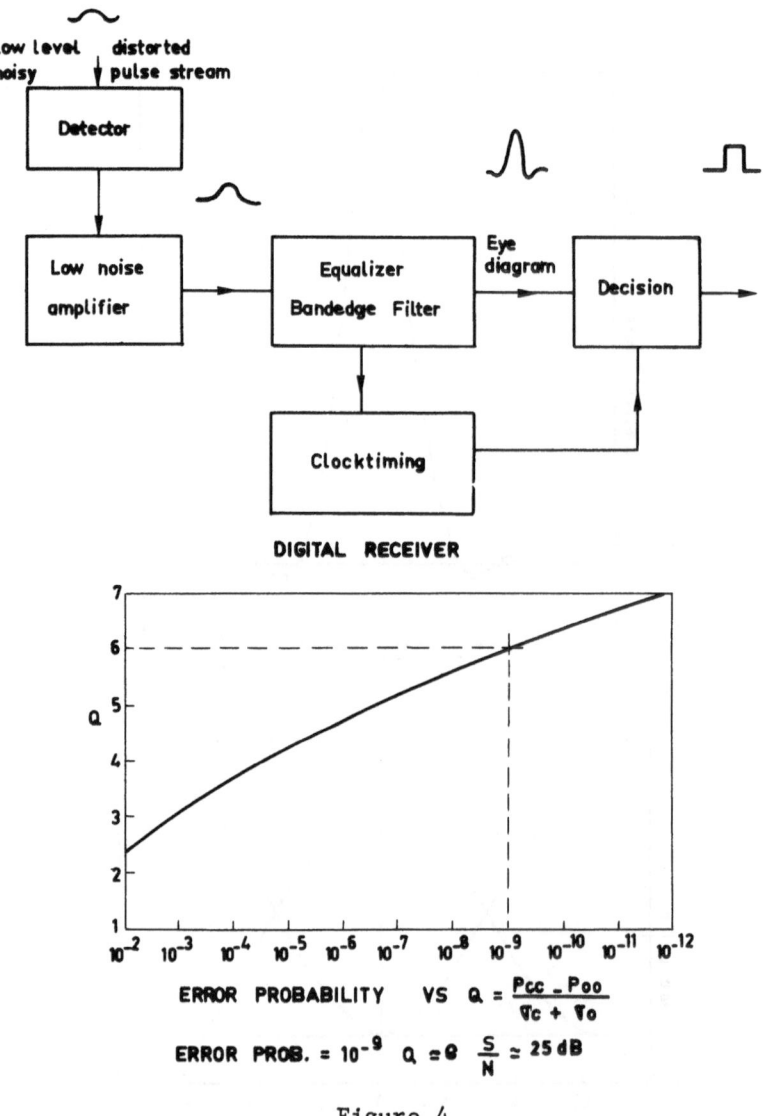

Figure 4

into a counter. For any code of this kind one can derive the theoretical relationship between the error rate measured from the counter and the effective error rate per symbol.

If we call P_{cc} the peak power of a received equalized pulse and P_{oo} the minimum power (darkness) and if σ_c and σ_o are the noise variances for P_{cc} and P_{oo}, respectively, a 10^{-n} error rate means that the decision threshold lying between P_{cc} and P_{oo} faces erratic pulses represented by σ_c and σ_o such that:

$$(\sigma_c + \sigma_o)/(P_{cc} - P_{oo}) = 1/Q$$

$$1/2\pi \int_Q^{+\infty} \exp[-x^2/2]dx = 10^{-n}$$

Assuming that $P_{oo} = 0$ and $\sigma_c \approx \sigma_o$, (signal/noise) = 25 dB corresponds to 10^{-9}.

I.2.5. <u>Components requirements</u>. The pulse shape may also be bound to the manner in which the light source and receiver operate in conjunction with the fiber. As we will see later (and as has been mentioned in a previous course), several types of light sources are available with respect to current/light power curve, linearity spectrum width, and speed of response, leading to totally different aspects of the system (pulse width modulation, pulse position modulation, etc.).

Among many "two state" (light and dark) modulations proposed earlier in many contributions and publications, we have shown some of the simplest ones in Fig. 1. Their spectral densities are shown in Fig. 2.

Thus far we have surveyed the fundamentals of digital and PCM techniques. We will discuss analog transmissions along optical fibers later in this course. Since this aspect is very much dependent upon component linearity and fiber frequency response, let me first introduce these in the frame of pulse transmission which presently seems to be of great interest. The next section will be devoted to the fiber itself, the ways systems people are working with fiber today; the receiver design will be developed in Section III and we will discuss transmitter design in Section IV.

II. FIBER TRANSMISSION CHARACTERISTICS (Refs. 1-5)

Some of the smartest experts in silica and glass materials have already presented and discussed all the troubles they faced in manufacturing long length fibers with a given uniform structure and low total spectral attenuation. Propagation theoreticians have also told you that fiber core and cladding dimensions and index profiles are of greatest interest as far as pulse dispersion is concerned. I

intend to summarize their contributions and concentrate on two
parameters (aside from geometrical tolerances and fiber homogeneity)
required by the system experts for repeater spacing calculation and
transmitter-receiver designs.

II.1. Static Spectral Attenuation Per Unit Length

Unmodulated light is attenuated along its path through the
fiber. The attenuation factor is highly dependent on the radiation
wavelength, spectral width, source radiation pattern and mode properties, fiber structure and material. Many results have been
reported by L. Jeunhomme (CGE Marcoussis) and some of them will be
presented here. From these results one can easily understand that
it is difficult to predict the total static attenuation with high
accuracy. However, since the common tendency is to assume that no
fiber bandwidth limitation will occur before lack of sensitivity
in the receiver, we will need more accurate data on the spectral
attenuation coefficient (or how to design efficient mode coupling
at any fiber transparency wavelength.

We could spend a long time discussing the different measurement techniques developed for the unit length attenuation of long
length fibers, but we will restrict it to a brief description of
the backscattering worked out by Barnoski and Jensen at Hughes
Research Laboratories (Fig. 5).

Light is first coupled into a tapered fiber coupler from a
pulsed GaAs injection laser and the waveform of the return light
pulse is detected by a photodetector, the output of which is amplified and put into a boxcar integrator operating in the scanning
mode. This approach, which does not require the fiber to be clearly
cut or the launching conditions to remain absolutely stable, allows
determination of the steady state attenuation coefficient and also
provides an estimate of the scattering and mode mixing characteristics of the fiber. If P(o) is the light power launched into fiber
input and P(x) the power at a distance

$$P(x) = P(o)\exp\{-[\int_o^x \alpha'(x)dx]\}$$

Assuming a scattering process independent of modal characteristics
and constant at all points along the fiber, the power scattered in
the opposite direction at position x is given by

$$P_B = SP(x)$$

and will be affected by the loss coefficient $\alpha'(x)$ on its way to the
photodetector. The outcoming backscattered power is

$$P_{Bout} = P_B(x)\exp\{-\int_o^x \alpha_B(x)dx\}$$

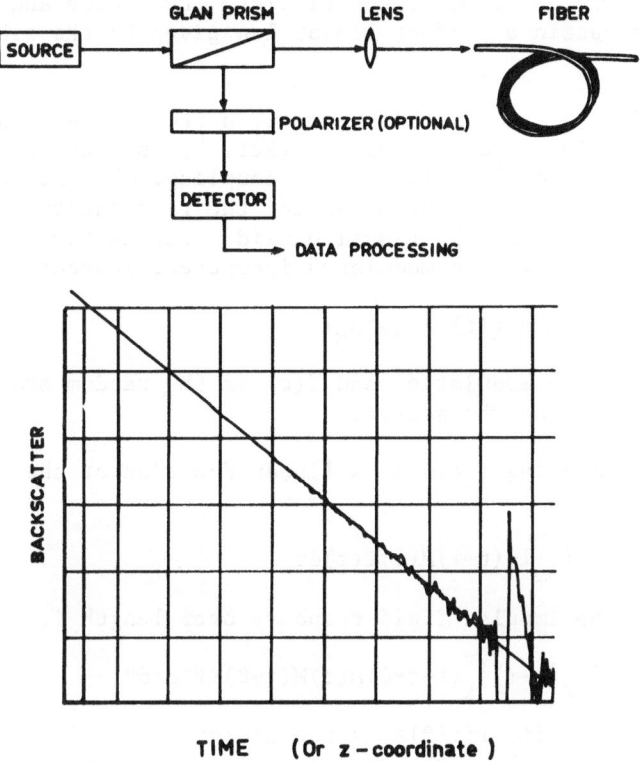

Fig. 5.   Fiber backscattering: experimental set-up and results
          (from D. Franzen, NBS, Boulder).

and finally

$$P_{Bout}(x) = SP(x) \exp\{-\int_0^x [\alpha'(x)+\alpha_B(x)]dx\}$$

$$\bar{\alpha} = -\frac{1}{2(x-x_1)} \{Log\ P_{Bout}(x) - Log\ P_{Bout}(x_1)\}$$

II.2.   Experimental Results

     The second parameter is also very much dependent on fiber
structure and material, radiation wavelength and source spectral
width, and source radiation pattern (the fiber transfer function
modulus, or Fourier transform of the output light pulse against
input light pulse). Once again the variations of this -3 dB bandwidth with length are difficult to predict with high accuracy for
multimode fibers; the curve slopes beyond the bandwidth upper limit
are also difficult to predict. The only case that fits theory well

is the association "broad spectral width LED source and graded index fiber" – we obtain a perfect linear law given by the fiber material dispersion.

In order to confirm this approach a few theoreticians worked on the fiber linear power response (Ref. 2) and derived different formulas, provided there is no mode coupling. "We may then calculate the response functions by adding the response functions of the individual modes," as Ch. Vassallo said. Let us take the input field in mode "k" from a modulated incoherent source:

$$(E,H) = M(t) \cdot f(t) \cdot (e_k^o h_k^o)$$

where $M(t)$ is the modulation and $f(t)$ is the random stationary function describing the source.

The propagating field is a linear function of the input field:

$$C_k(z,t) = \int_{-\infty}^{+\infty} I_k(t-\tau)M(\tau)f(\tau)d\tau$$

where I_k is the impulse field response over length ℓ. Then

$$<C_k^2> = \iint |_k(t-\tau)|_k(t-\tau+\theta)M(\tau)M(\tau-\theta)<f>d\tau d\theta,$$

the correlation $<f(\tau)f(\tau-\theta)>$ is the energy (optical frequency domain) spectrum of the light source $g(\nu)$ (to be consistent with Vassallo's writings).

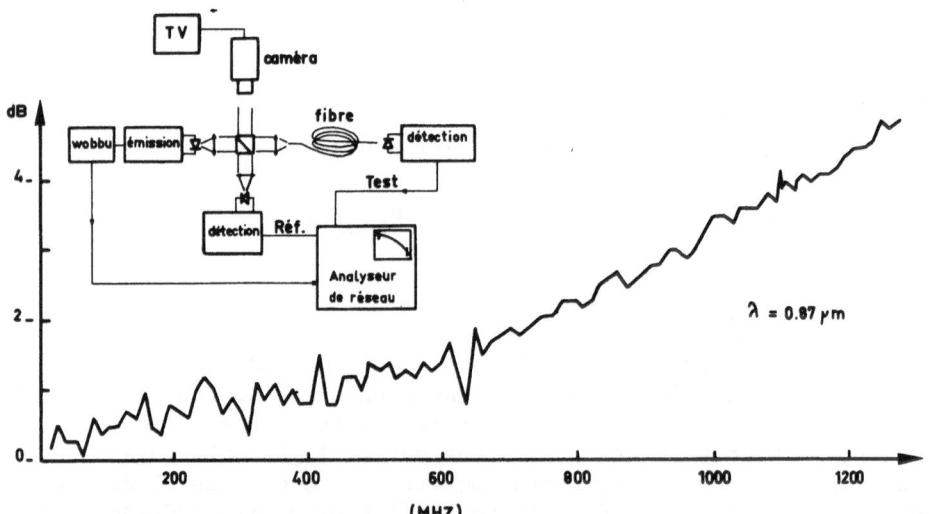

Fig. 6. Graded index (CNET, 6/21/77) bandwidth measurement.

If we set

$$\Gamma(\theta) = <f(\tau)f(\tau-\theta)>$$

the fiber pulse response for the kth mode is

$$L_k(t) = [1/\Gamma(o)]\iint_{-\infty}^{+\infty}g(\nu')\exp i[2\pi\nu t-\beta_k t+\beta_k(\nu'-\nu)td\nu d\nu']$$

and can be shortened when Z gets higher than Z_o and Z_1 as given in Ref. 3. Table 1 gives some values for silica-silica doped fibers and injection III-V compound sources.

Table 1

$\lambda = 0.85$ μm

$\beta'' = 0.5 \cdot 10^{-25}$ $m^{-1}s^2$

Δλ (Å)	300	30	3	0.3
ℓ_o(m)	$2 \cdot 10^{-3}$	$6 \cdot 10^{-2}$	2	60
ℓ_1(m)	1	10^2	10^4	10^6
Δν	10^4	10^3	10^2	10

Assuming mode coupling, this becomes quite different. I am not willing (and am even unable) to go through the whole calculation. Professor Someda is the expert. However, let me give rapidly the fundamental results for mode coupling: the domain of linearity is broader than it was assuming no mode coupling and independent travelling modes.

All this means we have the right to consider our new transmission medium as a regular linear low filter characterized by its transfer function in the frequency domain up to at least 1 GHz, whatever our light source is. We must be suspicious and careful about equilibrium length mode mixing and light source characteristics (Figs. 6 and 7).

If we go back to Section I.2 on digital transmission, our first definition was devoted to the baseband spectrum. From these considerations we know how to select the fiber whose structure leads to frequency characteristics most appropriate to the message band and the related "most appropriate" transmitter.

Many authors will not like this simple and crude type of presentation or accept the fiber bandwidth concept. When we look at fiber length association, they are right to say it does not have any physical meaning in the fiber case.

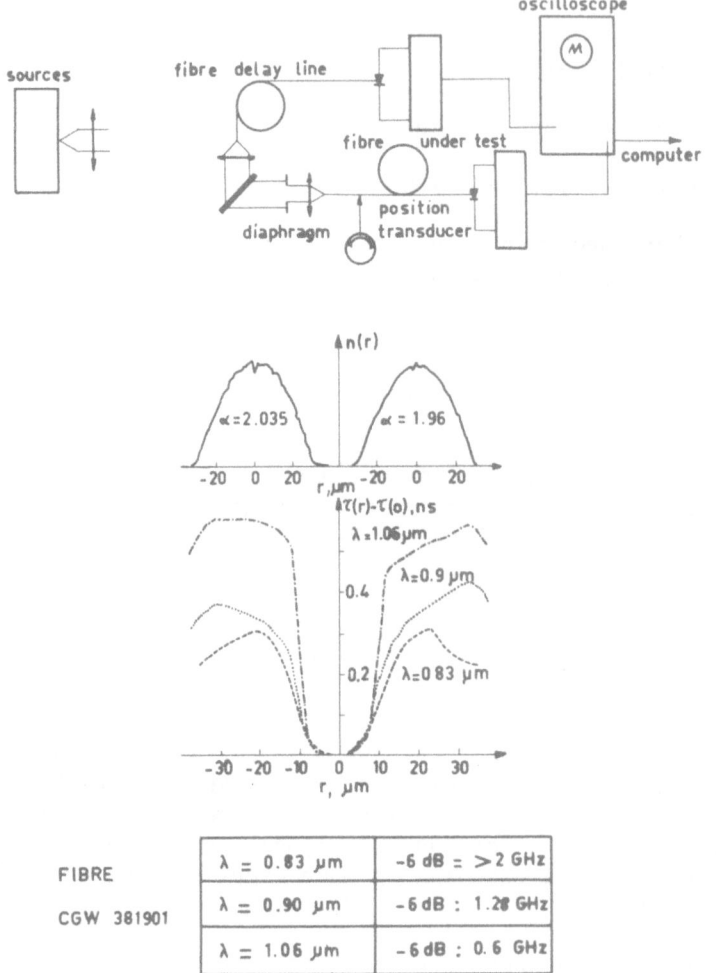

Fig. 7. Index profiles, differential mode delay, bandwidth of 1 km,
         CGW 381901 fiber (from L. Jeunhomme).

    Relying on Prof. Someda's presentations, let us now look at
the photoreceiver and its associated amplifier.

III. THE PHOTORECEIVER (Refs. 6-12)

III.1. Introduction

    For many years, different laboratories and semiconductor indus-
tries have been developing and manufacturing very convenient PIN

photodetectors in the visible and near infrared, starting from silicon substrates. These components are currently designed to absorb
almost 100% of incident light and convert all the absorbed photons
into electron-hole pairs. Their structures lead to a high efficiency
conversion and are also compatible with a high speed of response;
production costs are low. A drawback is that, unlike photomulti-
pliers or photoconductive detectors, they provide no internal gain
and in most cases require use of an amplifier to provide a large
enough signal current.

We will see in the next paragraph that amplifier noise will be
the limiting noise source in the case of PIN detectors. Therefore
it would be desirable to have an internal gain mechanism inside the
detector to multiply with minimum noise both light detected signal
and photodetector noise until the latter becomes greater than the
electronic amplifier noise.

III.2. PIN Diode: Structure and Operation (Figs. 8 and 9)

III.2.1. <u>Structure</u>. The PIN junction can be used as a photon
counter which converts light power into electric current. The intrinsic zone (high resistivity) is the middle layer of the junction
where:

- all the photons will be absorbed, depending on their wavelength, if the layer is thick enough;

- the electron-hole pairs are generated and electrons and holes
  are accelerated to their respective electrodes - speed of
  response depends on electric field in this depleted layer.

The first fundamental parameter is the spectral absorption of
the material in the desired wavelength range; it immediately leads
to the 100% absorption thickness required.

The second parameter is the field-dependent electron velocity
in the material, which leads to the reverse bias necessary to collect
the photon-generated carriers in the minimum time.

In the language of circuit designers, the PIN photodetector is
a reverse-bias junction, very linear current generator over several
orders of magnitude:

$$I_{ph} = (\eta q/h\nu)P_{opt}$$

where I_{ph} is the current delivered by the diode receiving a light
power P_{opt}, η is the wavelength-dependent external quantum efficiency,
q is the electron charge, and $h\nu$ is the photon energy. When no light
signal is applied, the diode delivers a "dark" leakage current function of temperature and applied reverse bias.

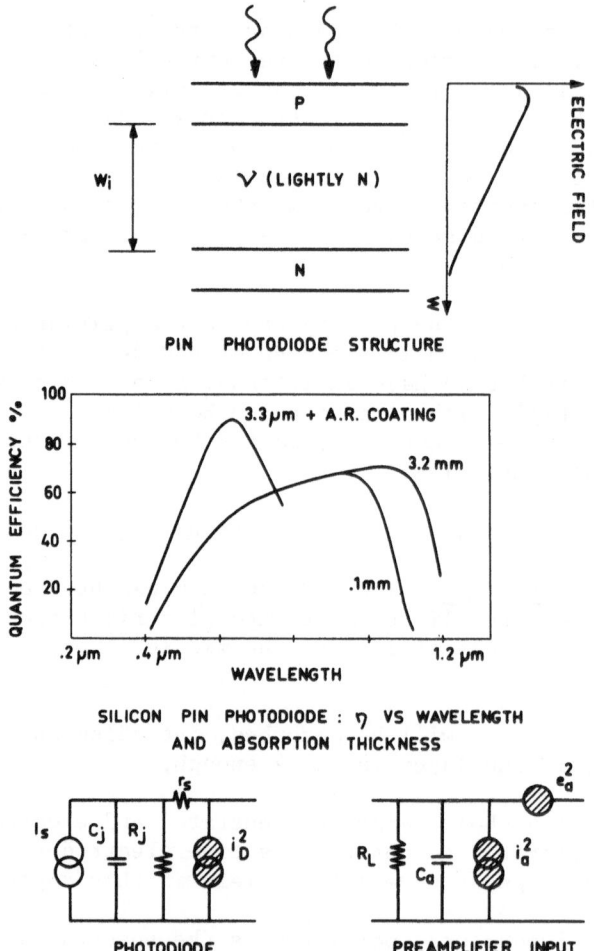

PIN   PHOTODIODE   STRUCTURE

SILICON PIN PHOTODIODE : η VS WAVELENGTH
AND ABSORPTION THICKNESS

NOISE EQUIVALENT CIRCUIT OF FIRST STAGE

Figure 8

The third important parameter is the two pole filter:

r_s is the series resistance of junction and wires, usually
    negligible except at very high frequencies;

C_j is the function capacitance, proportional to the inverse
    square root of applied voltage.

This capacitance will have a very important effect on the noise
performance one can achieve from a given component:

$$C_j = 2 \cdot 10^{-8} (S/\sqrt{\rho \upsilon}) \qquad \text{(silicon case)}$$

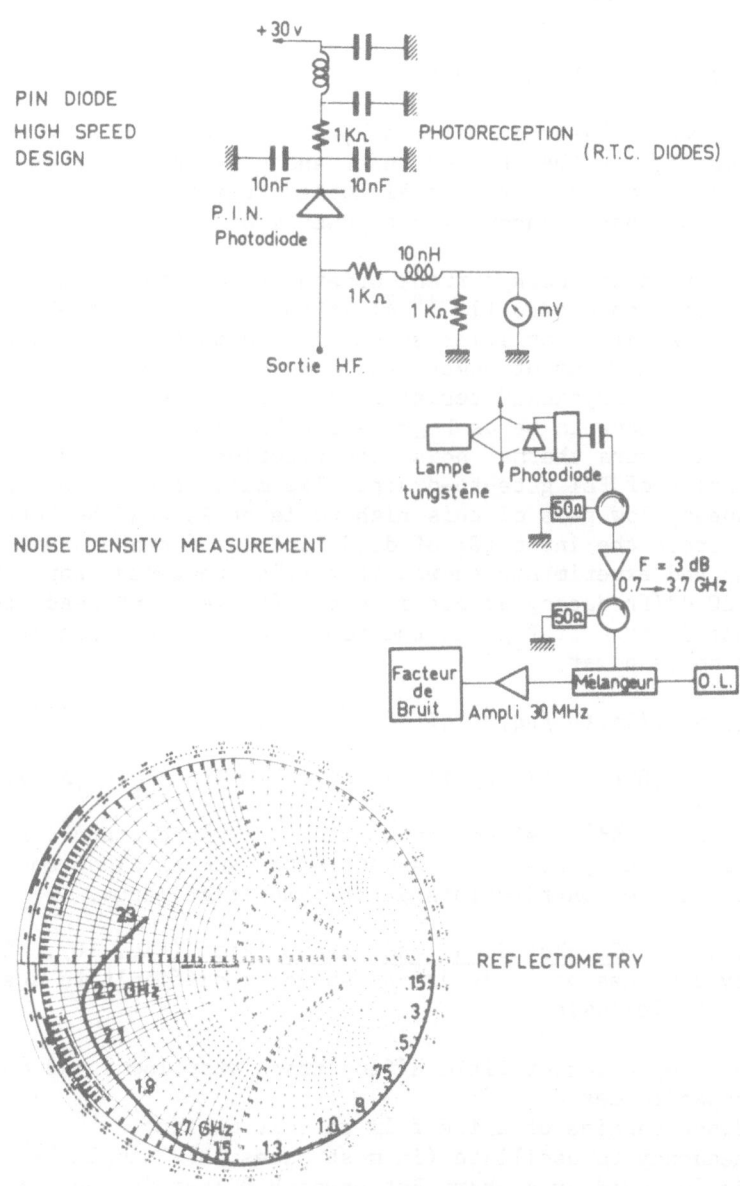

Figure 9

where S is the diode area and ρ is the Ω-cm material "I" resistivity.

III.2.2. <u>Noise sources</u>. The photodiode is a photon counter; it follows the classical Poisson process that leads to

$$E(i^2)\Delta F = [I_o + E(I_{ph})]2q\Delta F$$

where E( ) means "average value of", ΔF is the bandwidth under consideration, I_{ph} is the photocurrent, and q is the electron charge. This relation says that the total noise current source is proportional to the instantaneous light power input.

Looking at the dark current orders of magnitude commonly obtained on silicon diodes (10^{-10} Å) it appears that a field effect transistor as first amplifier stage will be most appropriate at least up to some tens of megahertz bandwidth. Since the diode will be loaded by a "physical" resistor somewhere in the circuit, this resistor will have to be as high as the FET accepts it in order to not bring an extra thermal noise contribution or the correlation noise sources of FET gate-to-drain. The major trouble we face is the frequency low pass of this high resistor R_2 and the total capacitance across the input (C_j of diode and C_t of transistor). Calculations and experiments showed that this frequency drop compensation of 20 dB/frequency decade is very efficient and leads to the best sensitivity. If B_{tot} is the root mean square value of electric to optic noise power

$$B_{tot} = (1/2\pi)(h\nu/\eta q)^2[2qI_o+2qE(I_{ph})+2qI_t+(2kT/R_L)]\int_{-\infty}^{+\infty}|A\omega|^2d\omega$$

$$+(h\nu/\eta q)^2(2kT/g_m)(1/2\pi)\int_{-\infty}^{+\infty}|(1/R_L)+jC_{tot}\omega|^2|A(\omega)|^2d\omega$$

where g_m is the field effect transistor transconductance, $A(\omega)$ the amplifier frequency response, and C_{tot} the sum of diode and amplifier capacitance contributions at the amplifier input.

III.2.3. <u>PIN photodiode and preamplifier</u> (Fig. 10). The high-frequency compensator we described above is difficult to design because of the following:

- discrepancies of field effect transistor input and output
  capacitances;
- discrepancies of thick film circuit design;
- tendency to oscillate (in most cases this simple "equalizer"
  is followed by a sharp Butterworth filter for better optical
  receiver band-edge definition which might overlap with field
  effect transistor roll off frequency);
- amplification of high frequency noise components from field
  effect transistor.

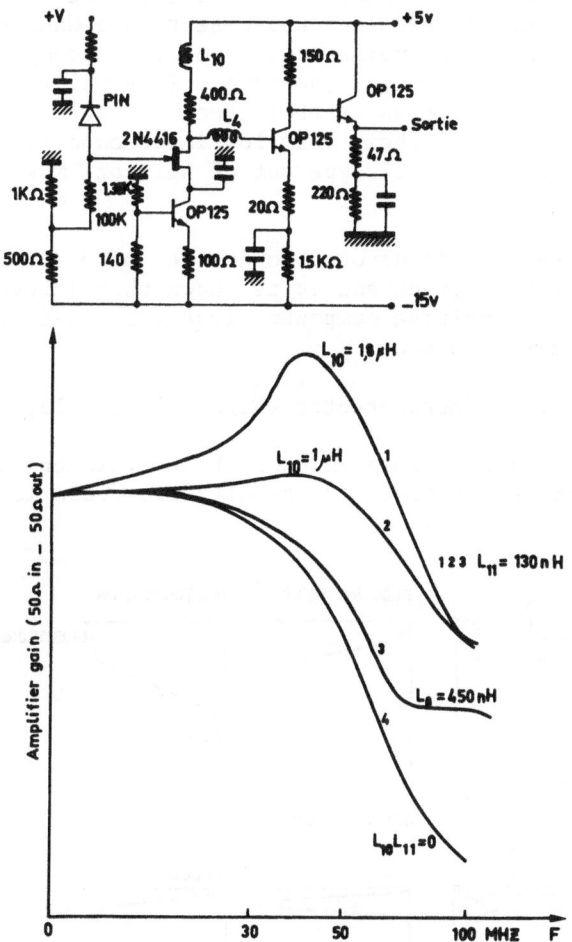

Fig. 10. Computer aided design on diode + hi-z input preamplifier
(APD or PIN) and bandwidth adjust.

All these limitations lead to a maximum compensation range of
100 dB. Since the compensator gain (in dB) is linearly dependent
on frequency, this maximum range gives the upper frequency limit
directly from the knowledge of low frequency start.

Many photodetector suppliers propose their own diode + amplifier module on the electronic components market. At CNET we directed
some efforts on computer-aided design and showed that thick film
hybrid circuitry was a requirement for overcoming the difficulties
listed above and improving silicon FET high frequency weakness. Our
other conclusion was that in the case of a silicon FET, for frequen-

cies higher than 20 Mc/s, preamplifier noise sources become the first
limitation. The next possibility would be to use GaAs FETs, but as
Rokos said five years ago, noise sources of these components remain
quite high in a broad frequency spectrum. They will probably be
used in high bit rate systems if their technology leads to more
reproducible components – their very low input capacitance (10 to
20 times lower than any N–FET type out of silicon) make them very
attractive.

The next paragraph is devoted to devices more sophisticated
than our simple PIN junction and it is shown that internal multiplication in the photosensitive component improves sensitivity when
high bandwidths are required.

III.3.   The Avalanche Photodetector (Figs. 11, 12, 13, and 14)

The basic physical mechanism upon which avalanche gain depends
is that of impact ionization. If the electric field is sufficiently

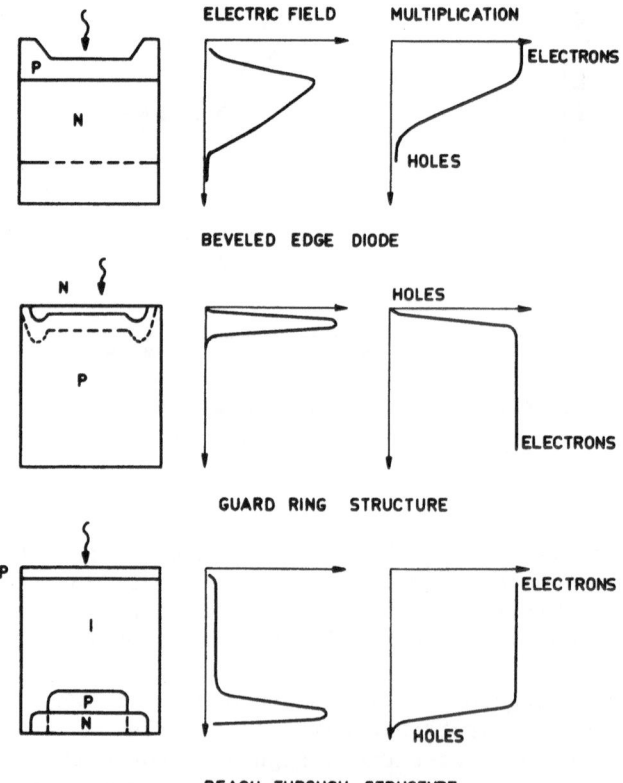

Figure 11.   (from Webb and McIntyre).

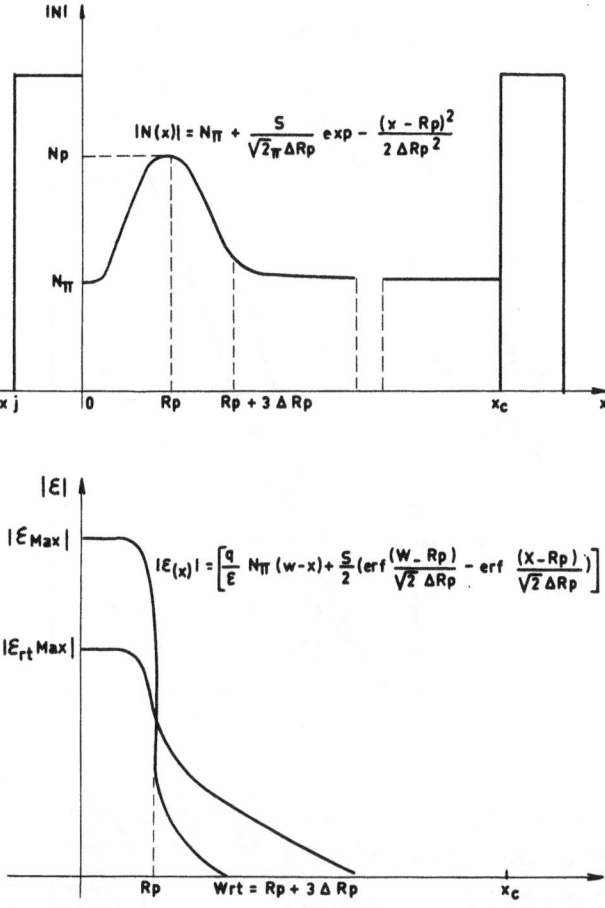

Fig. 12.   Ion implanted NΠPΠP structure for low noise applications
(from Ratsira, University of Montpellier).

high, an electron or a hole can collide with a bound electron with
sufficient energy to ionize it, thereby creating an extra electron
and hole pair. These additional carriers can gain enough energy
from the high field to cause further impact ionization until an
avalanche of carriers has been created. From the understanding and
calculations of these phenomena, P.P. Webb, R.J. McIntyre, and
J. Conradi derived the fundamentals of noise in avalanche photodiodes. As Tom Pearsall already said in his course, their derivation shows that good knowledge of ionization coefficients is required
to optimize the avalanche diode structure and material.

    III.3.1. <u>Signal-to-noise ratio</u>. Experts from RCA have cal-

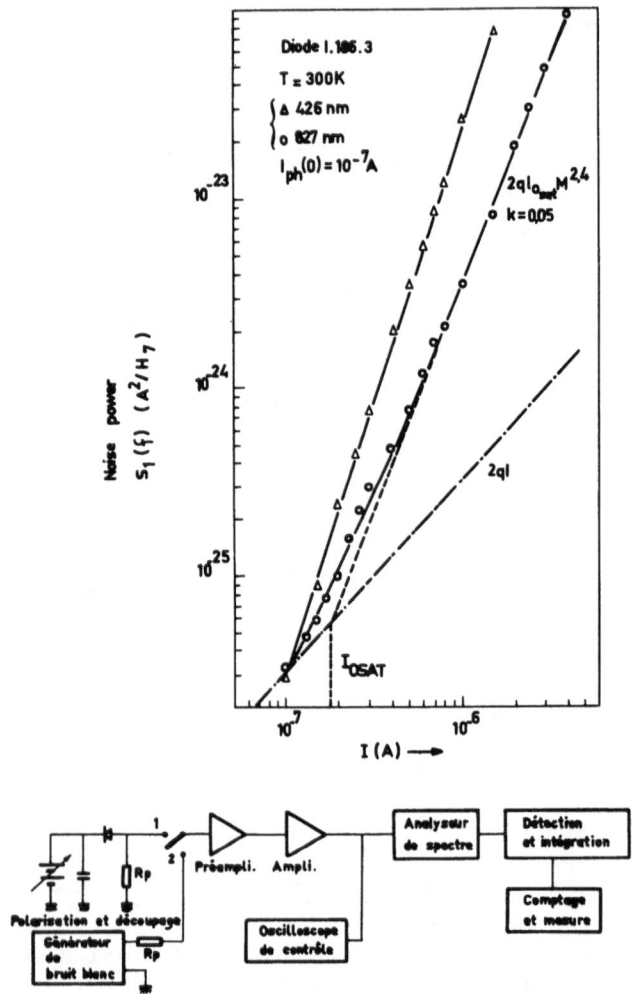

Fig. 13.   Low frequency noise measurement.

culated:

$$\frac{S}{N} = \left(\frac{i_s}{i_n}\right)^2 = \frac{1}{2} \frac{(P_o m R_o M)^2}{\{2q[I_{ds}+(P_o R_o F_s+I_{db}F_d)M^2]B+i_{na}^2\}}$$

where

      P_o is the average value of incident light;
      m is the modulation depth;
      R_o is the unity gain responsivity;

M   is the gain;
q   is the electron charge;
I_{ds} is the surface leakage current which does not go through
     multiplication;
I_{db} is the bulk leakage current which will be multiplied;
B   is the noise bandwidth;
i_{na}^2 is the amplifier input noise source;
F_d is the detector effective excess noise factor appropriate to
     the multiplication of the dark current;
F_s is the detector effective excess noise factor appropriate to
     the multiplication of the photocurrent.

We now look at F_d and F_s, starting with the gradients on carrier
current densities:

$$dJn/dx = -\alpha Jn - \beta Jp - g(x) = -dJp/dx$$

where α and β are the electron and hole ionization coefficients and
g(x) the optical generation; previous authors have derived the multiplication factor:

$$-M(x) = \frac{G(x)}{1 - \int_{x_2}^{x_1}\alpha G(x)\,dx}$$

$$G(x) = \exp[-\int_x^{x_2}(\alpha-\beta)\,dx]$$

which can be rewritten as follows:

$$M(x) = \frac{G(x)[1-k_1]}{G(x_1)-k_1}$$

$$G(x_1) = \exp[-(1-k_o)\delta] \qquad \delta = \int_{x_1}^{x_2}\alpha\,dx$$

where

$$k_o = \frac{\int_{x_1}^{x_2}\beta\,dx}{\int_{x_1}^{x_2}\alpha\,dx} \qquad k_1 = \frac{\int_{x_1}^{x_2}\beta M(x)\,dx}{\int_{x_1}^{x_2}\alpha M(x)\,dx}$$

In a first step this leads to the average multiplication factor an electron hole pair undergoes through the avalanche region:

$$\overline{M} = \int_o^\omega g(x)M(x)\,dx / \int_o^\omega g(x)\,dx$$

But what we are really looking for is some sort of evaluation
of the multiplication discrepancies between all the electron hole
pairs which truly give the noise associated with multiplication.

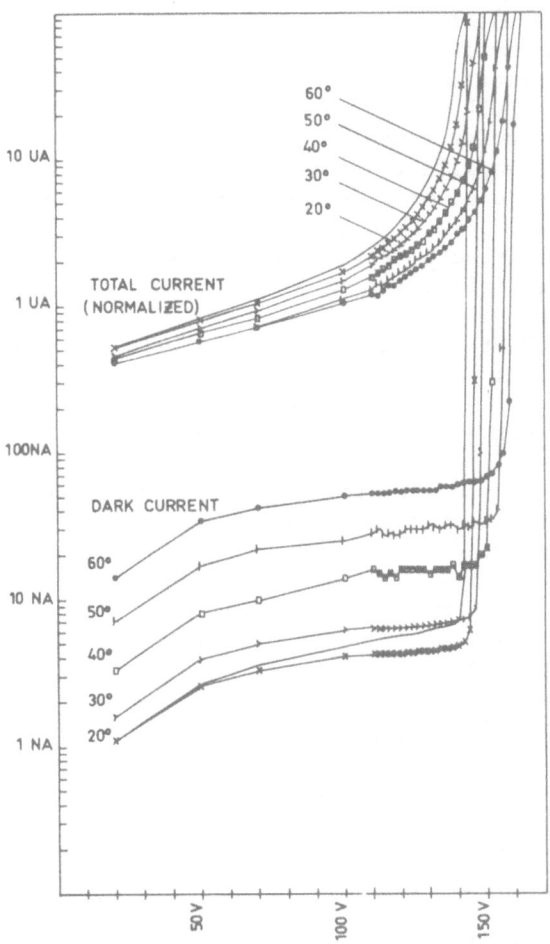

Fig. 14. 20 µm thick PΠPΠN diode: temperature dependence of dark
          current and gain vs. voltage.

We therefore need to derive the mean square gain and compare it to
M^2. If we set

$$k_2 = \frac{\int_{x_1}^{x_2} \beta M^2 dx}{\int_{x_1}^{x_2} \alpha M^2 dx}$$

we obtain

$$F_e = k_{eff} M_e + [2-(1/M_e)][1-k_{eff}]$$

$$F_n = k'_{eff}M_n - [2-(1/M_n)][k'_{eff}-1]$$

$$k_{eff} \sim k_2 \qquad k'_{eff} \sim k_2/k_1^2$$

When both carriers participate in the multiplication we obtain the most important equation [18] from McIntyre, Webb, and Conradi:

$$F_{eff} = \frac{fM_e^2 F_e + (1-f)M_n^2 F_n}{[fM_e + (1-f)M_n]^2} \qquad f = \frac{I_{nopt.}}{I_{popt.} + I_{nopt.}}$$

$I_{nopt.}$ and $I_{popt.}$ are the photon generated currents. Now we can go back to the signal-to-noise ratio given above and calculate it.

Before giving details on the reach-through structure and others derived from it, let me summarize speed of response and linearity aspects.

III.3.2. <u>Dynamic range and bandwidth of multiplication</u>. Space charge effects and temperature are the two major limiting factors on linearity and gain constancy. Mobile charges in the drift region increase the electric field in the drift region and decrease it in the avalanche region, causing a gain reduction; bad thermal dissipation also causes a gain reduction due to temperature increase. Aside from these two phenomena, and if the voltage drop across the load resistor does not rise too much, avalanche photodiodes remain linear over at least 7 orders of magnitude.

As we said in the case of PIN photodetectors, the first limiting term in the speed of response is the transit time of the carriers across the depletion layer. The multiplication time is the next limiting term. We can only talk about a gain bandwidth product when the diode is very thin and when the multiplication is the only time limiting phenomenon.

III.3.3. <u>Reach through structure and others</u>. Let us start from a PIN structure. The intrinsic layer is such that roughly 100% of the good wavelength photons are absorbed in it; under reverse bias conditions the photon generated pairs are dissociated and each carrier drifts to its proper electrode across constant field. Carrier multiplication will occur if we build up an ionization field inside the structure. In the first structure (RCA Canada) a P^+ layer is inserted between the surface N^+ layer and the high resistivity drift region. This internal junction is depleted at the high resistivity boundary just before reaching avalanche breakdown. In this case, multiplication occurs in a very thin part of the structure.

The second structure is more sophisticated. The extra internal P^+ layer is buried in the high resistivity layer. The avalanche

region is broader and the high field is more homogeneously spread
across it. Compared to the first structure a lower field is required
to achieve the same multiplication, leading to a higher α/β ratio and
consequently smaller excess noise factor. The multiplication time
is so short that both structures have very similar responses to
sharp pulses of proper wavelength radiation.

III.3.4. <u>Associated circuits</u>. The circuit design we developed
above in the PIN diode (Section III.2) remains unchanged: the avalanche photodiode is a current generator whose current results from
photon absorption and carrier multiplication. The internal multiplication is followed by the expression of all excess noise in terms
of an excess noise factor. Gain and excess noise may be adjusted
with diode bias so that diode noise and electronic amplifier noise
contributions are the same. The optimum detectivity can then be
derived as Personick showed in his first "Repeater Design" contribution in BSTJ.

The avalanche diode is quite sensitive to temperature variations:
quantum efficiency, leakage currents, and multiplication will change
under different thermal conditions. Different feedback and control
circuits have been designed and tested, the most common solution
consisting of a twin diode arrangement, the first diode as the light
sensitive element and the second one as the bias feedback reference.
Diode noise control is a second solution which has also been investigated. All these techniques regulate the high voltage bias.

III.4. Long Wavelength Photodetection

Wavelength radiations of 0.8 to 0.9 µm are efficiently absorbed
by reasonably fast responding silicon structures which are available
from many suppliers. Much research is being undertaken on III-V
materials for 0.9 to 1.5 µm photodetection. Tom Pearsall already
told you what his feelings were about diode structure and technology:
bulk and surface leakage, lattice mismatch traps, and high resistivity layers. As long as his components behave as photon counters
and current generators, nothing will be changed in the receiver
design.

Let us now turn to optical transmitter design.

IV. TRANSMITTER DESIGN (Refs. 9, 10, 13-15)

IV.1. Introduction

In this section, we will evaluate light sources which emit
light within the 0.8-1.4 µm wavelength range and are candidates for
use in fiber optic transmission systems. We will essentially review
the properties and characteristics of III-V semiconductor functions

starting from GaAlAs alloys. We will skip over the light sources requiring external modulation even though some of them might be of first importance due to their specific properties (YAG and ultra-phosphate Nd).

The laser diode structures will be described in Section IV.3, but first let us discuss the incoherent light emitting diode (LED).

IV.2.   The Electroluminescent Diode (Figs. 15, 16, and 17)

IV.2.1.   Diode structure. The structures most suitable for fiber optic transmitters are the carrier confinement junctions where recombinations take place in a very thin region of the PN junction.

Matérial	Bandgap type	Recombination Coef^t cm^3/sec	
Si	Indirect	1.79	10^{-15}
Ge	Indirect	5.25	10^{-14}
GaP	Indirect	5.37	10^{-14}
Ga As	Direct	7.21	10^{-10}
Ga Sb	Direct	2.39	10^{-10}
In As	Direct	8.5	10^{-11}
In Sb	Direct	4.58	10^{-11}

CALCULATED RECOMBINATION FOR DIRECT AND INDIRECT BANDGAP
SEMICONDUCTORS   (from Y.P. VARSHNI)

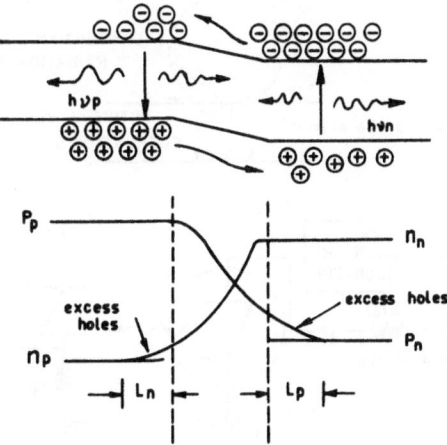

ELECTROLUMINESCENT p-n JUNCTION OPERATION

(from H. KRESSEL)

Figure 15

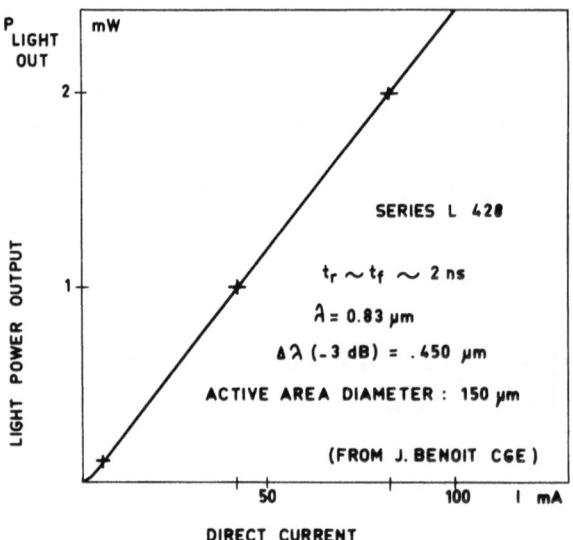

Figure 16

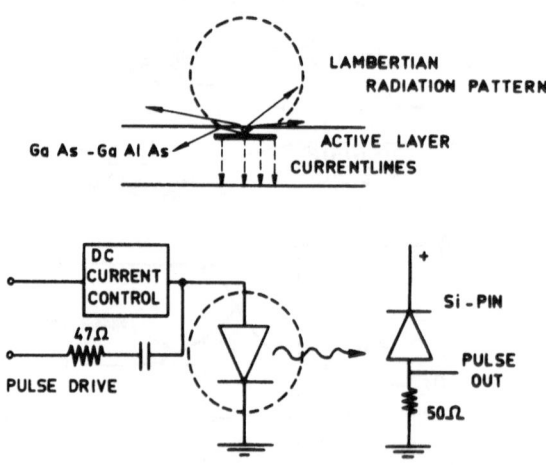

Fig. 17.   Rise and fall time measurement.

These structures can be grown by liquid phase epitaxy (epitaxial junctions), deep P diffusion down to GaAlAs N layer, or from GaAs or GaAlAs epitaxially grown layers, chemically etched through a circular mask to reduce some of the P or N layer thickness.

IV.2.2. Operation. The injected minority carriers will recombine at the vicinity of the P-N junction after application of a forward bias V. If they recombine radiatively, the photon that escapes has an energy nearly equal to the band gap energy E_q, assuming no deep level traps. If they do not, they transfer their energy into heat. The three basic equations are

$$I = I_s[\exp(qV/nkT)-1] \qquad\qquad 1 < n < 2$$

$$\eta_i = 1/[1+(\tau_r/\tau_{nr})]$$

$$1/\tau = (1/\tau_r) + (1/\tau_{nr})$$

where η_i is the internal quantum efficiency of the diode, τ_r and τ_{nr} are the minority carrier lifetimes for radiative and non-radiative recombination, and τ is the minority carrier lifetime. This radiative process is very efficient in direct bandgap semiconductors since direct recombination is possible without a third particle to conserve momentum.

From the above set of equations, we can derive the external quantum efficiency of the diode, that is to say, the ratio between the number of photons escaping through the surface and the number of injected carriers. Some experimental data are given in Fig. 16, showing linearity and speed of response to a square pulse.

The forward bias is around 1.5-2 V and the series resistance should be fractions of ohms: it behaves electronically as a usual forward bias PN abrupt junction and its optical response could be speeded up using current pulse shaping techniques. It is a lambertian light source: since fibers are of the same order of magnitude in dimensions and have limited numerical apertures, it will remain difficult to couple light into them from this plane surface source, unless one uses some sort of specific microoptics. Its spectrum width is broad and depends on the diode structure and confinement. However, this component offers low-cost expectation, high reliability, and good linearity in most cases.

IV.3. The Semiconductor Laser (Figs. 18-23)

As H. Kressel from RCA laboratories said, it is desirable to develop a light source with the widest possible system utility, even though its full potential will not be utilized. And this is where the laser diode stands. It combines high radiance, easy high frequency modulation up to some GHz, narrow-spectrum sharp radiation

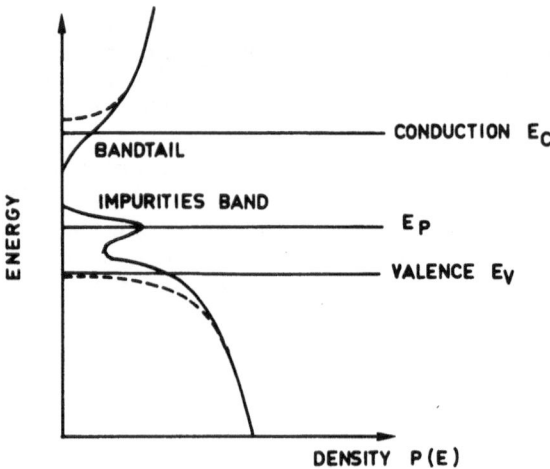

Fig. 18. State density distribution with conduction band tail and
high density of valence impurity.

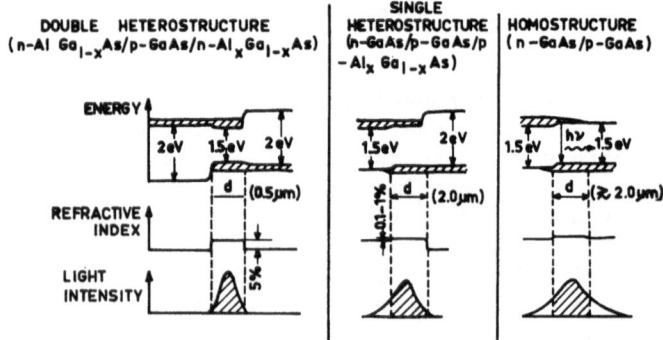

Fig. 19. Energy band configurations of some laser structures.

pattern, and small spot size. After a short discussion and critique
of the different available structures, we will concentrate on noise,
nonlinearities, and transients.

IV.3.1. Laser diode structures. A unique feature of the laser
diode not present in the other laser types is the ability to obtain
stimulated emission by minority carrier injection using a PN junction or heterojunction. The high packing density (10^{18} cm^{-3}) of
excited atoms in a semiconductor (compared to only 10^{10} cm^{-3} in gas
lasers) is advantageous because the optical gain coefficient is
relatively high, allowing for much shorter optical cavities. The

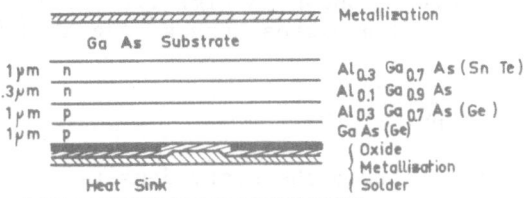

Oxide stripe isolation (from H. KRESSEL)

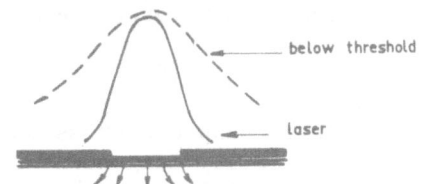

Near-field Pattern along junction plane

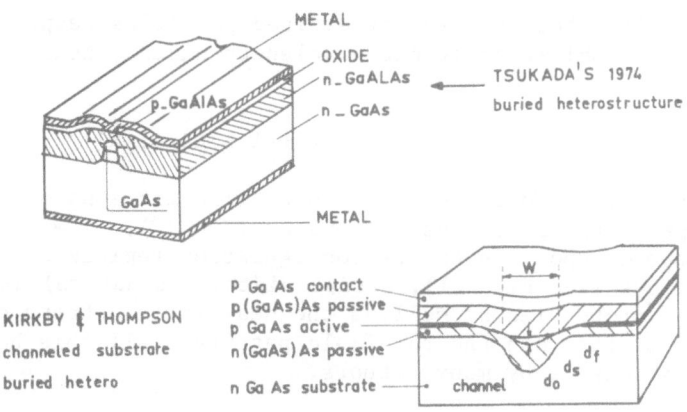

Figure 20

cavity is obtained by cleavage along natural crystal planes; since the material index of refraction is about 3.5, the reflection coefficient at these natural plane mirrors is about 30%.

Different laser structures are shown in Fig. 20. They show that improvements in the liquid phase epitaxy process and contact alloying lead to the stable stripe geometry double heterostructure. This last structure, which works C_w at room temperature, is the best suited for fiber transmissions up to now.

IV.3.2. <u>Operation and optical characteristics</u>. The junction
band structure is shown in Fig. 19. If τ_{sr} is the stimulated emission rate, one can express the gain g as a function of photon energy:

$$g(E) = \frac{\pi^2 c^2 h^3}{n^2(E)^2}\, \tau_{st}(E)$$

$\tau_{st}(E)$ obtained from the state density distributions are usually
assumed to be parabolic.

Since the number of free carriers is high (heavily doped material) these carriers will cause absorption losses under indirect band
to band transitions. Therefore, the overall gain will be

$$G(E) = g(E) - b(E) = g(E) - [n\tau_n + p\tau_p]$$

if τ_n and τ_p represent capture cross sections with electrons and
holes. If L is the cavity length and R the cleaved face reflection
coefficient, emission will occur when

$$G(E) \geq (1/L)\mathrm{Log}(1/R)$$

Let us call η the fraction of excited particles responsible for radiative recombination at the wavelength under consideration. Then
at 0°K

$$J_o = 8\pi \ell n^2 e(\nu_L/c^2)(\Delta\nu/\eta)[P(\nu)+(1/L)\mathrm{Log}(1/R)]$$

Above this threshold, the laser will transmit a narrow spectrum light
power proportional to the difference $(J - J_o)$. The modal distribution depends upon the cavity configuration (epitaxial layers of ter-
nary alloys (see Fig. 21), stripe width, and nature) and the
material properties as well as the near field pattern on the output
cleaved surface and the far field pattern. All this has been published in detail by many authors.

For a first approach to high bit rate applications, the systems
people need a set of four curves:

- the light power output VS current (pulsed or DC) in order
  to obtain a reasonably good knowledge of the threshold current and the differential external quantum efficiency
  $\Delta\eta_{ext} = \Delta P_{light}/\Delta I$ above threshold and the range of linearity;

- the modal distribution VS current (remember that for high bit
  rate transmissions the source spectral width has to remain
  small because of the bandwidth limitation from fiber material
  dispersion;

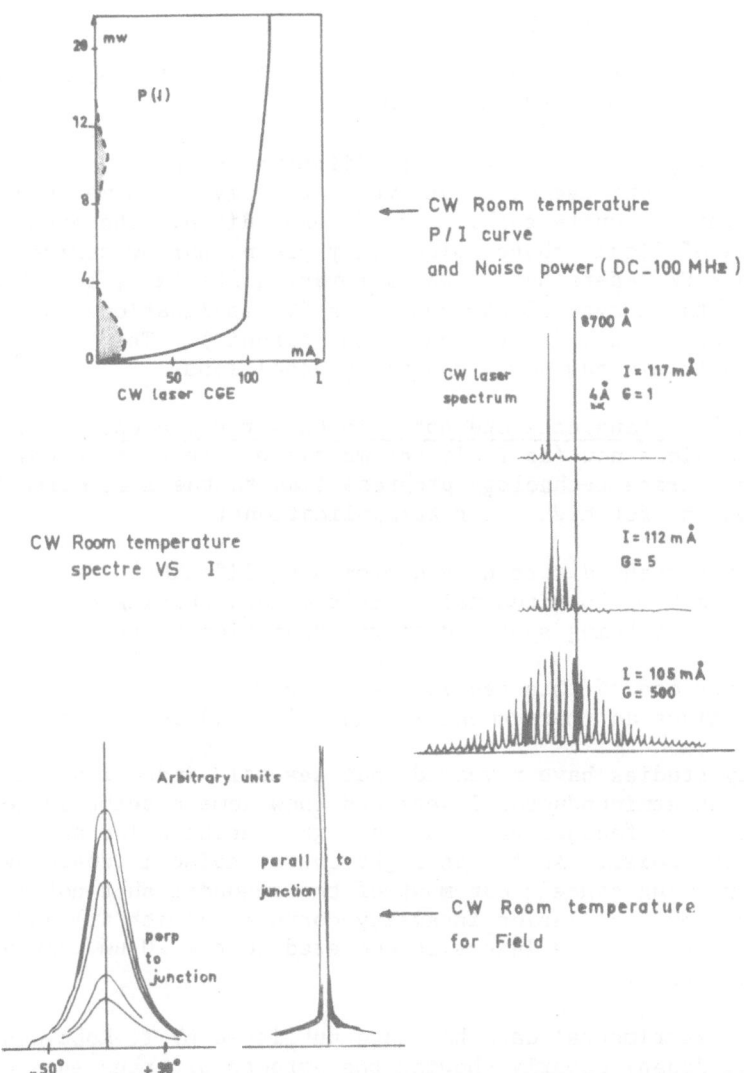

Fig. 21.   Typical curves of 20 μm stripe H⁺ implanted Ga–Al–As double Het laser (CW room temperature).

- the near field pattern, which gives the field intensity distribution across the emitting area and will be matched to the
fiber optical structure;

- the far field pattern which will also be matched to the fiber
optical structure.

All these data are currently gathered in different laboratories for
testing of lifetimes and long term drifts.

Actually, if we insist on the lifetime advantage of the incoherent electroluminescent diode, we cannot say the same about the
very similar but quite sophisticated laser diode. The optical
properties of liquid phase epitaxially grown, narrow striped room
temperature CW lasers have been improving quite fast, but they have
to remain stable over 10,000 hours for PTT applications as well as
the diode electronic input (equivalent circuit). They also have to
correspond to. minimum power consuming conditions.

IV.3.3. <u>Transients and noise in CW striped double heterolasers</u>.
It seems obvious now (as I already mentioned above) that epitaxial
growth and stripe technology progress lead to the most suitable
source concept for high bit rate applications:

- low threshold current and promising lifetime
- one set of longitudinal modes and good guiding conditions
- small emitting spot and narrow radiation pattern

The purpose of this section is to show how this type of structure also gives an improvement in the noise and transients aspects.

Early studies have revealed that several kinds of noise sources
exist in the semiconductor lasers and show some attempt to derive
the correlation factors between the light output noise and the junction current noise. As far as light output noise is concerned, the
stripe narrowing cancels out most of the resonant shot noise (and
the relaxation oscillation intensity correlated with it) and improves
the 6 μm narrow stripe lasers as compared to the 20 μm components
developed earlier.

Some experimental data has been published by T. Kobayashi et al.
(ECL, NTT, Japan) clearly showing the effects of transients on 3 μm
stripe width lasers and the usually obtained results for broader
stripe lasers. According to the analysis of Morgan and Adams, the
resonance frequency width of the shot noise is given by

$$\Delta f = \frac{1}{2\pi} \left(\overline{R}_{st} - \frac{1}{\tau_p} \right) - \frac{1}{2\pi} \left(\frac{\partial \overline{R}_{sp}}{\partial N_e} + \overline{N}_p \frac{\partial \overline{R}_{sr}}{\partial \overline{N}_e} \right)$$

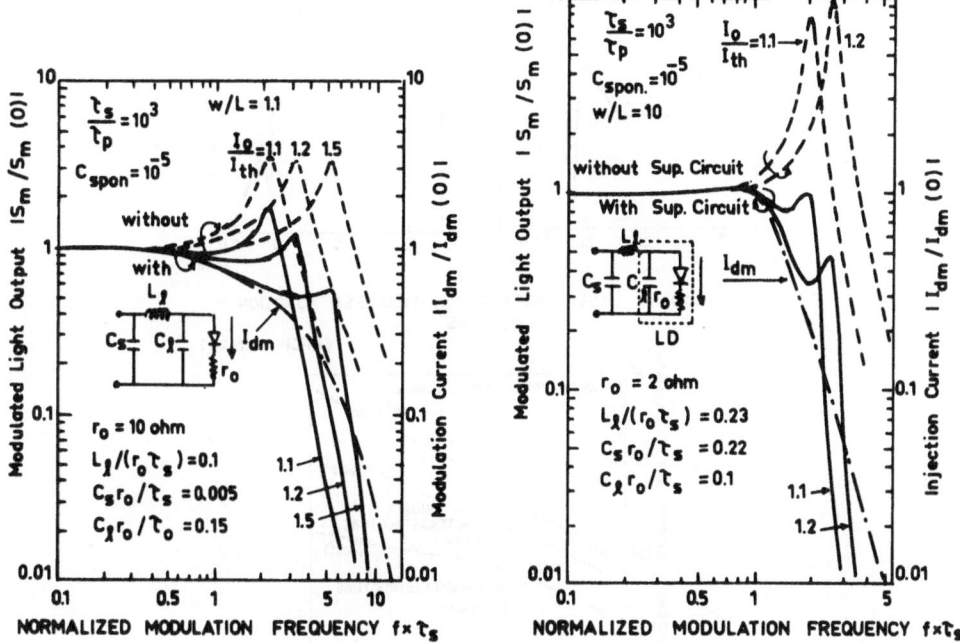

Fig. 22.   Modulation characteristics of injection lasers with π-type
           suppressor circuit (from Suematsu).

where $\overline{R}_{st}$, $\overline{R}_{sp}$, $\overline{N}_p$, and $\overline{N}_e$ are the averages of the stimulated emission rate, spontaneous emission rate, photon density, and electron
density, respectively, and τ_p is the photon lifetime. If we use
the Boers, Vlaardinger-Broek, and Danielsen equations, we have

$$\overline{R}_{sr} - \frac{1}{\tau_p} = \frac{\overline{R}_{sp}}{\overline{N}_p} = \frac{\overline{N}_e}{\overline{N}_p} \cdot \frac{\tau_s}{\beta}$$

where τ_s is the spontaneous electron lifetime and β is the fraction
of the spontaneous emission going into the lasing modes. Under
lasing conditions of any narrow stripe double hetero laser, the
spontaneously emitted light will be confined in the laser waveguide
and will participate in the stimulated emission modes. This can be
experimentally observed on the change in nearfield pattern. This
larger participation of the spontaneous emission plays an essential
role in the reduction of the shot noise resonance in narrow stripe
lasers.

     Going into further detail on current noise in GaAs CW laser
diodes, G. Guekos et al. found that the junction current noise is

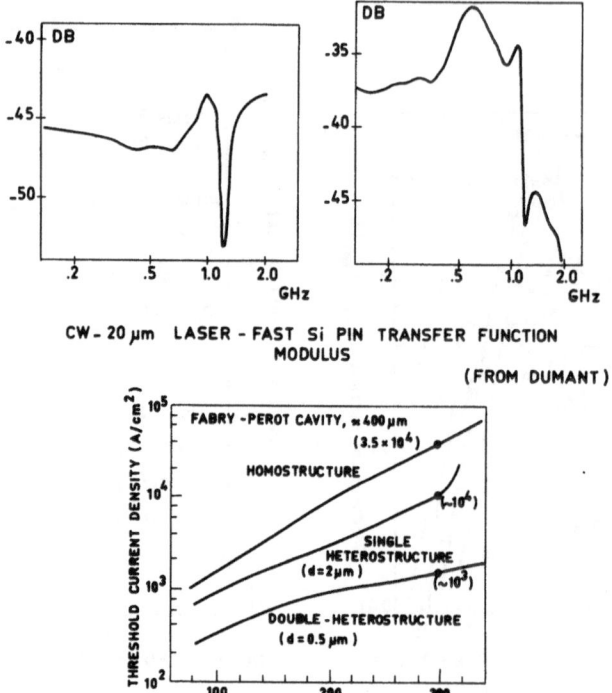

CW‑ 20 μm   LASER ‑ FAST Si PIN TRANSFER FUNCTION
MODULUS

(FROM DUMANT)

LASER THRESHOLD TEMPERATURE DEPENDENCE

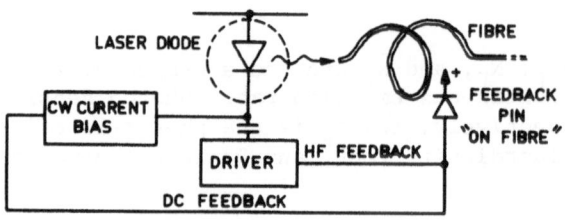

LASER MODULATION AND CONTROL

Figure 23

larger than shot noise and agree with Haug, who proposed:

- a term due to the fluctuations in emission and absorption
  rates;

- a second term due to light field fluctuations as pointed out
  by van der Ziel.

These considerations have a very important impact on transmitter circuit design for CW laser optimum operation. If proper DC
bias current is derived from the optical static curves for optical
power output and spectral distribution (as well as DC feedback from
a fraction of light output vs. temperature and launching conditions),
optimum pulse drive will require accurate knowledge of linearity
range, noise, and modal behavior.

IV.4. Conclusion

Without excess calculations on laser waveguiding properties and
transient response analysis, but with a lot of end-of-the-course
references, we have thus far presented a review of light source properties in terms of digital systems applications.

Now we will discuss new ideas on signal treatment which might
be helpful in digital transmissions through optical fibers.

V. SIGNAL PROCESSING IN DIGITAL FIBER OPTIC TRANSMISSIONS

The two preceding eminent speakers, Profs. Arnaud and Someda,
developed theoretical calculations and experimental results for
pulse propagation along optical fibers. As I have already said in
Section II, these calculations and experimental evaluations can be
converted into sharp pulse distortion (in time domain investigations)
or into frequency domain bandwidth and group delay distortions. I
have also said that this last concept, applied to multimode fibers,
is limited by the length of fibers, and its long length dependence
could not be derived properly with high accuracy.

Nevertheless, multimode fibers are low pass filters and we might
need to know how to compensate for the distortions they cause and
the receiver penalty resulting from those compensations. Several
authors have published work on that subject.

V.1. Personick's Approach (Ref. 7)

Starting from his first "Repeater Design," Personick adds a
delay distortion equalizer Heg_2 (f) in series with the "high input
impedance" equalizer to compensate for the incoming light pulses.
This leads to a trade off between extra noise and less intersymbol
interference and also requires more information about the fiber
impulse response than is available from time domain Fourier transforms or frequency domain measurements. However, as Personick says,
"repeaters which can accommodate small amounts of intersymbol interference increase the yield of useable filters in real cables where
the index gradings are not ideal."

V.2. Tamburelli-Luvison and Dogliotti's Approach (Ref. 16)

Tamburelli-Luvison and Dogliotti build up their non-linear
equalization from Gloge's mathematical model of baseband pulse response of multimode fibers and from the Poisson noise behaviour of
avalanche photodetectors. Equalizers can be designed to keep intersymbol interference within tolerable bounds with respect to a desirable and achievable goal of error probability. Figs. 24-27 show
comparative results on the signal-to-noise ratio required for error
probability of 10^{-9} vs. pulse normal dispersion (Gloge's pulse shape).

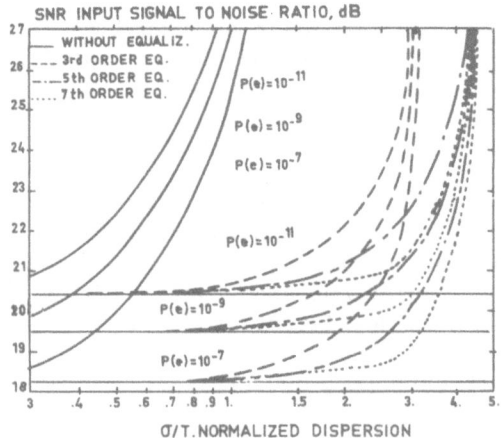

Fig. 24. MMSE equalizer of different orders.

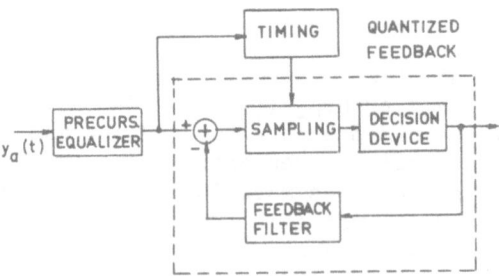

Fig. 25. Decision feedback equalizer and linear compensation of
          precursors (from A. Luvison, CSELT).

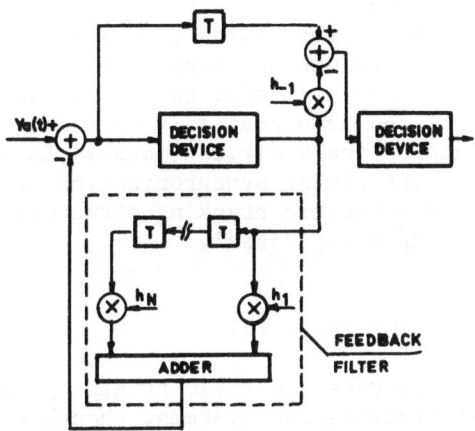

Fig. 26. Nonlinear equalizer with decision feedback and precursor correction (from A. Luvison, CSELT).

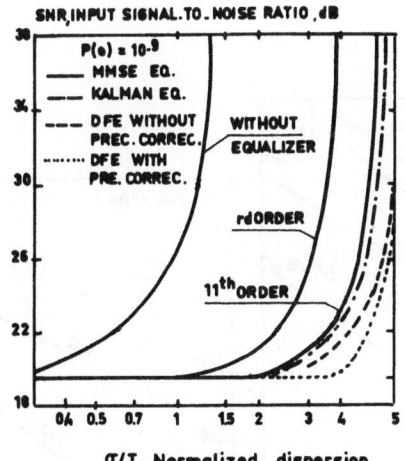

Fig. 27. (From A. Luvison, CSELT.)

V.3.   Conclusion

The selection of one of these equalization techniques will
obviously come from the logical speed available at the given bit
rate and the system requirements (maximum repeater spacing, total
cost, and power consumption). Aside from these final optimizations,
some results about low reduncancy binary pulse format seem to be
promising. In particular, we refer to the two level alternate mark
inversion code (AMI) and more generally to coding plans that convert
m binary digits into n binary digits (mBnB). In connection with the
use of line coding, the timing synchronization at the receiver could
be carried out by an automatic tracking structure which operates by
forming a zero-seeking error signal.

VI.   ANALOG TRANSMISSION (Fig. 28)

Most system prototypes and field trials operating on fiber optic
cables are digital transmission systems, though the very first demonstrations were very simple direct videotransmissions of TV pic-
tures. The characteristics of analog transmission are tightly
related to the modulation; in the following paragraphs, we will study
and compare three kinds of modulation which can be applied to optical
fiber transmission.

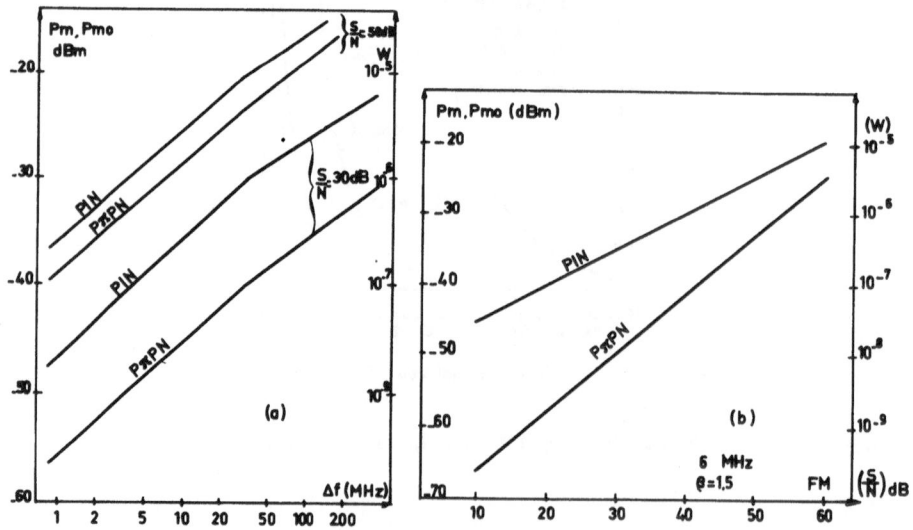

Fig. 28.   Optimum minimum power vs bandwidth: (a) videotransmission;
           (b) from Rousseau, CGE.

VI.1. Amplitude Modulation

This is very easy and straightforward: the signal to be transmitted is directly applied to the optical source DC bias in its
linear mode of operation (class A). We have already seen that the
fiber output photodiodes, PIN or avalanche, have a linear dynamic
range at low levels of about three orders of magnitude. From these
considerations:

- care must be taken to eventually compensate for light source
  non-linear distortions (feedback or double diode);

- receiver design is almost identical to what has been seen in
  Section III about the digital receiver.

The signal-to-noise ratio calculation remains the same except
that in the case of analog transmissions the ratio required for good
quality might be much higher. If S/N stands for the signal-to-noise
ratio, ΔF the frequency bandwidth, and m the modulation ratio, let
us call

$$A = (S/N)^2 \; \Delta f/2m^2$$

$$B = NI_r + NI_a + (d/3)\pi^2 C_{tot}^2 \Delta F^2 NE_a$$

where all the other parameters have been defined in Section III.
In the case of PIN diode the average power will be

$$P_m = \frac{h\nu}{\eta q} \{qA + \sqrt{q^2A^2+2AB}\}$$

In the case of the avalanche diode, it will be

$$P_m = \frac{h\nu}{\eta q^2} \frac{2B}{x} \{\frac{q^2A(x+2)x}{2B}\}^{\frac{2+x}{2+2x}}$$

From the curves shown, the avalanche diode is first, as one
would expect, but not by much for high signal-to-noise ratio since
its internal gain raises Poisson quantum noise.

VI.2. Frequency Modulation

Let us take the example of TV transmission with a videoband of
6 MHz and a modulation ratio of 1.5 at maximum frequency. From
Carson's formula we derive the modulated signal bandwidth:

$$B_{FM} = 2\Delta F(1+\beta) = 30 \text{ MHz}$$

where β is the modulation ratio and ΔF (6 MHz) the videoband. In order to avoid second order cross term interference with the signal, the subcarrier can be

$$F = 45 \text{ MHz}$$

Thus far

$$\left(\frac{S}{N}\right)^2_{FM} = \frac{i^2 m}{4\Delta F} \cdot \frac{1}{NI_q + NI_r + NI_a + 4\pi^2 c^2 F^2 NE_a}$$

$$\left(\frac{S}{N}\right)^2_{video} = 3\beta^2 \left(\frac{S}{N}\right)^2_{FM}$$

and if we set

$$A = \frac{2}{3} \frac{\Delta F}{\beta^2} \left(\frac{S}{N}\right)^2_{video}$$

$$B = NI_r + NI_a + 4\pi^2 c^2 F^2 NE_d$$

we obtain

$$\text{PIN diode} \qquad P_m \sim \frac{h\nu}{\eta q} \frac{2}{\beta} \left(\frac{S}{N}\right)_{video} \sqrt{\frac{B\Delta F}{3}}$$

$$\text{Avalance diode} \qquad P_m \sim \frac{h\nu}{\eta q^2} \frac{2B}{x} \left(\frac{q^2 A (x+2) x}{2B}\right)^{\frac{2+x}{2+2x}}$$

Curves are given on the previous page for this particular TV case. The triangular "noise density vs. frequency" and the center frequency shift improve the signal-to-noise ratio by a factor greater than 15 dB compared to direct amplitude modulation.

VI.3. Pulse Position Modulation

In this case, a super linear light source is not required and the amount of light in the fiber input will be much higher since we can use higher efficiency laser diodes. The transmitter delivers pulses whose delays with respect to a clock signal depend on video signal amplitude. At the receiver stage the video signal is restored by going through pulse duration modulation followed by a simple low pass filter.

The results obtained reveal an extreme sensitivity to transmitted pulse widths. For a pulse width around 25 ns the results show
that this type of modulation is less favorable than amplitude modulation and frequency modulation. However, for sharper pulses using
an avalanche diode, it leads to an improvement of 8.5 and 20.5 dB
compared to frequency and amplitude modulations, respectively.

VI.4. Conclusion

The system design methodology is not a strict and 100% logical
series of operations. Many users' requirements may force the designer
to go back and forth to look very carefully at many possible solutions. However, according to rules that have been well established,
the starting point is the receiver design. The next steps are given
in the last section of this course.

VII. GENERAL CONCLUSION

As a very common rule for all kinds of systems, design must
begin with analysis of the user's needs. In the case of our transmission system, the following data must be listed:

- the signal nature (digital or analog, bandwidth or bit rate);
- the quality (signal-to-noise ratio or error rate, distortion
  differential phase, and gain or regenerated clock jitter);
- the length of the link and if possible the unit cable lengths;
- input and output signals that will be available;
- the network configuration (in the case of a whole network);
- equipment reliability;
- environmental conditions (thermal, pressure, humidity, aerial
  or underground);
- delays for real installation and field trials - cost.

From these data, the "countdown" must be undertaken as described
below:

- first selection of modulation and simplest code;
- receiver definition (calculation of the minimum input power
  for the required signal-to-noise ratio or error rate; even-
  tual modification of coding schemes);
- selection of the light source vs. bandwidth, reliability,
  linearity, spectrum width, etc.;
- total loss calculations and selection of fiber structure and
  composition.

This last point will have to be fed back into the others and as
usual, one will have to arrive at many exciting compromises. I
thank you for your attention and all my colleagues and "fiber optic"

friends who in one way or another helped me on this work. I am
sincerely delighted to have met you here. Finally, I thank Dan and
DAINO and all the committee members.

REFERENCES

(1) B. Danielson, G. Day, and D. Franzen, Fibre Metrology at NBS,
    URSI Conference Measurement in Telecommunications, Lannion,
    1977, France.
(2) C. Vassallo, Linear power responses of an optical fibre, IEEE
    Trans. Microwave Theory and Techniques, Vol. MTT 25, n° 7,
    July, 1977.
(3) C.Y. Boisrobert, A. Cozannet, and C. Vassallo, Sweep frequency
    measurement applied to optical fibre, IEEE Trans. Inst. Meas.,
    December, 1976.
(4) L. Jeunhomme, A. Cozannet, R. Bouillie, and J.P. Hazan, Mesure
    des caractéristiques de transmission de conducteurs optiques,
    Second European Conference on Optical Fibre Communications,
    Paris, September 27-30, 1976.
(5) L. Jeunhomme, J.P. Pocholle, and J. Raffy, Wavelength depen-
    dence of modal dispersion in graded index optical fibres,
    Electronics Letters, June 8, 1978, Vol. 14, n° 12.
(6) G.H.S. Rokos, Optical detection using photodiodes, Optoelec-
    tronics 5, n° 4, Juillet, 1973.
(7) S.D. Personick, Receiver design for digital fibre Optic Com-
    munication systems, BSTJ, July-August, 1973, Vol. 50.
(8) M.K. Barnoski, Fundamentals of Optical Fibre Communications,
    Academic Press, New York, 1976.
(9) J.M. Dumant, C.Y. Boisrobert, and J. Debeau, Modulation rapide
    d'une diode laser GaAlAs en bande de base, Onde Electrique,
    Vol. 56, n° 12b, December, 1976.
(10) J. Conradi, F. Kapron, and J. Dyment, Fibre Optical transmis-
    sion between 0.8 and 1.4 μm, IEEE Transactions on Electron
    Devices, Vol. ED 25, n° 2, February, 1978.
(11) P.P. Webb, R.J. McIntyre, and J. Conradi, Properties of
    avalanche photodiodes, RCA Review, Vol. 35, June, 1974.
(12) T. Kaneda, H. Takanashi, H. Matsumoto, and T. Yamaoka,
    Avalanche buildup time of silicon reach through photodiodes,
    JAP, Vol. 47, n° 11, November, 1976.
(13) M. Maeda, K. Nagano, I. Ikushima, M. Tanaka, K. Saito, and
    R. Ito, Buried-heterostructure lasers for wideband linear
    optical sources, Central Research Laboratory, Hitachi, Ltd.
(14) M. Yano, K. Seki, Y. Yanai, and T. Kamiya, Opto-electronic
    digital devices modulation characteristics of laser diodes,
    Annual Research Report of Electron Device Laboratory, n° 15,
    University of Tokyo.

(15)   T. Hong and Y. Suematsu, Suppression of resonance-like pheno-
       mena in the light output of directly modulated injection lasers
       by π-type suppressor circuit, Transactions of the IECE of
       Japan, Vol. E61, n° 3, March, 1978.
(16)   R. Dogliotti and A. Luvison, Signal processing in digital fibre
       communications, Annales des Télécommunications, Volume 32,
       n° 9-10, September-October, 1977.

PHOTODETECTORS FOR INTEGRATED OPTICS AND FIBER OPTICAL COMMUNICATION

T. P. Pearsall

Thomson. CSF - L.C.R.

BP. 10 - Orsay, France

1. Introduction

The rapid progress toward telecommunication by optical fibers and optical information processing using integrated optics has raised considerable scientific interest in appropriate photodiodes for these applications.[1-4] While there is universal appreciation that silicon photodiodes have the best performance characteristics of any available photodiode, other features intervene, notably the properties of the transmission medium,[5] which give a considerable advantage to optical signal transmission at wavelengths in the infra-red where silicon is not sensitive. It is difficult to appreciate the true performance of a photodiode without also considering the conditions under which it must operate. A detector can be characterized by three properties : spectral response, rise-time, and noise. In even an average photodiode, the quantum efficiency and rise-time are usually quite good and near the maximum possible. Noise is the crucial feature in determining photodiode performance, because detection in optical communication is concerned with very weak signals. In this case the detector sensitivity and thus the system performance is always limited by the noise. Noise arises from three different origins : shot noise in the optical signal, shot noise in the detector and excess noise generated by avalanche gain in the detector, and amplifier thermal noise. In the first section we will show how these three sources determine the signal-to-noise ratio for the communication system, and in the second section we will see the conditions under which each of these noise sources will be the limiting factor for high bit-rate communications by optical fibers.

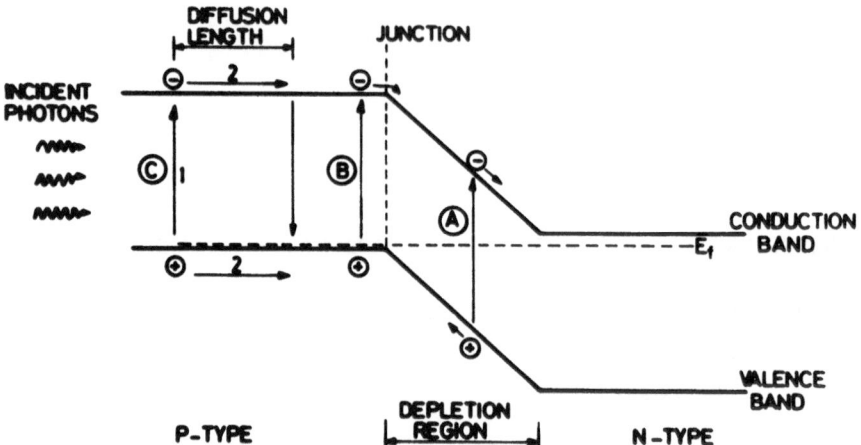

Fig. 1 Diagram of photon absorption in a p-n junction photodiode.

2. Characteristics of Photodiodes Sensitive between 0.5 and 1.5 eV

2a. Spectral Response

The fundamental aspect of photodiode performance is the
spectral response. This response is usually presented as a
function of photon wavelength :

$$\eta_Q = \frac{\text{Number of electrons collected}}{\text{Number of incident photons}} \qquad (1)$$

The calculation of the quantum efficiency does not involve the energy of the photon. Sometimes the spectral response is given in
terms of the absolute sensitivity which is related to the quantum
efficiency :

$$s = \frac{\text{electrons/sec}}{\text{incident photon power}} = e\eta_Q/\hbar\omega \;\; (\text{amperes/watt}) \quad (2)$$

The absolute sensitivity is a more useful expression because it
gives the transfer characteristic of the photodiode : photocurrent
per unit optical power. We note that it takes a greater optical
power at 0.8 μm than at 1.3 μm to achieve the same photon flux and
thus the same signal photocurrent. The quantum efficiency and
absolute sensitivity are numerically the same at λ = 1.24 μm.

Photon absorption in a photodiode is diagrammed in Fig. 1
where we show a simple energy level versus distance representation
for a p-n junction. Photons, incident on the semiconductor are
absorbed within a few absorption lengths creating electron-hole
pairs. Photo-excited carriers created in the depletion region are
separated by the electric field and are collected as majority

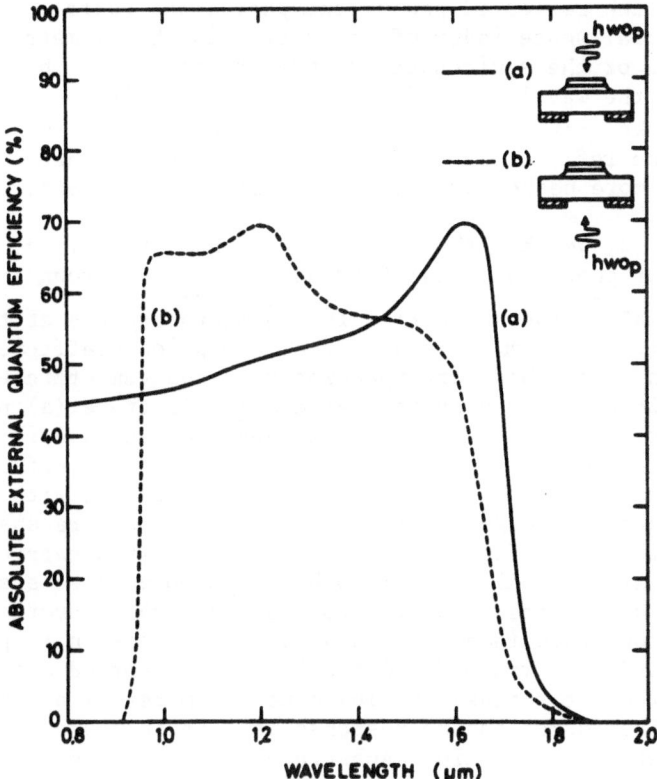

Fig. 2 Photoresponse of $Ga_{0.47}In_{0.53}As$, a direct bandgap photodiode with an absorption edge at 1.7 μm. The spectral
response is shown for direct incidence (a) and through
the substrate (b).

carriers on each side of the depletion region. Photons absorbed
within a diffusion length of the junction also create electron
hole pairs. Only the minority carrier of this pair can diffuse
into the high-field depletion region. The majority carrier is
automatically excluded by the electric field. For a photodiode
of only moderate quality, the quantum efficiency is nearly 100 %
for the photons absorbed in the semiconductor. A significant
fraction of the photon flux, however, is reflected at the surface
because of the discontinuity in the index of refraction between
the air (usually), n = 1, and the semiconductor, n $\simeq$ 3.5. The
reflection coefficient for normal incidence is :

$$R = \frac{(n_1 - n_2)^2}{(n_1 + n_2)^2} \simeq 30\% \qquad (3)$$

To decrease the reflected power, the photodiode surface is coated
with a material whose index of refraction is the geometric mean
of the index of the semiconductor and air, and whose thickness is
1/4 that of the wavelength of interest. Si_3N_4 with an index
n = 1.97 works very nicely and is easily deposited. This coating
increases the reflectivity at other wavelengths so that the overall
response is more peaked around the wavelength of interest.

   In Fig. 2, we show the spectral response of a direct band-gap
semiconductor photodiode made from $Ga_{0.47}In_{0.53}As$ grown lattice-
matched on InP.[5] The sharp rise in the photoresponse at the
bandedge λ_g = 1.7 μm reproduces the absorption coefficient until
a saturation is reached corresponding to a quantum efficiency of
65-70 %, (90-100 % for a coated detector). In trace (a) photons
are incident on the detector surface, and the decrease in photoresponse at higher photon energies is the result of surface
recombination. The absorption coefficient increases with
increasing photon energy meaning that more photons are absorbed
very close to the surface so that the photo-excited carriers
have a greater possibility of recombination on surface states.
This effect of surface recombination may be largely avoided in a
heterojunction photodiode. In this case the junction is protected
by a wide band-gap "window" and the light is absorbed deep within
the device when it reaches the low band-gap material at the
junction. The corresponding response is seen in trace (b) for
light incident from the diode substrate side, and the response is
somewhat more uniform over the 1.0-1.6 μm range.

2b. Detector Response Time
   The speed of response is determined by the larger of the
carrier transit-time across the depletion region, or the product

λ_{OPT} = 1.06 μm

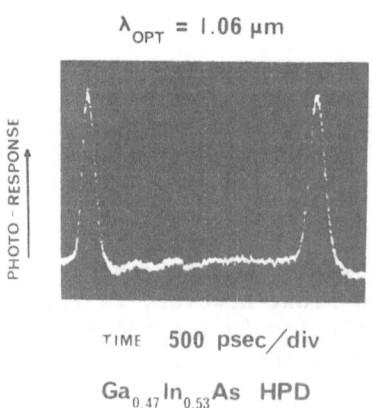

TIME 500 psec/div

$Ga_{0.47}In_{0.53}As$ HPD

Fig. 3 Response time of a $Ga_{0.47}In_{0.53}As$ photodiode to a pulsed
    Nd-YAG laser at 1.06 μm.

of the diode load resistance and capacitance. In most direct-gap
photodiodes the depletion region width is $\sim$ 2 μm and the drift
velocity is $\sim$ 10^7 cm.sec^{-1}. The transit-time limitation on
response is $\sim$ 10 psec. This is so short that these photodiodes
are almost response-limited by the circuit parameters instead.
For a small photodiode diameter (∅ = 100 μm) typical for a fiber-
optical photodiode, and a depletion region 2 μm wide, the
capacitance is on the order of 1 pf. The capacitance of the
package is on the same order, and the R C -limited rise-time
is about 100 psec. Direct-gap photodiodes of this dimension
routinely show a sub-nanosecond rise-time. In Fig. 3 we show the
response of a $Ga_{0.47}In_{0.53}As$ photodiode to a 1.06 μm Nd-YAG
laser.[6] A response time of 65 psec has been measured using
$Al_xGa_{1-x}Sb$ avalanche photodiodes sensitive between 1.0 and
1.4 μm. [7]

2c. Photodiode Sensitivity : Noise and the Minimum Detectable
    Power
      The photodiode sensitivity is determined by two factors. One
is the quantum efficiency; the other is the noise which determines
how weak a signal can be detected. In optical communications
applications, system design is based on the weakest detectable
signal. It is therefore not surprising that the noise properties
play a major role in detector selection. Noise comes from two
sources : the detector and the pre-amplifier. The signal-to-
noise ratio after the pre-amplifier can be expressed :

$$S/N = \frac{\text{Signal power}}{\text{(shot noise poser) + (amplifier noise power)}} . \quad (4)$$

The signal power is given by the square of the photocurrent.
The photocurrent is proportional to the product of the optical
power and the quantum efficiency :

$$\langle i_s \rangle^2 = \frac{1}{2}\left[2 P_{OPT} \, e\eta_a/\hbar\omega \right]^2 \quad (5)$$

P_{opt} = optical power.

The signal also carries shot noise with it. Noise power in a
good semiconductor laser is about 10-5 or 50 dB below the signal.[8]
The source shot noise is :

$$\langle i_{NS} \rangle^2 = 4e\left[(P_{opt})\left(e\eta_a/\hbar\omega\right) \right]B \quad (6)$$

The shot noise generated in the photodiode is proportional to the
dark current :

$$\langle i_{ND} \rangle^2 = 2e i_b B \quad (7)$$

In Fig. 4, we show the measured dark current density in a number
of photodiodes whose band-gaps range from 0.3 eV to 1.1 eV.

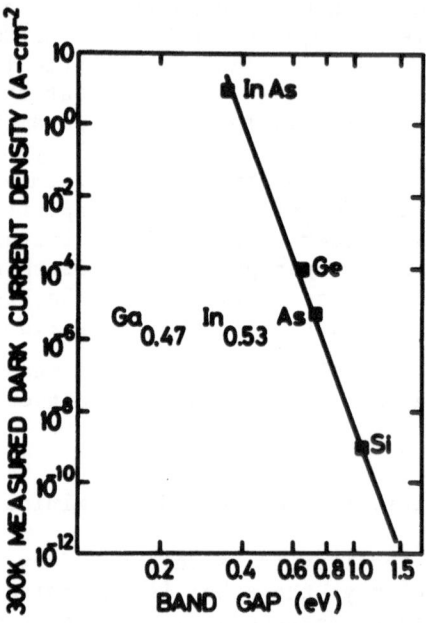

Fig.4 Measured dark current density in several semiconductor
       photodiodes. The straight line through the points is
       empirical.

        The shot noise power in the amplifier is characterized by our
effective temperature, T_{eff}. A good noise figure for a state of
the art F.E.T. Amplifier is 5 dB, corresponding to a noise
temperature of 1000K.

$$\langle i_{NA} \rangle^2 = \frac{4 \; k \; T_{eff} \; B}{R_L} \qquad (8)$$

where R_L is the photodiode load resistance. Noise is minimized
further if R_L is chosen as large as possible consistent with the
bandwidth and the detector capacitance :

$$R_L = \frac{1}{2 \pi B C} \qquad (9)$$

A quick calculation using Eqns 7 and 8 shows that the amplifier
noise level always dominates, even for Ge photodiodes. If the
photodiode is biased near breakdown the resulting avalanche gain
increases the signal power and the shot noise power, but not, of
course, the amplifier noise. The maximum useful gain is that

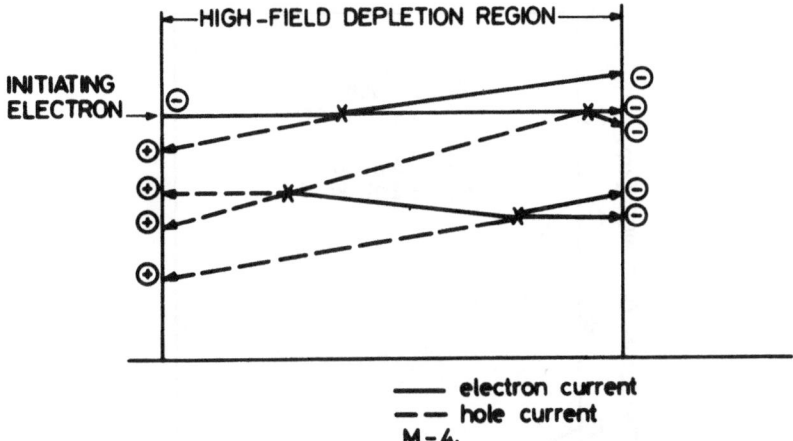

Fig. 5 Avalanche gain in a semiconductor photodiode has feed-
       back which accelerates the multiplication process and
       adds to the noise.

which increases the shot noise power to the amplifier noise level.
It is clear that the maximum useful gain will depend both on the
shot noise level and how that noise is multiplied. With this
possibility in mind, the signal-to-noise power ratio may be
written :[8]

$$\frac{S}{N} = \frac{2\left[P_{OPT}\ e\eta_a/\hbar\omega \right]^2 M^2}{\left[2ei_b + 4e\left(P_{OPT} \right)\left(e\eta_a/\hbar\omega \right) M^2 F(M) + 4k\, T_{eff}/R \right]B} \qquad (10)$$

where : M is the multiplication from avalanche gain
        and
        F(M) is the excess noise factor which depends on the
        ionization rates.

2d. Avalanche gain: Impact Ionization and Excess Noise
     Electrons and holes traversing the high-field region of a
semiconductor photodiode gain kinetic energy. In a sufficiently
high electric field they can acquire enough energy to create an
additional electron hole pair through an inelastic collision in
which the energy lost is used to promote an electron from the
valence band to the conduction band. This collision event is
called impact ionization, and the subsequent increase in junction
current is called avalanche gain. Impact ionization is
characterized by the ionization rates α , for electrons and β ,
for holes. The ionization rates give the number of secondary

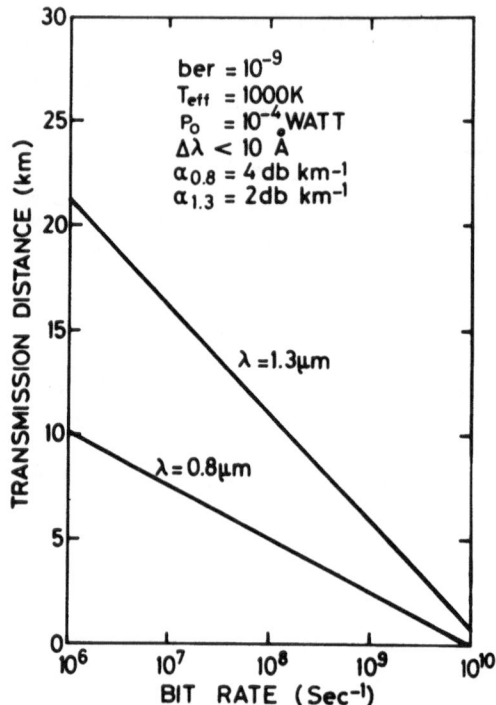

Fig. 6 Transmission distance using optical fibers and sources
at 1.3 μm and 0.8 μm as a function of bit-rate for a
10^{-9} bit-error rate.

electrons created by a single initiating carrier per cm of travel
in an electric field of magnitude $\mathcal{E}$. In general $a \neq \beta$
reflecting the different dynamic properties of electrons and
holes as summarized in the band structure.[9] Unlike the gain in
a photo-multiplier, avalanche gain has feedback, as shown in
Fig. 5. At each stage where multiplication occurs, two carriers
are created - an electron and a hole. If the primary carrier is
an electron, as is the case in Fig. 5, then the holes created by
impact ionization are the feedback carriers, because the holes
are accelerated in the direction from which the primary carrier
entered the junction. These holes may undergo ionizing collisions
and generate new electrons which can be swept back across the
high-field region to ionize additional carriers. It is this
process which leads to avalanche.

     The excess noise in avalanche gain is directly related to
the amount of feedback and can be expressed in terms of the
ionization rates a and β . As the ratio a/β becomes
more different only the carrier with the larger ionization rate

Table 1. Measured ionization rate ratios for some semiconductors whose bandgaps lie in the 0.5–1.5 eV range.

Material	Bandgap at 300K	Ionization Rate Ratio α/β Carrier Concentration (cm^{-3})		Electric Field Orientation	Ref.
		$N_D = 10^{15}$	$N_D = 10^{16}$		
GaAs	1.43 eV	0.12	1	$\langle 100\rangle$	9
		1	1	$\langle 110\rangle$	
		0.01	1	$\langle 111\rangle$	
Si	1.12 eV	8.0	4.5	$\langle 111\rangle$	16
		60.0	4.5	Unknown	
$Ga_{0.86}In_{0.14}As$	1.15 eV	-	0.25	$\langle 111\rangle$	12
$GaAs_{0.88}Sb_{0.12}$	1.15 eV	-	2.5	$\langle 100\rangle$	13
$Ga_{0.47}In_{0.53}As$	0.75 eV	-	5.0	$\langle 111\rangle$	6
Ge	0.6 eV	-	0.5	Unknown	16

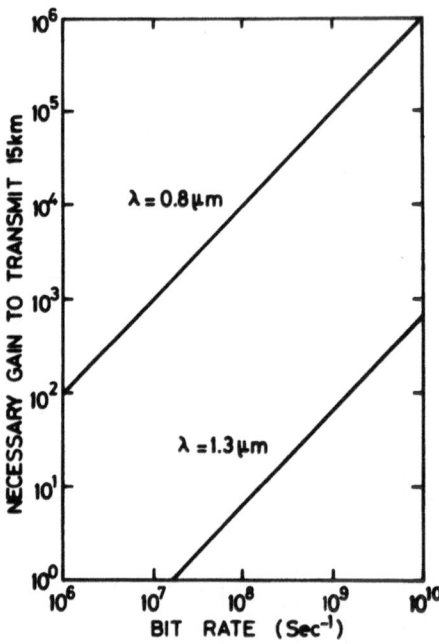

Fig. 7 Avalanche gain required to transmit 15 km at a 10^{-9} b.e.r.
at 0.8 and 1.3 μm.

contributes to impact ionization, and the excess noise factor
becomes less tending toward its lower limit of 1, provided, of
course, that the carrier with the higher ionization rate initiates
impact ionization. If the ionization rates are equal, the excess
noise is at its maximum and F is at its upper limit of M. The
detailed dependence of avalanche noise on the ionization rates
can be expressed as : [10]

$$F = M \left[1 - (1-k) \left[\frac{(M-1)}{M} \right]^2 \right] , \qquad (11)$$

for the case where electrons have the higher ionization rate and

$$k = \beta / \alpha < 1.$$

An equivalent expression exists for hole-initiated avalanche
gain.

     The dynamic properties of electrons and holes vary with
crystal orientation, and so the ionization rates and their ratio
α/β is also orientation dependent.[11] It has been shown that
this orientation dependence can be quite large in GaAs.[11]   Very
little ionization rate data including the electric field orienta-

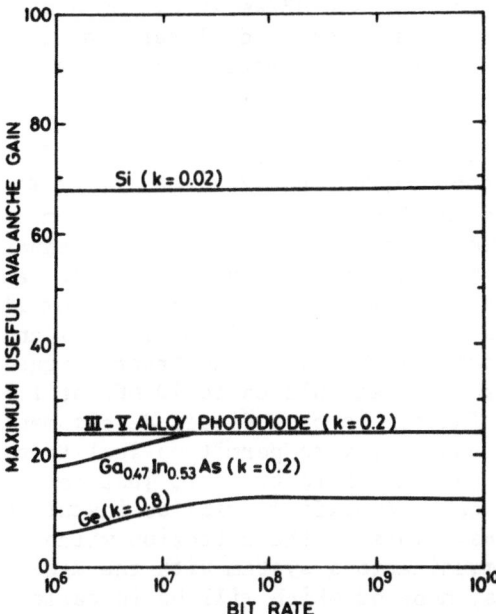

Fig. 8 Maximum useful avalance gain as a function of bit rate.
Avalanche gain is limited by laser shot noise and diode
dark current.

tion is available so that the present picture of the ionization
rates in well-studied materials like Si and Ge is somewhat
incomplete. A summary of ionization rate measurements is given
in table 1 for semiconductors of interest for optical communications. The basic feature seen in all measurements is that the
ionization rates become increasingly different as the carrier
concentration in the lower doped depletion region becomes smaller.
The excess noise factor subsequently becomes less as the doping
level is lowered : Silicon, whose excess noise factor is small,
is not unique among the semiconductors in table 1. Measured
ionization rates in GaAs indicate an excess noise factor lower
than that in Si for low doping in the high field region. Similar
behavior may be shown by III-V alloy avalanche detectors at
1.3 μm;[12-13] but sufficiently pure material has not been produced
to verify this picture.

3. An Example : Photodiode performance in Optical Fiber Communication Systems

The master equation for system performance is Eqn. 10, given
in the previous section. We will now show how this Eqn. can be
used to specify the performance of an optical fiber unit. In
this example, the transmission medium is a mono-mode fused silica

fiber whose attenuation is 4 dB/km at 0.8 μm and 2 dB/km at 1.3 μm.
The optical source is a single-mode laser, and the optical power
coupled into the fiber is 100 μWatts. First we will consider
transmission without avalanche gain in the detector. In this case,
the detector sensitivity is limited by amplifier noise. The
minimum detectable power for a bit error rate of 10^{-9} requires
that the signal photocurrent be 12 times larger than the noise
current. By equating Eqns. 5 and 7, the minimum detectable optical
power is determined. Then using the fiber attenuation characteristic, the transmission distance as a function of bandwith can be
calculated. This is shown in Fig.6. For 0.8 μm light, the
maximum transmission distance is 10 km (10^6 Hz) and at 100 MHz the
transmission distance is 5 km. By contrast, transmission distances
greater than 15 km are feasible up to 10 MHz at 1.3 μm, even using
Ge photodiodes. The aim in photodiodes development for optical
communication is not really to permit optical transmission "as
far as possible" but rather to transmit just far enough so that
the necessary repeater amplifiers will be installed in existing
switching stations. This is the criterion which will allow the
design of the communications system with the minimum number of
repeater stations, none of which will be installed in remote
locations. In this example we will take 15 km as a realistic
estimate of maximum inter switching office distance. For the case
of 0.8 μm transmission, recall that with a fiber attenuation of
4 dB/km, each order of magnitude of gain extends the transmission
distance 2.5 km. At 1 GHz it can be seen clearly that the fiber
attenuation makes severe demands on the performance of detectors
at 0.8 μm, such as Si, for 15 km transmission even with avalanche
gain. In fig.7 we show how much avalanche gain is required to
transmit 15 km at 0.8 μm and 1.3 μm. Even at moderate bit rates,
100 MHz, huge gains are required at 0.8 μm while at 1.3 μm a gain
of less than 10 will suffice. Avalanche gain is not very stable
against small changes in temperature or bias voltage near breakdown.[14-15] At a gain of 200, bias fluctuations of 0.1 V can
produce variations of ± 50% in the gain. It is unlikely that
any practical system could be built requiring avalanche gain in
excess of 500. Therefore it is clear that Si photodiodes are not
capable of resuscitating the signal at 0.8 μm, at bit rates higher
than 10 MHz in a 15 km link.

We have shown the avalanche gain needed to transmit 15 km
as a function of bit-rate, and now in Fig.8 we show the possible
avalanche gain calculated by equating the multiplied shot noise
and the amplifier noise :

$$M^2 F(M) = \frac{\text{amplifier noise}}{\text{shot noise}} \qquad (12)$$

There are two possibilities :
(a) the gain is limited by the detector dark current shot noise, or
(b) the gain is limited by the laser shot noise as is the case

Table 2. Parameters used in the calculation of maximum transmission distance by optical fibers at 0.8 and 1.3 μm using avalanche gain.

Optical Wavelength λ	Fiber Attenuation α_L	Quantum Efficiency η_Q	Amplifier Noise Temp. T_{eff}	Ionization Rate Ratio α/β
0.8 μm : GaAs Si	4 dB/km	90 %	1000K	0.02 50
1.3 μm : Ge $Ga_{0.47}In_{0.53}As$ III-V Alloy, $\lambda_g =$ 1.3 μm	2 dB/km	60 % 90 % 90 %	1000K	1.25 5.0 5.0

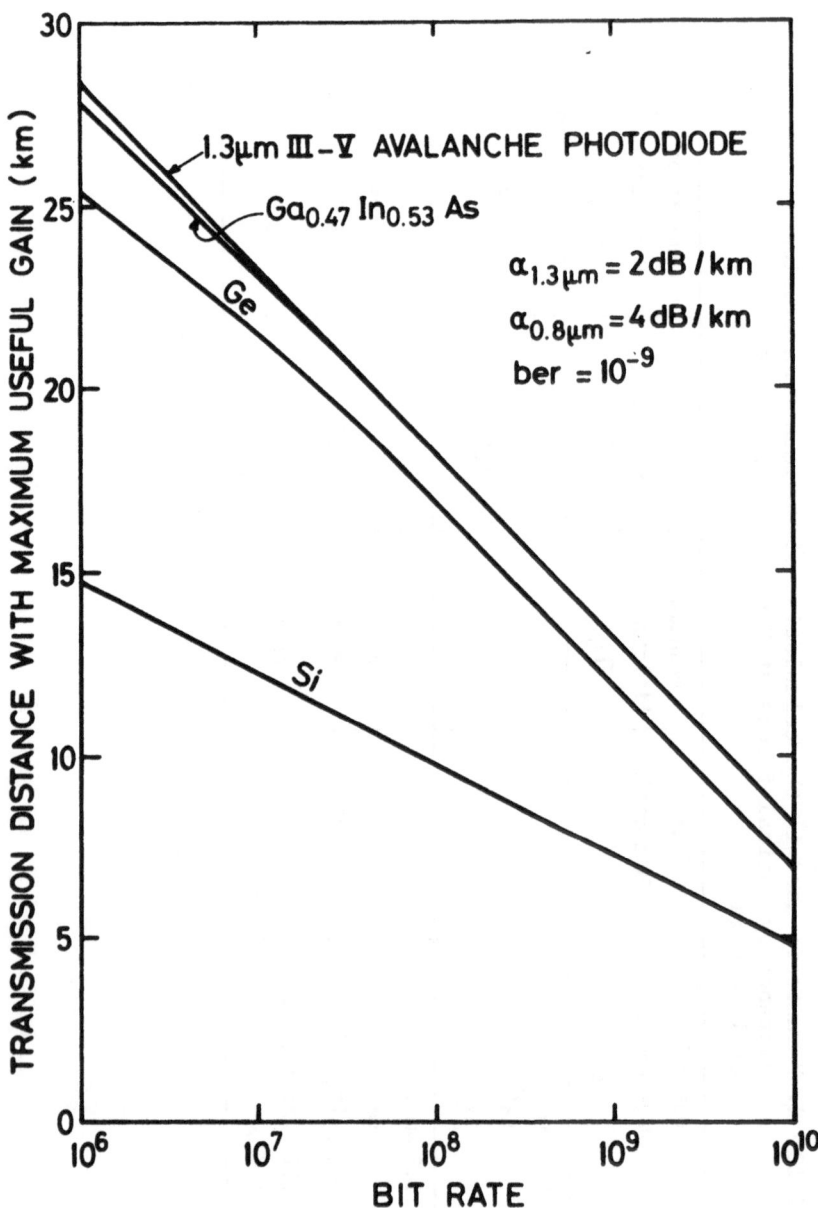

Fig. 9 Transmission distance at a 10^{-9} b.e.r. as a function of bit
rate using the maximum possible avalanche gain.

in Si whose dark current is quite low. $Ga_{0.47}In_{0.53}As$, with a dark current intermediate between Si and Ge is limited at low bit rates by detector noise and at high bit rates by laser shot noise. It is estimated that a 1.3 μm photodiode, made for example from $Ga_{0.27}In_{0.73}As_{0.60}P_{0.40}$ will be limited by laser shot noise over the bandwith range of interest to optical communica- tion. The laser shot noise power increases with bandwidth because it is proportional to the signal power. The signal power seen at the detector must increase with bandwidth to maintain the signal to noise ratio. The laser noise limits the gain in Si to 65 In $Ga_{0.47}In_{0.53}As$, the same laser noise limits the gain to $\sim$ 24. This lower noise-limited gain occurs because the excess noise is higher in $Ga_{0.47}In_{0.53}As$ than in Si.

However in optical communications at 0.8 μm, this extra factor of 2.5 favoring Si detectors extends the transmission distance less than 1 km. Using the avalanche gain shown in Fig. 8, the transmission distance as a function of bit-rate can be deduced and is shown in Fig. 9. The effective ionization ratio for Si is 0.02, Ge : 0.8, $Ga_{0.47}In_{0.53}As$: 0.2, and a III-V avalanche alloy photodiode is estimated to have an ionization rate ratio of 0.2 as well. The dark current is taken from Fig. 4 assuming a 100 μm diameter diode. These properties are summarized in Table 2. Using these results it is now possible to draw some conclusions about detector performance in optical fiber communications systems.

4. Summary

The example of the previous section illustrates the most important principle regarding photodiodes and optical communica- tions. The detector must be chosen to maximize the system performance. The converse : choosing the best detector (Si) and designing the communications system around it leads in this case to the worst system performance. This difference illustrates the point that in many applications, such as optical fiber communications, the system performance is not determined by the detector. It is difficult to imagine a better avalanche photo- diode than Si at 0.8 μm. The quantum efficiency is good, the speed sufficient, the noise very low. Yet the performance of the system at 0.8 μm is much worse than at 1.3 μm using a Ge photodiode. The culprit is the fiber attenuation which pushes Si detector really to the limit of its performance. To compete with Ge (at 1.3 μm) the silicon photodiode would have to show gains on the order of 10^4. Thus it is necessary to know both the properties of the detector and the system in which it is to be used. We have shown that the system performance can be limited by pre-amplifier noise, light source shot noise or detector dark current noise. It is these sources of noise which determine the minimum detectable signal. Which of these is largest depends on the amount of ava-

lanche gain used and the physical noise properties of both source
and detector. In the development of a high sensitivity detector one
should bear in mind the intended application. In our example we
have shown that if the majority of optical fiber links are less
than 15 km, there may be little motivation to search further than
Ge for an avalanche photodiode in optical communication. If the
majority of links will cover longer distances, then a new detector
from the III-V semiconductor alloys such as $Ga_{0.47}In_{0.53}As$ or
$Al_xGa_{1-x}Sb$ will be required.

References

1. S. Hata, K. Kajiyama, Y. Mizushima, Electron. Lett. 13, p 668
    (1977).
2. J. Conradi, F. P. Kapron, and J. C. Dyment, I.E.E.E. Trans.
    Elect. Dev. ED-25, 180 (1978).
3. J. R. Grierson, and S. O'Hara, Sol. St. Electron. 18, 1003
    (1975).
4. T. P. Pearsall, "Photodetectors for Communication by Optical
    Fibres", in Optical Fibre Communication, ed. D. V. Morgan
    and M. J. Howes, (New York, John Wiley and Sons, 1979).
5. M. Horiguchi, Electron. Lett. 12, 310 (1976).
6. T. P. Pearsall, and M. Papuchon, Appl. Phys. Lett. 33, to be
    published (1978).
7. H. D. Law, L. R. Tomasetta, K. Nakano, and J. S. Harris,
    Proc. Conf. on Integ. and Guided Wave Optics, Salt Lake City,
    (1978).
8. L. K. Anderson, and B. J. Mc Murty, Proc. I.E.E.E. 54, 1335
    (1966).
9. T. P. Pearsall, F. Capasso, R. E. Nahory, M. A. Pollack, and
    J. R. Chelikowsky, Sol. St. Electron. 21, 297 (1978).
10. P. P. Webb, R. J. Mc Intyre, and J. Conradi, R.C.A. Review 35,
    235 (1974).
11. T. P. Pearsall, R. E. Nahory, and J. R. Chelikowsky, Phys. Rev.
    Lett. 39, 295 (1977).
12. T. P. Pearsall, F. E. Nahory, and M. A. Pollack, Appl. Phys.
    Lett. 27, 329 (1975).
13. T. P. Pearsall, R. E. Nahory, and M. A. Pollack, Appl. Phys.
    Lett. 28, 403 (1976).
14. A. M. Burd, Y. A. Leichenko, B. N. Motenko, and A. S. Popov,
    Prib. Tekh. Eksp. (USSR) 18, 176 (1975).
    Trans. : Instr. and Exp. Tech. 18, 1224 (1976).
15. J. Conradi, Sol. St. Electron. 17, 99 (1974).
16. G. E. Stillman, and C. M. Wolfe, "Avalanche Photodiodes", in
    Semiconductors and Semimetals, Vol. 12, p 291 (New York,
    Academic Press, 1977).

FIBER RAMAN LASERS

R. H. Stolen

Bell Telephone Laboratories

Holmdel, New Jersey

INTRODUCTION

Optical fibers can exhibit wavelength conversion and other nonlinear optical effects at powers less than one watt. The reason has little to do with the glass in the fiber core, which is one of the least nonlinear of all materials, but rather with the tremendous interaction length possible in low-loss fibers. The combination of low power and well-controlled geometry makes fibers a particularly useful medium for the study of nonlinear optics. There will also be device implications any time nonlinear effects can be produced at low powers. Of course, the presence of such effects has implications for fiber communications since fiber transmission lines may not always be completely passive devices.

We will first deal in some detail with stimulated Raman scattering.[1] This case introduces most of the essential differences between fibers and plane wave interactions in more conventional samples as well as serving as a good example of nonlinear interactions generally. We then turn to the fiber Raman oscillator which utilizes these principles in device applications. Stimulated Raman scattering is only one of many nonlinear effects which occur in fibers, often in complicated and bewildering combinations. Several of these processes will also be described with particular emphasis on the experimental conditions which isolate one process to the exclusion of others.

STIMULATED RAMAN SCATTERING

Raman Gain

We will take an operational approach to the Raman interaction as illustrated schematically by the optically pumped amplifier of Fig. 1. A weak signal frequency ν_s is tuned around a fixed pump at ν_0. If absorption and surface reflections are neglected, the signal amplification is:

$$P_s(\ell) = P_s(o)\exp\left[g(\Delta\nu)\frac{P_p\ell}{a}\right] \tag{1}$$

where $\Delta\nu = \nu_0 - \nu_s > 0$, P_p/a is the pump intensity and ℓ is the sample length. The gain coefficient $g(\Delta\nu)$ is a measure of the strength of the Raman interaction and is plotted for fused silica[1] in Fig. 1. If $\Delta\nu < 0$ the form of $g(\Delta\nu)$ is the same but the sign is reversed so the signal will be absorbed rather than amplified.[2]

Classically, the Raman amplifier is a three wave parametric process in which pump and Stokes signal are electromagnetic waves and the idler is a highly damped vibrational wave. In the quantum treatment, a pump photon is destroyed and a signal photon and idler phonon are created. The process could be stimulated by either the signal or idler but, because of damping, the phonon population is negligible.

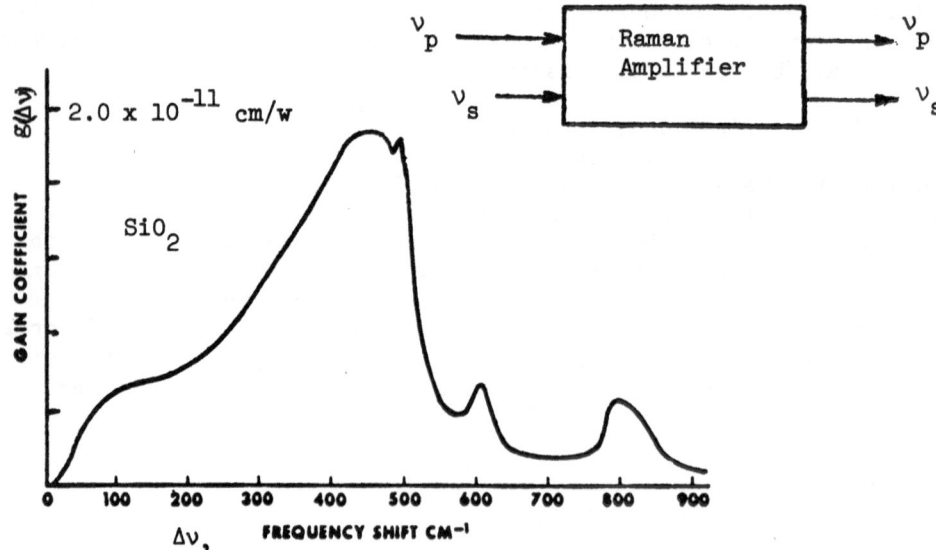

1. Schematic Raman amplifier and Raman gain curve for silica

The Raman gain coefficient is related to the Raman cross section as measured by conventional spontaneous scattering:

$$g(\Delta \nu) = \frac{\sigma_0(\Delta \nu)c}{h \varepsilon \nu_s^3} \tag{2}$$

where c is velocity of light, ε the dielectric constant, h is Planck's constant and ν_s the Stokes frequency. $\sigma_0(\Delta \nu)$ is the Raman cross section which would be measured at low temperatures. The usual Raman cross section increases with temperature because of thermal excitation of vibrational states while the Raman gain is temperature independent. Note that Eq. 2 contains the factor ν_s^{-3} while σ_0 varies approximately as ν_s^4 so the Raman gain is linear in Stokes frequency. An important feature of the gain curve in glasses is the possibility of wideband tuning. The Raman spectrum in glasses is about two orders of magnitude larger than in crystals because the usual wave vector selection rules break down and the spectrum reflects the entire density of states of the glass.[3]

Raman Gain in Fibers

In a fiber, the spot size a, in Eq. 1, is approximately the core area and ℓ is limited only by the fiber absorption. Fiber cores can be as small as 2μm and lengths can be kilometers, so enhancements of 10^8 are possible in comparison with the focal region of a gaussian beam. The peak Raman gains in glasses are more than two orders of magnitude less than the strongest Raman scatterers, but the length factor more than overcomes this reduction in cross section and, as can be seen from Eq. 1, this increase in length lowers the power requirements for amplification. This trade-off between length and power is in fact the central element of nonlinear optics in fibers and applies to a wide range of nonlinear processes in addition to the stimulated Raman effect.

In a long fiber, it is seldom possible to ignore the loss of either the pump or the Stokes wave. For simplicity we assume that the loss, α, is the same at the two wavelengths and that there is no pump depletion. The differential equation for a forward traveling Stokes wave is then:[4]

$$\frac{dP_s(z)}{dz} = \left[g \frac{P_p(z)}{a} - \alpha \right] P_s(z)$$

$$P_p(z) = P_p(o)e^{-\alpha z} \tag{3}$$

Integrating Eq. 3 from 0 to ℓ produces the generalized form of Eq. 1.

$$P_s(\ell) = P_s(o)\exp\left[-\alpha\ell + g\frac{P_p(o)}{a}L\right]$$

$$L = \frac{1 - e^{-\alpha\ell}}{\alpha}$$

(4)

where L is an effective fiber length[5] and incorporates the fact
that most of the amplification in a long fiber takes place closer
to the input end where the pump intensity is strong. For a short
fiber L = ℓ, while for a long fiber, L = $1/\alpha$.

The spot size is also modified in fibers because the gain
varies across the transverse intensity profile of the guide modes.
This gain variation can be accounted for by using an effective
core area which is derived from an overlap integral. This integral
can be derived from Eq. 3 written in terms of intensities which
depend on r and θ as well as z.

$$\frac{dI_s(r,\theta,z)}{dz} = gI_s(r,\theta,z)I_p(r,\theta,z)$$

(5)

$$I(r,\theta,z) = const \left|E_m(z)\right|^2 \psi^2(r,\theta)$$

where $\psi(r,\theta)$ is the mode field and E_m is the peak field. Integration
of Eq. 5 leads to the fiber version of Eq. 3:

$$\frac{dP_s(z)}{dz} = gP_p(z)P_s(z)\frac{\langle\psi_s^2\,\psi_p^2\rangle}{\langle\psi_s^2\rangle\langle\psi_p^2\rangle} = \frac{gP_pP_s}{A_{eff}}$$

(6)

where
$$P(z) = \int_o^{2\pi}\int_o^{\infty} I(r,\theta,z)rdrd\theta$$

(7)

$$\langle\psi^2\rangle = \int_o^{2\pi}\int_o^{\infty} \psi^2\, rdrd\theta$$

Any variation in r of the gain coefficient g has been neglected
which is a good approximation in weakly doped silica based fibers.

The ratio of the effective to actual core area for a single-
mode step-index fiber can be calculated as a function of normalized
frequency V. For V = 1.5, 2.0, 2.5, and 3.0, the ratios are 2.43,
1.47, 1.10 and 0.93. If the mode is approximated by a gaussian:

$$\psi_s^2(r,\theta) = \psi_p^2(r,\theta) = e^{-r^2/w^2}$$

(8)

$A_{eff} = 2\pi w^2$ where w is the 1/e intensity radius. A gaussian fit
to a step index mode produces an A_{eff} which is 10–15% too large.
The gaussian approximation may actually be better for highly

graded cores[6] and most real fibers probably lie somewhere between the cases of step-index and gaussian modes.

Most of the present considerations deal with fibers that support one or a small number of transverse modes. All these nonlinear effects will, of course, appear in large-core multimode fibers at correspondingly higher powers. For such highly multi-moded guides, mode mixing and the distribution of power between the modes become important.[7] For rough estimates, the core can be regarded as uniformly filled so the spot size is just the core area.

The effect of polarization on the Raman gain can be different in fibers than in bulk samples because strains and defects scramble polarization in most fibers. Raman gain coefficients are usually quoted for parallel linearly polarized pump and Stokes fields[1], but the cross section[8], and hence the gain, is typically reduced by an order of magnitude for perpendicular polarizations. Both pump and Stokes waves see the same strains and will for some distance experience the same depolarization but, eventually, because their wavelengths differ, the relative polarizations will become completely random. Actually, maximum gain occurs not just for linear polarizations but whenever pump and Stokes polarizations are the same[9], so the distance over which the gain decreases is not the depolarization distance but rather the distance in which the pump and Stokes polarizations get out of step. This length will depend inversely on the frequency separation and on the magnitude of the strains in the fiber itself.

$$\ell_p \sim (\Delta\nu\,\delta n)^{-1} \qquad\qquad\qquad (9)$$

If $\ell \gg \ell_p$ polarizations are completely scrambled and the best procedure seems to be to reduce the gain coefficient by a factor of two. It is not known what to use for δn but threshold measurements in various single-mode fibers suggest values of ℓ_p around 1-10 m. Similar considerations apply to multimode fibers except that complete polarization scrambling takes place in much shorter lengths.

There can also be differences between cw and pulsed gains. Fortunately, the Raman gain band is always much broader than the pump linewidth so the gain coefficient is the same for pulses and cw.[2] The problems arise because of pulse spreading which reduces the peak intensity and by physical separation of pump and Stokes pulses due to group velocity dispersion.

The walk-off between pump and Stokes pulses can be corrected by group velocity matching using waveguide modes. The approach is illustrated in Fig. 2 with a plot of normalized group delay[10] vs. V for the two lowest order modes. The group delay per unit length between a pulse at λ and a pulse at $\lambda+\Delta\lambda$ from material dispersion is:

$$\frac{\Delta t_{MA}}{\ell} = M(\lambda)\Delta\lambda \qquad\qquad\qquad (10)$$

where $M(\lambda) = (-\lambda/c)(d^2n/d\lambda^2)$ is the material dispersion and ℓ is the fiber length. Δt is about 700 ps for typical Raman shifts in 100 m silica fibers using argon laser pump lines. The Stokes pulse travels faster so it must propagate in the mode with greater group delay. In the example, V = 4.0 so compensation occurs with pump in the LP_{01} and Stokes in the LP_{11} mode. The difference between the group delays in Fig. 2 is:

$$\frac{\Delta t_{MO}}{\ell} = \Delta n\ (\tau_{11} - \tau_{01})/c \qquad\qquad\qquad (11)$$

where Δn is the core–cladding index difference. Group matching occurs at that $\Delta\lambda$ where $\Delta t_{MA} + \Delta t_{MO} = 0$. So far, group velocity matching has not been observed in single pass stimulated Raman scattering but has occurred in repetitively pulsed Raman oscillators.[11]

Stimulated Raman Threshold

It is not necessary to insert the Stokes signal to see amplified Stokes output because there will always be some spontaneous Raman scattering. If the power is large or the fiber long, the spontaneously scattered light can be amplified to intensities comparable to the pump. Figure 3 shows a basic experiment of non-linear optics in fibers. Pulsed light is coupled into a long fiber

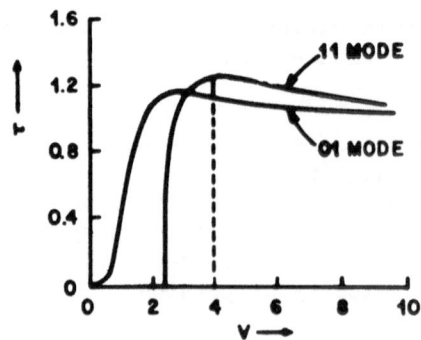

2.   Normalized group delay vs. V.

and the output light is dispersed by a prism. The output consists
of residual pump light and several Stokes orders showing that
sequential Stokes conversion takes place in a long fiber. The
color change along the fiber shows up clearly in the side-scattered
light. Examination of the time structure of the various pulses
shows that conversion is almost complete near the center of each
pulse. Even in this basic form, the fiber Raman laser has proved
a useful tool for generating a series of pulses in the 1.1-1.5 μm
range for fiber dispersion studies[12].

The Stokes output is the sum of all the spontaneous emission
weighted by the net gain. This output expressed as the number
of Stokes photons is:

$$N_s(\ell) = N_s(0)\exp\left[-\alpha\ell + g\frac{P_p(o)}{A}L\right]$$

$$N_s(0) = 1 - \exp\left[-g\frac{P_p(o)}{A}\ell\right]$$

(12)

Linear loss has been included but pump depletion is a much harder
problem and is usually ignored with the argument that one is most
interested in calculating the threshold power. In a long fiber
$N_S(0) \approx 1.0$ showing that adding up all the spontaneous emission
is equivalent to assuming an input of one photon. In the limit
$\ell \to 0$ the Stokes output goes to zero and at low powers the Stokes
output is the ordinary spontaneous Raman scattering.

There is no exact stimulated Raman threshold but it is useful
to define a critical power for which the Stokes power increases
from noise to equal the pump power.[4] This condition is satisfied
for:

$$\gamma^{3/2}e^{-\gamma} = \beta = \frac{\sqrt{\pi}}{2}h\nu_s g_o \frac{\Delta\nu_{1/2}}{A}L$$

$$\gamma = g_o\frac{P_p(0)}{A}L \approx 16$$

(13)

where g_0 is the peak Raman gain and $\Delta\nu_{1/2}$ is the Raman bandwidth
(FWHM) and L the effective length. For a typical long fiber with
a 10^{-7} cm^2 core area and a 20 dB/km loss at 500 nm, β is 4×10^{-6} and
γ is approximately 16. γ varies only slowly with changes in fiber

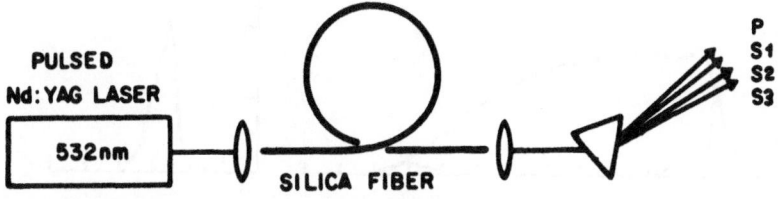

3. Single pass stimulated Raman scattering in a fiber.

parameters – a factor of 100 change in β changes γ by only about
25%. There are still questions concerning thresholds in highly
multimoded fibers. β will be increased by the number of modes
but decreased when overlap integrals are considered. It appears
that γ will again be close to 16.

     Some caution is necessary in measuring threshold powers using
pulses whose bandwidth is much larger than the transform limit.
For example, a long pulse consisting of a train of mode-locked
pulses will have a higher apparent Stokes gain in a short fiber
than in a long fiber where group velocity dispersion begins to play
a role. Similar behavior would be expected from non mode-locked
wide band pulses because of the sharp microstructure in their
time behavior. For fine time structure and long fibers, one would
expect to see gain characteristic of the overall pulse.

Backward Stimulated Raman Scattering

     The gains for forward and backward Raman amplification will be
the same at low pump powers but the two configurations differ
greatly in the pump saturation regime. In forward amplification
the Stokes power can be no greater than the original pump power
while in the backward direction it is possible to extract all of
the pump energy by the leading edge of the Stokes pulse.[13]

     This is shown in Fig. 4 which shows a 200 ns pulse before and
after transmission through a 90 m 6 μm core fiber.[14] The sharp
spike on the leading edge of the pulse produced forward stimulated
Raman scattering in a 20 ns pulse. An external mirror was used to
feed this pulse back into the fiber. This backward pulse then
swept out the rest of the pump energy and produced a 5 ns Stokes
pulse. The holes in the pump pulse from both forward and backward
Stokes passes can be clearly seen.

     The limit to the usable fiber or pump pulse length is set by
conversion of backward amplified Stokes to second Stokes. The
combination of pump depletion and higher order Stokes conversion

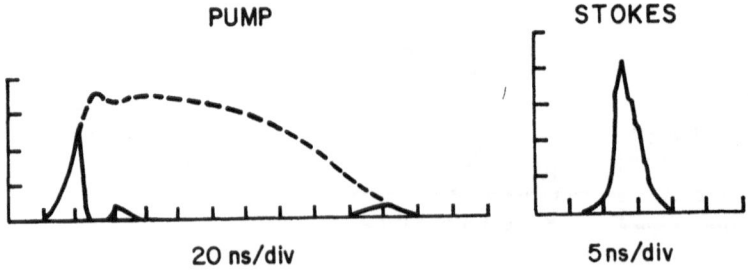

4. Backward stimulated Raman scattering showing initial and
    depleted pump and backward Stokes output.

is a complicated problem. It should be pointed out, however, that the process in a single-mode fiber is simpler to deal with than the same process in bulk samples with focused beams.

Practical Problems

There are a few practical details involving the handling of small core single-mode fibers and accuracy of measurement that bear mentioning. For example, the tilt of the input end of such fibers is much more critical than for large-core multimode fibers. It is helpful to first couple light from the output end and use this as a reference to adjust position and tilt of the fiber input with the input lens removed. The back-coupled beam also facilitates initial positioning of the input lens and provides a check on the match between input beam, lens magnification and mode size. The best lenses for many applications are 20X uncoated microscope objectives since expensive objectives seem more prone to damage. Where low insertion loss is required, long-working-distance coated 20-40X objectives work well. The mode quality of the pump laser is extremely important. Coupling efficiency can vary between 20% and 90% for different lasers, all of which have at least nominal TEM_{oo} outputs.

Power in the fiber core is measured from the output power. This requires an accurate value for the loss, and mode strippers to eliminate spurious light guided in the fiber substrate. These are standard techniques. Loss is determined from the ratio of output powers before and after breaking the fiber near the input end. Mode stripping is accomplished with index oil such as glycerine near both ends.

The greatest uncertainty lies in the value of the core size. This is complicated by the fact that the core-cladding interface is usually graded and not only is the diameter ill-defined, but the mode shape itself is uncertain. Fortunately, the diameter

5. Photograph of the end of a birefringent fiber.

obtained by looking at a short section of fiber with a microscope
is usually reliable to +10%. Improvement in accuracy can be
obtained by comparing the far field pattern with the far field
transform of the estimated guide mode.[1]

Recently, single-mode fibers which maintain linear polarization
have been developed.[15] Figure 5 is a photograph of the end of such
a fiber and shows the round silica core and substrate tube and
the elliptical borosilicate cladding. The fiber maintains linear
polarization by virtue of its birefringence which removes the
degeneracy between the two orthogonal polarizations of the fundamental mode. This birefringence is due to the anisotropic strain
introduced by the elliptical cladding. The polarization axes
which lie along the principal axes of the cladding ellipse can
be identified for alignment purposes by viewing the input end of
the fiber.

RAMAN OSCILLATORS

cw Oscillators

When mirrors are used to feed back Stokes light into the
fiber, the Raman amplifier becomes an oscillator. The addition
of frequency selective elements permits tuning over the broad
Raman gain curve shown in Fig. 1.

A simple Raman oscillator is illustrated in Fig. 6. Pump
light is coupled through mirror M1 which is highly reflecting at
the Stokes wavelength and M2 is the output mirror. Usually, M1
is the output mirror of the pump laser.[16] The feedback with high
reflectivity mirrors can be better than 90%.

At the threshold power the round trip gain equals the loss.

$$2g \; \frac{P_p(o)}{A} \; L \;\; = \;\; \frac{(\text{loss in dB})}{10} \; \ln(10) \qquad\qquad (14)$$

The factor of 2 comes from adding forward and backward gain.

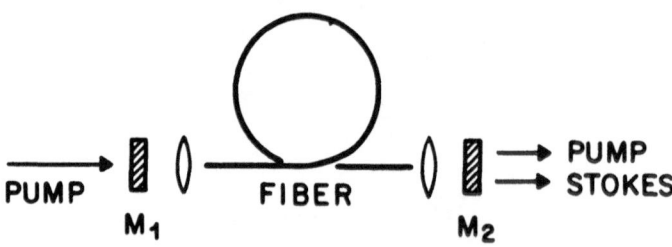

6. Basic fiber Raman oscillator.

We can estimate the threshold power for a 100 m single-mode polarizing fiber. We choose a 2.5 μm core diameter and an index difference such that V = 2.5 at a 514.5 nm pump wavelength. The loss is 20 dB/km (α = 4.6 x 10^{-5} cm^{-1}) so L from Eq. 4 is 80.1 m. A is 1.10 times the core area from the overlap integral, and g is 1.90 x 10^{-11} cm/w from Fig. 1 and taking $g \sim \nu_s$. We estimate the feedback loss on each end as 1 dB so the total loss in one round trip is 6 dB which amounts to a net transmission of 25%. There is, perhaps, a $\pm$ 1 dB uncertainty in the round trip coupling loss but this is only a 15% uncertainty in threshold power. Threshold power in this example is 245 mw. This power is not a very sensitive function of length. For example, threshold is 293 mw in a 50 m length of the same fiber and 251 mw in a 200 m length. The actual Stokes output power will be greater for the shorter fiber since the primary loss is absorption and scattering in the fiber.

Threshold powers in real fibers are very close to the predicted values. Thresholds around 200 mw have been measured using 2.5 μm core polarizing fibers pumped by 488 or 514.5 nm. The good agreement is a little surprising since there are several problems with these simple oscillators. Feedback from the fiber and output mirror affects the output power of the pump laser and there is competition from stimulated Brillouin scattering which will be discussed later. Also the use of the pump laser output mirror to provide feedback means that the position of the input lens must be a compromise between optimum coupling and feedback.

Figure 7 shows a fiber Raman oscillator with prism tuning.[17] At a power twice the threshold power, the tuning range from Fig. 1 will be 240 cm^{-1}. The limit for the tuning range is set by feedback from the fiber end. At first sight, the effect of this 4% reflection would appear negligible since it amounts to a 14 dB loss as compared to the estimated 1 dB loss using the mirror. However, the loss part of Eq. 14 has only increased from 0.6 ln10 to 1.9 ln10 which corresponds to an increase in threshold power by a factor of 3.2.

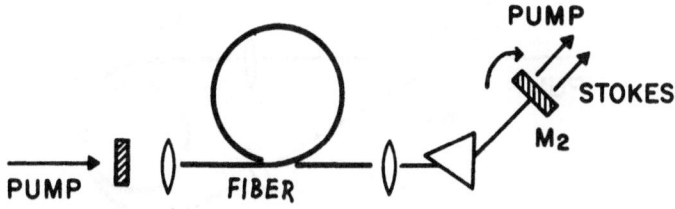

7. Fiber Raman oscillator with prism tuning.

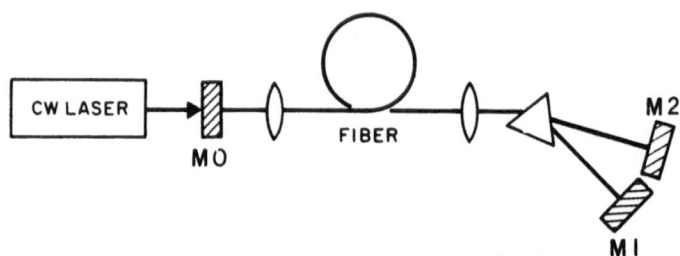

8.   Multiple Stokes cavity.

Well above threshold, the conversion of pump to Stokes light
is almost complete. The Stokes power within the cavity is in
fact high enough to act as pump for tunable second Stokes as
illustrated in Fig. 8. By adding more feedback mirrors, four orders
of tunable Stokes have been obtained from a 100 m 3.3 μm core
fiber pumped with 8 w at 514.5 nm.[18] The total tuning range was
350Å.

We can estimate the Stokes intensity in the cavity but it
is easier to work with the ring configuration shown in Fig. 9.
We assume for simplicity that Stokes is only amplified in the
forward direction and that at the fiber output all the pump has
been converted to Stokes power. If P_o is the pump power and T
is the transmission of the coupling loop exclusive of fiber loss,
the Stokes power at the fiber input end will be:

$$P_s = P_o Te^{-\alpha\ell}(1 + Te^{-\alpha\ell} + T^2 e^{-2\alpha\ell} + \ldots) = \frac{P_o Te^{-\alpha\ell}}{1 - Te^{-\alpha\ell}} \qquad (15)$$

If the feedback coupling loss is 1 dB and the fiber loss is 2 dB,
then $Te^{-\alpha\ell} = 0.5$. P_s is then equal to the initial pump power.

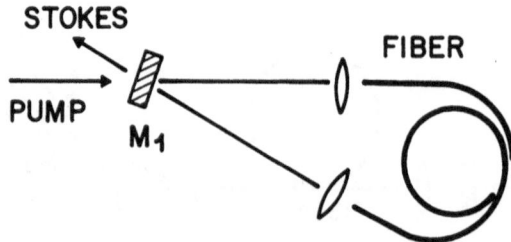

9.   Riber Raman ring laser

Pulsed Oscillators

Fiber Raman oscillators can be pumped with repetitively pulsed as well as cw lasers. The condition for pulsed operation is that of synchronous pumping - the pump pulse spacing must be an integral multiple of the cavity round trip time which is typically around a μs. This is illustrated in Fig. 10 where the cavity length is adjusted by moving the output mirror so the fed-back Stokes pulse lines up in time with a succeeding pump pulse. Equivalently, synchronization could be achieved by varying the pump pulse spacing.

In pulsed as opposed to cw operation, the backward traveling Stokes wave sees no gain and the ring configuration of Fig. 9 is more efficient than that of Fig. 10. Such a fiber Raman ring laser has been pumped by a cavity-dumped argon laser and tuned with a piezoelectric intracavity interferometer.[19] The cavity-dumped laser appears to be quite stable with respect to fiber-pump cavity interactions and as will be discussed later, pulsed operation discriminates against stimulated Brillouin scattering.

The cavity length (or pump spacing) required for synchronization will depend on Stokes wavelength because of group velocity dispersion. The oscillator can in fact be tuned this way - the technique is called "time-dispersion-tuning" to differentiate it from spatial dispersion using prisms or gratings. Time-dispersion-tuned oscillators similar to Fig. 10 have been pumped by mode-locked argon and Nd:YAG lasers.[20] Mode-locked pulses are spaced much closer (∿10 ns) than the fiber round-trip time (1-5 μs) so the Stokes pulse is synchronized, not with the next pump pulse, but with a much later pulse.

Optimum time-dispersion tuning requires enough group-velocity dispersion to just about separate the pump and Stokes pulses in one pass through the fiber. This imposes constraints on pulse and fiber length over and above those set by gain and linear

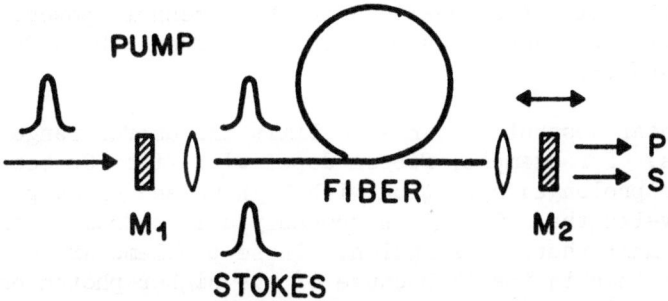

10. Synchronization in a repetitively pulsed fiber oscillator.

absorption. If the pulses are too short, they will separate in space long before the end of the fiber and the Stokes pulse will experience no gain while continuing to see the fiber loss. On the other hand, if the pulses are too long, oscillation will take place at the peak of the gain curve and there will be no tuning.

The time delay in a single pass through the fiber between two Stokes pulses separated in wavelength by $\Delta\lambda$ is given by Eq. 10. By translating the output mirror $\Delta\ell = c\Delta\tau$ the synchronized Stokes wavelength is changed by $\Delta\lambda$ so the slope of the tuning curve is:

$$\frac{\Delta\lambda}{\Delta\ell} = \frac{1}{c\,\ell M(\lambda)} \tag{16}$$

This slope was 6.9Å/cm in a 100 m oscillator pumped at 514.5 nm and 18Å/cm in 600 m with a 1.06 µm pump.[20]

Reflected pump light from the tuning mirror tends to broaden the mode-locked pulse so this was often decoupled by means of a prism after the fiber output end. This required some tilting of the mirror while it was translated but in this way tuning was achieved over the Raman gain curve of Fig. 1 from 20 to 600 cm^{-1}. As long as the fiber output end does not form a cavity in synchronism with the pump, the limit in pump power is set by single pass stimulated Raman scattering. Once this threshold is exceeded, the stimulated first Stokes can be used to pump time-dispersion tuned second Stokes.

Materials Considerations

The spectral region of operation of fiber Raman oscillators is controlled by the transmission properties of the glass in the fiber core. Fiber loss generally follows the ν^4 Rayleigh scattering limit so fiber Raman oscillators should work even better in the 1 µm region than in the visible. Because the gain varies as ν and the mode area will be larger at longer wavelengths, the full factor of ν^4 will not be realized in the threshold power, but one might expect lower thresholds in the infrared by a factor between ν and ν^2.

Nonlinear absorption can also limit the useful range of fiber oscillators. For example, fibers doped with Ti or Ge become lossy after prolonged operation with high pulsed or cw powers at visible wavelengths. This loss probably arises from color centers created by multiphoton absorption. These problems are more severe in the blue than in the IR because of the higher photon energy. This was seen in pure GeO_2 core fibers which were damaged by 532 nm light but withstood high peak powers at 1.06 um.[21]

From the point of view of oscillator characteristics, silica is not necessarily the best material. It would be desirable to have both larger cross sections and flatter gain curves. Figure 11 shows a rough comparison between the gain curves of three glasses as derived from spontaneous Raman spectra. The peak gain in germania core fibers is a factor of 10 larger than in fused silica and this was verified by measuring the stimulated Raman threshold at 1.06 μm.[21] Unfortunately, the loss was more than an order of magnitude larger than in corresponding silica fibers. The lead silicate glass is Schott F2 and shows that desirable gain curves are possible, at least in principle, although once again such glasses have not been used in low-loss fibers.

OTHER NONLINEAR PROCESSES

Stimulated Brillouin Scattering

There are many nonlinear optical processes which occur in fibers. Of these, perhaps stimulated Brillouin scattering appears at the lowest power.[22] Brillouin scattering can be viewed as Raman scattering from acoustic modes with the difference that wave vector considerations limit Brillouin amplification to the backward direction in fibers while both forward and backward Raman gains are possible.

The peak Brillouin cross section in fused silica is about 300 times that for Raman scattering so at first sight, it is surprising that stimulated Raman scattering can be seen at all.

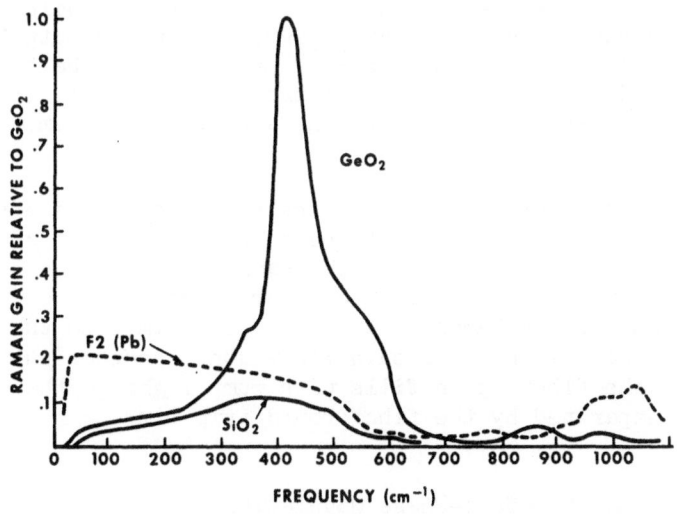

11. Comparison between the Raman gains of SiO_2, GeO_2 and Pb glasses.

By using pulses, however, stimulated Brillouin scattering can be easily suppressed since the forward Raman interaction length is determined by pulse walk-off while the Brillouin interaction is limited to the pulse length.

In cw Raman oscillators, one can take advantage of the fact that the narrow Brillouin linewidth is inefficiently pumped by typical lasers. For example, the linewidth at 514.5 nm is 145 MHz[23] so using the factor:[24]

$$G = G_0 \frac{\Delta\nu(\text{Brillouin})}{\Delta\nu\ (\text{pump})} \qquad\qquad (17)$$

the Brillouin gain would be reduced by a factor of 31 for a 4.5 GHz pump linewidth. This factor alone would not suppress Brillouin oscillation in the simple oscillator of Fig. 6 where feedback is provided at both Raman and Brillouin frequencies. In the tunable oscillator of Fig. 7, however, the comparison is between Brillouin oscillation from Ml and the fiber end vs. Raman oscillation with the feedback mirror. As previously discussed, the difference in coupling efficiency amounts to about 13 dB or a factor of 20. This, combined with the linewidth factor, should be more than sufficient to favor Raman over Brillouin in the prism-tuned oscillator.

In many oscillators, however, Brillouin oscillation is seen even with the prism in the cavity. Examination of the pump spectrum shows that interactions between the fiber and pump cavity act to reduce the number of longitudinal pump modes from around 30 to about 5 which then favors Brillouin oscillation. By using an etalon in the pump laser to produce single frequency output, Brillouin oscillation is observed[22,25] at powers far below the Raman threshold. In some sense, the presence of Brillouin Stokes power is not too serious since it will also pump the Raman Stokes line.

Brillouin scattering is not entirely a deleterious effect. Recently Brillouin oscillation has been used to convert cw laser light into 10 ns pulses.[26] Here, the backward Brillouin light sweeps out the pump power producing a strong sharp pulse similar to backward stimulated Raman scattering. The limit to the pulse shortening is set by the Brillouin linewidth. Another pulse cannot appear until the fiber again fills with pump light so these pulses will appear separated by the fiber round trip time.[22]

Self-Phase Modulation

One particularly ubiquitous nonlinear effect is the intensity

dependent refractive index which usually leads to self-focusing.
In fibers any additional confinement caused by self-focusing is
negligible but the resultant phase shifts, even though small, can
add up in a long fiber and have significant effects.

This is illustrated by Fig. 12, which shows the electric
field before and after the fiber. Because of the intensity dependent refractive index, the phase at the peak has been delayed with
respect to the wings. The phase modulated output pulse will then
be broadened in frequency. Measured spectra from a 100 m 3.35 μm
silica core fiber are shown in Fig. 13.[27] In this fiber, a peak
power of 3 w results in a factor of 10 frequency broadening. Such
spectra can be used to obtain the pulse width by the relation[28]

$$\Delta\tau \ (FW \ 1/e) = 1.72 \ \Delta\phi_{max} \ /\pi\Delta f \ (FW \ 1/e) \tag{18}$$

and in this example, the value was 155 ps which was in excellent
agreement with measurements using a fast diode.

Self-phase modulation will limit the ultimate linewidth of
any pulsed fiber Raman oscillator. Frequency broadening will
also appear in single pass stimulated Raman scattering experiments.
In this connection, it is worth noting that the pulses from most
Q-switched lasers are not transform limited and the value of Δτ
which governs the broadening is not the total pulsewidth, but
more likely, the width of sharp microstructure within the pulse.
A useful consequence of this is the smooth continuum produced
when the pump is the broad-band superfluorescent output of a
nitrogen laser pumped dye laser.[29]

The self-phase modulation results illustrate another useful
aspect of nonlinear measurements in fibers - the use of low
powers and well-controlled geometry to measure nonlinear
coefficients.

$$\delta n = \frac{\lambda\Delta\phi}{2\pi L} = \frac{4\pi n_2 P \times 10^7}{ncA} \tag{19}$$

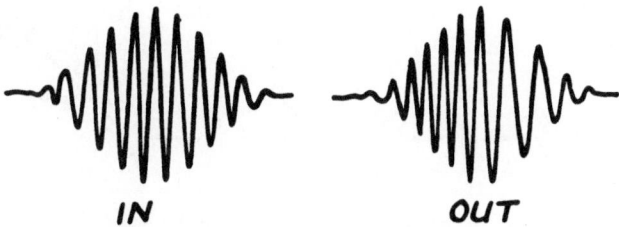

IN   OUT

12. Effect of the intensity dependent index on the phase of the
     electric field of an optical pulse.

so knowledge of the fiber parameters such as **effective length**
and area and the power in watts gives n_2 which is the self focusing
coefficient. From these experiments, a value of 1.1×10^{-13} was
obtained which is in excellent agreement with values obtained
by other techniques.[30]

Similar experiments can be done to measure the Kerr index,
which is the index change induced by a different frequency,[31] and
to measure the Raman cross section directly by the Stokes
amplification.[1] Since spontaneous spectra from silica and benzene
are easily compared, the gain method can be viewed as a completely
independent technique for measuring the standard Raman cross
section of benzene.

Four-Photon Mixing

Most of the nonlinear processes observed in fibers are based
on the $\chi_3 E^3$ term in the polarization expansion. The $\chi_2 E^2$ term
which produces frequency doubling and three photon parametric

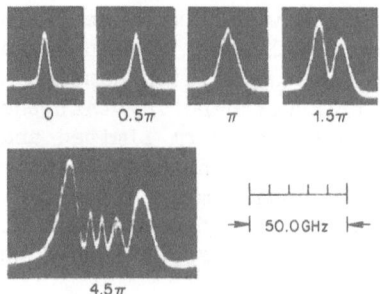

13. Frequency broadening of a mode-locked argon laser pulse in
    a 100 m fiber. The maximum phase shift in units of π is
    proportional to peak intensity.

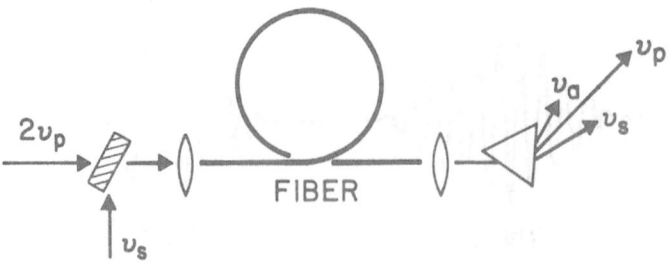

14. Four-wave mixing in which two pump photons at ν_p are mixed
    with a signal at ν_s to produce ν_a.

interactions is zero in glasses because of inversion symmetry. Raman and Brillouin come from the imaginary part of χ_3 and the real part gives rise to the nonlinear index and four-photon mixing.

A four-photon mixing experiment is illustrated in Fig. 14. Here, two green pump photons at ν_p are mixed with a tunable red signal at ν_s to generate a blue idler at ν_a. In the limit of no pump depletion or significant Stokes amplification Maxwell's Eq. becomes:

$$\frac{dE_a}{dz} = \text{const} \quad \chi_3 E_p E_p E_s^*$$
(20)

and energy conservation requires that $\nu_a = 2\nu_p - \nu_s$.

The central problem in four-photon interactions is wave vector matching. Because of index despersion, the wavelength of the polarization produced by mixing $2\nu_p - \nu_s$ is slightly less than that of the free running wave at ν_a. This is no problem if distances are short but in a long sample, the electric field at ν_a will get out of phase with the polarization wave and energy will start to flow back from ν_a to ν_p. The distance for this to happen is called the coherence length, L_c.

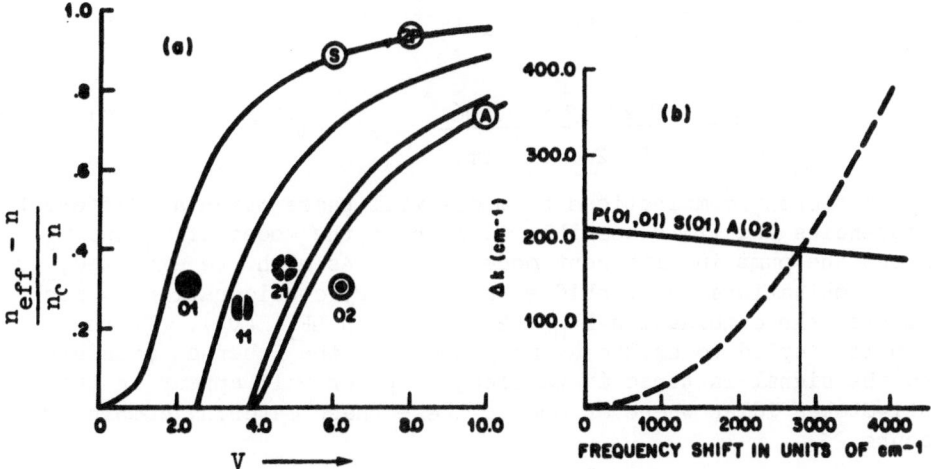

15. (a) Plot of normalized effective index vs. V showing one combination of modes for phase matching. (b) Plots of wave vector mismatch vs. $\Delta\nu$ in silica for a 532 nm pump (dashed line) and the correction from the combination of modes in 15a.

$$L_c = 2\pi/\delta k$$

$$\delta k = k_a + k_s - 2k_p$$

(21)

This distance can be surprisingly long and at a frequency difference of 1 cm^{-1} in silica at 500 nm (which is the Brillouin shift) L_c = 2 km. $L_c \sim (\nu_p - \nu_s)^2$ so that at typical Raman shifts of 400 cm^{-1} $L_c \approx 1$ cm which is small when dealing with fibers.

Four-photon mixing at large frequency separations requires phase matching which is usually accomplished by using birefringent crystals or by mixing the beams at an angle. An important feature of fibers is that the waveguide modes in multimode fibers can be used for phase matching.[32] This is illustrated by the plot of effective refractive index vs. normalized frequency[10] in Fig. 15a. The problem is to correct the wave vector mismatch $k_s + k_a > 2k_p$. Because k is proportional to the effective index, what is required is to get ν_p and ν_s into a high index mode and ν_a in a low index mode as shown by the circles on the curves.

We can estimate the frequency at which phase matching occurs for the example in Fig. 15. We choose $n_c - n$ = 0.006 and a 532 nm pump wavelength. In Fig. 15b, we first plot δk vs. $\Delta\nu$ from index dispersion. Fig. 15a can then be used to work out the correction to δk that would be obtained using the waveguide modes and this correction as a function of $\Delta\nu$ is also plotted on Fig. 15b. The two lines intersect at $\Delta\nu$ = 2800 cm^{-1} and this is where phase matching will occur. It is particularly easy to calculate δk at $\Delta\nu$ = 0 because here $\delta k = k_p - k_a$ at the same value of V. For the example of Fig. 15b:

$$\delta k = \frac{2\pi\delta n_{eff}}{\lambda} = \frac{2\pi \times 0.32 \times 0.006}{5.32 \times 10^5 \text{ cm}} = 213 \text{ cm}^{-1}$$

(22)

Different combinations of modes will phase match at different frequencies and it is possible to reverse the modes of ν_a and ν_s or put the pump in different modes, etc. As might be expected, some combinations are forbidden by symmetry considerations - for example, the combination LP_{01} ($2\nu_p, \nu_s$) with LP_{11} (ν_a). If the pump is coupled to excite as many modes of the fiber as possible and the signal is tuned in wavelength, power will appear in the anti-Stokes idler whenever one of these phase matching combinations occurs.[32]

Use of waveguide modes to phase match has not completely eliminated the coherence length problem. Fig. 16 shows idler intensity vs. wavelength as the signal was tuned for a 9 and a 25 cm length of the same 10 mode fiber. For ideal phase matching the frequency bandwidth decreases as 1/ℓ and the peak power

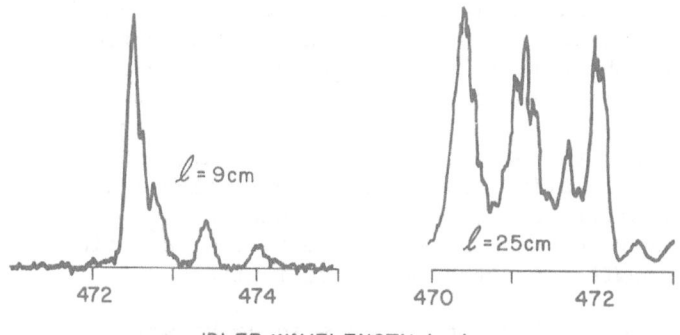

IDLER WAVELENGTH (nm)

16. Idler intensity vs. wavelength as signal is tuned for two
     lengths of the same 10 mode fiber.

increases as ℓ^2. The messy spectrum in the longer fiber occurs
because diameter variations along the fiber cause the phase matching
frequency to vary. In a short fiber such fluctuations are less
than the bandwidth, but in a long fiber, the phase matching frequency
will wander outside the narrower bandwidth. The important point
is that we have replaced the coherence length determined by index
mismatch with an effective coherence length governed by fiber
imperfections. In the fiber used for Fig. 16, this effective
coherence length was about 10 cm while L_c at $\Delta\nu$ = 3000 cm^{-1} would
be 0.02 cm so we have gained a factor of 500.

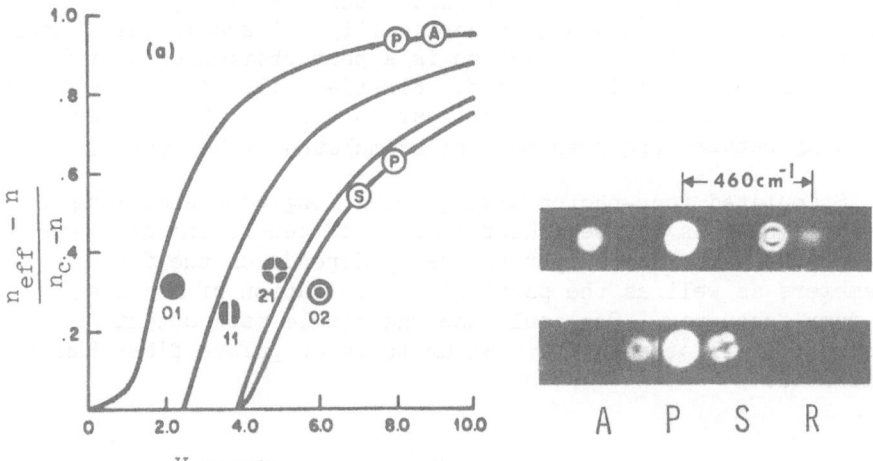

17. (a) Combination of modes for a long coherence length process.
     (b) Stimulated four-photon mixing from mode combinations
         similar to 17a.

There are some combinations of modes for which the phase matching frequency is less sensitive to guide imperfections.[33] In these combinations, the pump is split between the same two modes as the Stokes and anti-Stokes as shown in Fig. 17a. Note that at phase matching $k_{p1} + k_{p2} = k_{a1} + k_{s2}$ where 1 and 2 refer to the LP_{01} and LP_{02} modes. Now the effective coherence length is determined by changes in the relative slopes of the effective index curves with diameter variations rather than by changes of n_{eff} directly. Effective coherence lengths were 10 m in the same fiber that produced only 10 cm lengths for other mode combinations.

Four-photon mixing often occurs in combination with stimulated Raman or Brillouin scattering. For example, the Brillouin Stokes lines mix with the pump to produce anti-Stokes in the cw Brillouin oscillator.[25] Similarly, anti-Stokes output has been reported in single pass stimulated Raman scattering in multimode fibers.[34] In this case, the anti-Stokes was produced by mode combinations like Fig. 15a and the output was weak because the fiber was much longer than the effective coherence length.

The effective coherence length for the mixed-mode combination of Fig. 17 is long enough for stimulated four-photon mixing to occur. The Stokes and anti-Stokes output will appear in different modes as shown in Fig. 17b. In this case, the experimental setup was similar to Fig. 3 except that a 10 mode fiber was used. The frequency shifts for these processes are usually less than 500 cm^{-1} from the pump. At these frequencies, stimulated Raman scattering increases the Stokes and reduces the anti-Stokes intensity. Interactions between four photon and Raman type interactions can be viewed in two limits. The four photon susceptibility χ' in silica is about 10X the peak Raman susceptibility χ'' and in this limit ($\chi' >> \chi''$) the Raman interaction is a perturbation on a stimulated four-photon mixing process. For the Brillouin case, $\chi'' \approx 30\chi'$ so $\chi'' >> \chi'$ and it is best to treat the anti-Stokes as generated by mixing between the pump and the stimulated Stokes waves.

Stimulated four-photon mixing using long coherence length mode combinations may be a useful means of generating discrete wavelengths since the output frequency depends on the fiber parameters as well as the particular combination of modes and the pump frequency. One could use the single pass output or construct oscillators quite similar to cw or pulsed fiber Raman oscillators.

CONCLUSIONS

The nonlinear processes observed in fibers are listed in Fig. 18 along with some potential applications. The most advanced of the applications is the tunable fiber Raman oscillator and it

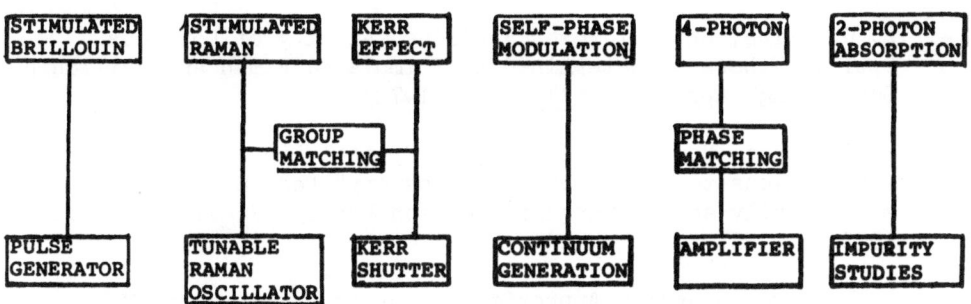

18. Nonlinear processes observed in fibers along with some
     potential applications.

is interesting to speculate on the future form such devices might
take. One attractive possibility is a combination of a fiber with
a pulsed diode laser pump. Present diode lasers have output
powers not far from the thresholds of Raman ring lasers pumped by
cavity dumped lasers although further progress in coupling such
diodes to small-core fibers is required. These lasers could be
tuned either with intracavity etalons or by varying the pulse
rate and would provide a compact and inexpensive source of tunable
polarized radiation throughout the near infrared.

     In conclusion, there are three important aspects of nonlinear
optics in fibers. First, the advantages of the fiber geometry for
the study of nonlinear optics; second is the power limit placed
on single-mode fiber communication systems by nonlinear effects;
and third is the device aspect as exemplified by the fiber Raman
laser.

REFERENCES

1. R. H. Stolen, E. P. Ippen, and A. R. Tynes, Raman Oscillation
    in Glass Optical Waveguide, Appl. Phys. Lett. 20, 62 (1972);
    R. H. Stolen and E. P. Ippen, Raman Gain in Glass Optical
    Waveguides, Appl. Phys. Lett. 22, 276 (1973).

2. N. Bloembergen, The Stimulated Raman Effect, Am. J. Phys. 35,
    989 (1967); W. Kaiser and M. Maier, Stimulated Rayleigh,
    Brillouin and Raman Scattering in "Laser Handbook" (ed.
    F. T. Arecchi and E. O. Schulz-DuBois) North-Holland Publishing,
    Amsterdam, 1077, (1972).

3. B. Shuker and R. W. Gammon, Raman-Scattering Selection-Rule
    Breaking and the Density of States in Amorphous Materials,
    Phys. Rev. Lett. 25, 222 (1970).

4.   R. G. Smith, Optical Power Handling Capacity of Low Loss
     Optical Fibers as Determined by Stimulated Raman and Brillouin
     Scattering, Appl. Opt. 11, 2489 (1972).

5.   E. P. Ippen, Nonlinear Effects in Optical Fibers, in "Laser
     Applications to Optics and Spectroscopy" (ed. S. F. Jacobs,
     M. O. Scully and M. Sargent) Addison-Wesley, Reading, Mass.,
     1975, p. 213.

6.   D. Marcuse, Gaussian Approximation of the Fundamental Mode
     in Graded Index Fibers, J. Opt. Soc. Am. 68, 103 (1978).

7.   F. Capasso and P. DiPorto, Coupled-Mode Theory of Raman
     Amplification in Lossless Optical Fibers, J. Appl. Phys. 47,
     1472 (1976).

8.   M. C. Tobin and T. Baak, Raman Spectra of Some Low-Expansion
     Glasses, J. Opt. Soc. Am. 58, 1459 (1968).

9.   P. D. Maker and R. W. Terhune, Study of Optical Effects Due
     to an Induced Polarization Third Order in the Electric
     Field Strength, Phys. Rev. 137A, 801 (1965); A. Owyoung,
     The Origins of the Nonlinear Refractive Indices of Liquids
     and Glasses. Ph.D. Thesis Cal. Inst. of Tech. (1971).
     Clearinghouse for Federal Scientific and Technical Information
     Report No. AFOSR-TR-71-3132.

10. D. Gloge, Weakly Guiding Fibers, Appl. Opt. 10, 2252 (1971).

11. Chinlon Lin, R. H. Stolen and R. K. Jain, Group Velocity
     Matching in Optical Fibers, Optics Lett. 1, 205 (1977).

12. L. G. Cohen and Chinlon Lin, Pulse Delay Measurements in the
     Zero Material Dispersion Wavelength Region for Optical Fibers,
     Appl. Opt. 16, 3136 (1977).

13. M. Maier, W. Kaiser, and J. A. Giordmaine, Backward Stimulated
     Raman Scattering, Phys. Rev. 177, 580 (1969).

14. Chinlon Lin and R. H. Stolen, Backward Raman Amplification and
     Pulse Steepening in Silica Fibers, Appl. Phys. Lett. 29,
     428 (1976).

15. R. H. Stolen, V. Ramaswamy and P. Kaiser, Linear Polarization
     in Elliptically-Clad, Birefringent, Single-Mode Fibers,
     Paper PD-1, IEEE/OSA Topical Meeting on Integrated and Guided
     Wave Optics, Salt Lake City (1978).

16. K. O. Hill, B. S. Kawasaki, and D. C. Johnson, Low Threshold
     cw Raman Laser, Appl. Phys. Lett. 29, 181 (1976).

17. R. K. Jain, Chinlon Lin, R. H. Stolen, W. Pleibel, and P. Kaiser,
     A High-Efficiency Tunable cw Raman Oscillator, Appl. Phys.
     Lett. 30, 162 (1977); D. C. Johnson, K. O. Hill, B. S.
     Kawasaki, and D. Kato, Tunable Raman Fiber-Optic Laser,
     Electron. Lett. 13, 53 (1977).

18. R. K. Jain, Chinlon Lin, R. H. Stolen, and A. Ashkin, A
     Tunable Multiple Stokes cw Fiber Raman Oscillator, Appl. Phys.
     Lett. 31, 89 (1977).

19. R. H. Stolen, Chinlon Lin, J. Shah and R. F. Leheny, A Fiber
     Raman Ring Laser (to be published).

20. R. H. Stolen, Chinlon Lin, and R. K. Jain, A Time-Dispersion-
     Tuned Fiber Raman Oscillator, Appl. Phys. Lett. 30, 340 (1977);
     Chinlon Lin, R. H. Stolen, and L. G. Cohen, A Tunable 1.1 μm
     Fiber Raman Oscillator, Appl. Phys. Lett. 31, 97 (1977).

21. Chinlon Lin, L. G. Cohen, R. H. Stolen, G. W. Tasker, and
     W. G. French, Near-Infrared Sources in the 1-1.3 μm Region
     by Efficient Stimulated Raman Emission in Glass Fibers,
     Opt. Comm. 20, 426 (1977).

22. E. P. Ippen and R. H. Stolen, Stimulated Brillouin Scattering
     in Optical Fibers, Appl. Phys. Lett. 21, 539 (1972).

23. J. Pelous and R. Vacher, Thermal Brillouin Scattering Measure-
     ments of the Attenuation of Longitudinal Hypersound in
     Fused Quartz from 77 to 300 K, Solid State Comm. 16, 279 (1975).

24. M. Denariez and G. Bret, Investigation of Rayleigh Wings and
     Brillouin-Stimulated Scattering in Liquids, Phys. Rev. 171,
     160 (1968).

25. K. O. Hill, D. C. Johnson, and B. S. Kawasaki, cw Generation
     of Multiple Stokes and Anti-Stokes Brillouin Shifted
     Frequencies, Appl. Phys. Lett. 29, 185 (1976); K. O. Hill,
     B. S. Kawasaki, and D. C. Johnson, cw Brillouin Laser, Appl.
     Phys. Lett. 28, 608 (1976).

26. B. S. Kawasaki, D. C. Johnson, Y. Fujii, and K. O. Hill,
     Bandwidth-Limited Operation of a Mode-Locked Brillouin
     Parametric Oscillator, Appl. Phys. Lett. 32, 429 (1978).

27. R. H. Stolen and Chinlon Lin, Self-Phase-Modulation in Silica
     Optical Fibers, Phys. Rev. A17, 1448 (1978).

28. C. H. Lin and T. K. Gustafson, Optical Pulsewidth Measurement
     Using Self-Phase Modulation, IEEE J. Quantum Electron. 8,
     429 (1972).

29.   Chinlon Lin and R. H. Stolen, New Nanosecond Continuum for
      Excited State Spectroscopy, Appl. Phys. Lett. 28, 216 (1976).

30.   D. Milam and M. J. Weber, Measurement of Nonlinear Refractive-
      Index Coefficients Using Time-Resolved Interferometry, J. Appl.
      Phys. 47, 2497 (1976).

31.   R. H. Stolen and A. Ashkin, Optical Kerr Effect in Glass
      Waveguide, Appl. Phys. Lett. 22, 294 (1973).

32.   R. H. Stolen, J. E. Bjorkholm, and A. Ashkin, Phase-Matched
      Three-Wave Mixing in Silica Fiber Optical Waveguides, Appl.
      Phys. Lett. 24, 308 (1974).

33.   R. H. Stolen, Phase-Matched Stimulated Four-Photon Mixing
      in Silica-Fiber Waveguides, IEEE J. Quantum Electron. QE-11,
      100 (1975); R. H. Stolen and W. N. Liebolt, Optical Fiber
      Modes Using Stimulated Four-Photon Mixing, Appl. Opt. 15,
      239 (1976).

34.   J. Botineau, F. Gires, A. Saïssy, C. Vanneste, and A. Azema,
      Laser Raman à Fibre Émettant dans un très Large Domaine
      Spectral, Appl. Opt. 17, 1208 (1978).

PLANAR OPTICAL WAVEGUIDES

H.-G. Unger

Institut für Hochfrequenztechnik der T.U.
Postbox 3329, D-3300 Braunschweig
Federal Republic of Germany

Recent advances in opto-electronics and electro-optics have opened
the infrared and visible part of the electromagnetic spectrum for
communications and general data processing applications. Planar
optical waveguides are needed in these applications to form distributed components and connect components and subsystems.

1. TOTAL REFLECTION

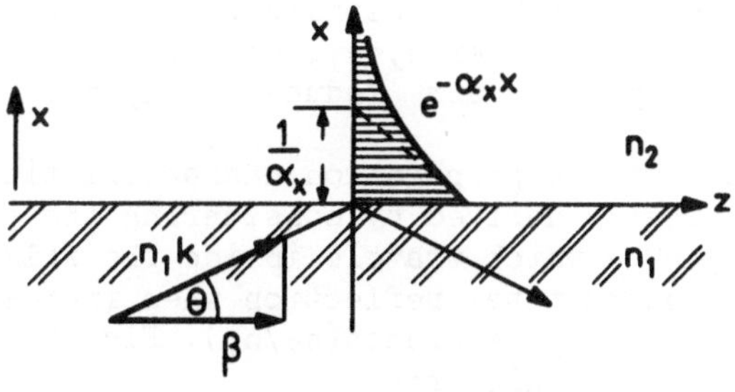

Of the various guiding principles for electromagnetic
waves the phenomena of total
reflection proves particularly effective at optical
frequencies and most of the
optical waveguides work on
this principle. When the
plane uniform wave in Fig. 1
impinges on a plane boundary to a region of lower
refractive index at an angle
Θ smaller than Θ_c
= arccos (n_2/n_1) the component

Fig. 1: Total reflection at a
plane boundary

β of its wave vector $n_1 k$ parallel to the boundary is larger than
the wave number $n_2 k$ of the region beyond the boundary. For the
space-periodic excitation with period $\lambda = 2\pi/\beta$ which this incident
wave provides, the region beyond with n_2 is, figuratively speaking,
too narrow and below cutoff. All fields therefore decay exponentially into this region and the boundary totally reflects the incident wave. The reflection coefficient as the ratio of reflected to

incident field phasor has unit amplitude and a phase that depends on
the polarization of the incident wave. For an electric field vector perpendicular to the plane of incidence, this phase angle is

$$\phi_e = 2 \arctan(jk_{2x}/k_{1x}), \tag{1.1}$$

while it is

$$\phi_m = 2 \arctan(jn_1^2 k_{2x}/n_2^2 k_{1x}), \tag{1.2}$$

when E is parallel to the plane of incidence.

$$k_{1x} = n_1 k \sin\theta = \sqrt{n_1^2 k^2 - \beta^2} \tag{1.3}$$

is the x-component of the incident wave vector and

$$jk_{2x} = \alpha_x = \sqrt{\beta^2 - n_2^2 k^2} \tag{1.4}$$

represents the decay constant of the evanescent field beyond the
boundary.

2. DIELECTRIC FILMS

Optical film waveguides consist of a thin dielectric film of low
optical absorption and an index of refraction n_1 deposited on a
transparent substrate of preferably likewise low absorption and
with an index of refraction n_2 which is somewhat smaller than n_1.
The region above the film remains usually free space or air with
refractive index $n_0 = 1$. Symmetrical films constitute the special
case in which the regions on both sides of the film have the same
refractive index $n_2 < n_1$, which, in case of a film suspended in
free space or air, reduces to $n_2 = 1$.

Most of the propagation characteristics of film modes are con-
veniently derived by considering the plane uniform wave in
Fig. 2.1 which travels inside the film on a zig-zag path and experiences total reflection when incident on the boundaries at an
angle $\theta_1 < \theta_{2c} = \arccos(n_2/n_1)$, Fig. 2.1 shows the typical ray
path ABCD which re-
presents this wave.
Also indicated by
dotted lines are
phase fronts be-
longing to the
uniform plane wave
while travelling
from A to B and
from C to D, re-
spectively. After
two successive

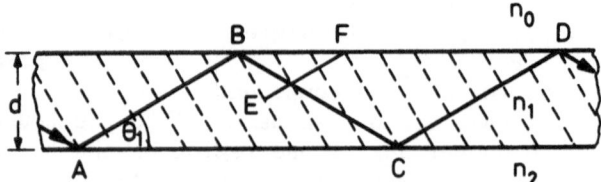

Fig. 2.1 Film guide with zig-zag path of a
guided mode

reflections, for example at B and C, the wave must repeat in phase
in order to yield a selfconsistent field distribution, which then
forms a film mode. The phase front containing B and E must change
by the same phase angle when moving directly from E to F into the
phase front through F and C, as when undergoing the two reflections
at B and C, and moving from B to C in between. Any phase differences
between these two paths may only amount to $2m\pi$, with m an integer
number. Introducing ϕ_{10} and ϕ_{12} as phase shifts associate with the
total reflection at B and C, respectively, the phase condition for
the ray between B and C in Fig. 2.1 obtains as

$$-n_1k\,(\overline{BC} - \overline{EF}) + \phi_{10} + \phi_{12} = -2m\pi \qquad (2.1)$$

where $\overline{BC} = d/\sin\theta_1$ and $\overline{EF} = \overline{BF}\cos\theta_1$ with $\overline{BF} = d/\tan\theta_1 - d\tan\theta_1$.
Substituting these expressions for the various distances into
eqn.(2.1) leads to

$$-2n_1kd\sin\theta_1 + \phi_{10} + \phi_{12} = -2m\pi \qquad (2.2)$$

This equation constitutes the characteristic equation of film modes.

The phase angles ϕ_{10} and ϕ_{12} take on different values depending on
the polarization of the plane uniform wave. If the magnetic field
is parallel to the plane of incidence, the film modes are desig-
nated as H-modes or TE-modes, because of its total field, only the
magnetic part has a component in the direction of propagation;
also these film modes have a transverse electric (TE) field. To
evaluate the characteristic equation for these H-modes, the phase
angle ϕ_e from eqn.(1.1) must be substituted for ϕ_{10} and ϕ_{12} in
eqn.(2.2), leading to

$$\arctan\frac{jk_{0x}}{k_{1x}} + \arctan\frac{jk_{2x}}{k_{1x}} = dn_1k\sin\theta_1 - m\pi. \qquad (2.3)$$

For a representation of this eigenvalue equation in terms of dimensionless quantities, we take the product of the transverse
component k_{1x} of the wave vector inside the film and the film
thickness d and denote it by

$$u = k_{1x}d = dn_1k\sin\theta_1. \qquad (2.4)$$

Similarly we take the transverse attenuation constants jk_{0x} and
jk_{2x} of the evanescent fields outside of the film times the film
thickness and denote them by

$$v = jk_{2x}d \qquad\qquad w = jk_{0x}d. \qquad (2.5)$$

With the wave vector components in eqn.(2.3) thus normalized the
eigenvalue equation appears as

$$\arctan (v/u) + \arctan (w/u) = u - m\pi. \tag{2.6}$$

Taking the tangent of this equation reduces it to the even simpler form

$$\tan u = u(v+w)/(u^2 - vw) \tag{2.7}$$

The eigenvalue equation for film modes of the E- or TM-type follows similarly as

$$\tan u = n_1^2\, u\, (n_o^2 v + n_2^2 w)\big/(n_o^2 n_2^2 u^2 - n_1^4 vw) \tag{2.8}$$

The parameters u, v, w representing the transverse phase and attenuation constants inside the film and in the space above as well as in the substrate below the film relate to the longitudinal component $k_z = \beta$ of the wave vector and to the wave number k_o, k_1, k_2 of the three regions as follows ($k_i = n_i k$)

$$\begin{aligned}
u/d &= k_{1x} = (k_1^2 - \beta^2)^{1/2}\\
v/d &= jk_{2x} = (\beta^2 - k_2^2)^{1/2}\\
w/d &= jk_{ox} = (\beta^2 - k_o^2)^{1/2}\,.
\end{aligned} \tag{2.9}$$

The longitudinal wave vector component β describes the z-dependence of field. If real values for β obtain from the solution of the eigenvalue equation, they are phase constants for the propagation of film modes.

In case of H-modes and their eigenvalue equation, a numerical solution of eqn.(2.7) can be represented in universal form by expressing the transverse attenuation parameter w in terms of u and v by way of eqn.(2.9)

$$w^2 = v^2 + a_H\,(u^2 + v^2)\,. \tag{2.10}$$

The parameter

$$a_H = (n_2^2 - n_o^2)/(n_1^2 - n_2^2) \tag{2.11}$$

measures the asymmetry of the struture. For a symmetric structure with $n_2 = n_o$, we have $a_h = 0$.

Substituting for w from eqn.(2.10) into eqn.(2.7) leads to a form of the eigenvalue equation for H-modes which, according to

$$\tan u = \frac{u\{v + \left[v^2 + a_H\,(u^2 + v^2)\right]^{1/2}\}}{u^2 - v\left[v^2 + a_H\,(u^2 + v^2)\right]^{1/2}}\,, \tag{2.12}$$

contains only the transverse wave parameters u and v, and in

Fig.2.2: Phase parameter B as a
function of film parameter
V for low order film waves.
Asymmetry parameter

$$B = \frac{(\beta/k)^2 - n_2^2}{n_1^2 - n_2^2}$$

$$a = \frac{n_2^2 - n_o^2}{n_1^2 - n_2^2} \quad \text{for } H_m \text{ -waves}$$

$$a = \frac{n_1^4 (n_2^2 - n_o^2)}{n_o^4 (n_1^2 - n_2^2)} \quad \text{for } E_m \text{ -waves}$$

(after Kogelnik et al. /1/)

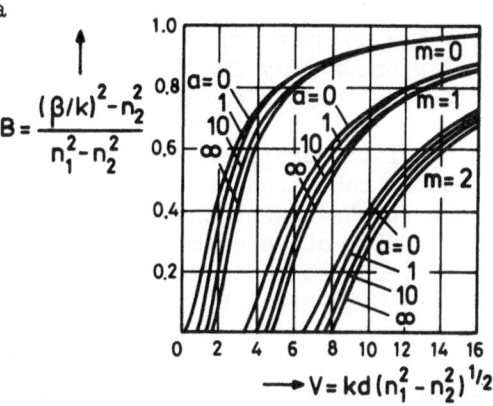

addition the asymmetry parameter a_H. The solution of eqn.(2.12)
together with eqn.(2.9) may be plotted in the form of the phase
parameter $B = v^2/V^2$ as a function of $V = \sqrt{u^2 + v^2}$ with a_H as a
parameter. Fig.2.2 shows the diagram for a number of the lower
order H_m-modes.

The universal representation for the phase parameter of H_m- modes
may also be applied to E_m-modes by considering eqn.(2.10) for
$n_1 \simeq n_2$ and hence large values of a_H. We then have approximately

$$w^2 \simeq a_H (u^2 + v^2) \tag{2.13}$$

and with this approximation, the eigenvalue equation for H-modes
reduces to

$$\tan u \simeq \frac{u \left[v + a_H^{1/2} (u^2 + v^2)^{1/2} \right]}{u^2 - a_H^{1/2} v (u^2 + v^2)^{1/2}} \tag{2.14}$$

while eqn.(2.12) for E-modes appears approximately as

$$\tan u = \frac{u \left[v + a_E^{1/2} (u^2 + v^2)^{1/2} \right]}{u^2 - a_E^{1/2} v (u^2 + v^2)^{1/2}} \tag{2.15}$$

with

$$a_E = (n_1/n_o)^4 \cdot a_H . \tag{2.16}$$

Eqn.(2.15) has the same form as eqn.(2.14). Its solutions may
therefore also be read from Fig.2.2 if only the new asymmetry

factor a_E for E-modes according to eqn.(2.16) is used in its evalua-
tion. The solution of eqn.(2.15) as represented with a = a_E in
Fig.2.2 holds only for $n_1 \approx n_2$ and should only be applied for
E-modes as long as $a_H >> B$.

For specific conditions only a finite number of film modes exist.
At cutoff the waves have the limiting angle θ_c of total reflection
at the substrate and their fields extend very far into the substrate. High above cutoff, the waves travel with very small θ and
their fields concentrate inside the film. In the symmetric film
with a = 0 in Fig.2.2, the two lowest order modes have zero cutoff.

3. FILM MODE SCATTERING

The guided modes of films suffer loss mainly from scattering at
deviations of their boundaries from a perfectly plane geometry.
Such deviations result at both boundaries from the fabrication
process. Usually, however, they occur more likely at the boundary
of larger index difference and also cause more scattering at this
boundary. Sputtered films on a glass substrate are a typical case,
where more imperfections must be expected at the boundary of the
film to the cover space with its larger index difference.

Fig.3.1 shows such an imperfect boundary with random displacements
f(y,z) of the actual boundary from its nominal plane at x = 0.With
n as the actual index near the boundary, and $\bar{n}$ as the index at the
same point for a perfect boundary, the displacement may be accounted for in the field equations by introducing the effective current
density

$$\vec{J} = j\omega\varepsilon_0 \ (n^2 - \bar{n}^2) \ \vec{E} \qquad\qquad (3.1)$$

Here $\vec{E}$ represents the electric field, which is associated with the
particular guided mode in presence of the perturbation. More spe-
cifically, we have

$$\vec{J}_{a;b} = j\omega\varepsilon_0 \ (n^2_{b;a} - n^2_{a;b}) \ \vec{E}_{ab;ba} \qquad , \qquad (3.2)$$

if the region with $n_{b;a}$ extends beyond the nominal boundary into
the region with $n_{a;b}$.

$E_{ab;ba}$ represents the actual field in the particular dent. To a
good approximation we have

$$E_{ab;ba} = E_{b;a} \ ,$$

with $E_{b;a}$ as the field of the guided mode of the perfect guide in
the region with index $n_{b;a}$ at the boundary, i.e. at x = ±0.

With the effective currents of the density $\vec{J}$ as given by eqn.(3.2),
any boundary element dydz which is displaced by f(y,z) has the

Fig.3.1 Nominally plane boundary
        between regions of re-
        fractive index n_a.

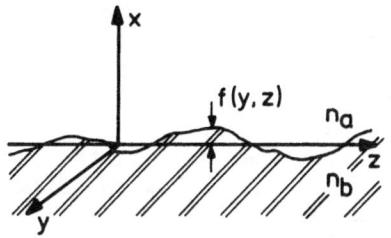

---

electric current element of moment

$$I\vec{\ell} = \vec{J}\, f(y,z)\, dydz \qquad (3.3)$$

associated with it. Depending on the direction of displacement,
and therefore the sign of $f(y,z)$, we have such an electric dipole
moment either on side a or on side b of the boundary, but always
directly adjacent to it.

For a more unified representation of these effective dipole moments
we replace any dipole moment in region b by an equivalent source
in region a. Consider the distribution of dipole moments in
Fig.3.2 with $(I\vec{\ell})_a$ and $(I\vec{\ell})_b$ directly above and below the boundary,
respectively, and $(I\vec{\ell})_p$ of any orientation located anywhere in
space. Denoting by E_{pa} the field component of source p in the direction of source a and at its location, we have with corresponding
notations for the other field components, and by law of reciprocity
$E_{pa}(I\ell)_p = E_{ap}(I\ell)_p$, as well as $E_{pb}(I\ell)_p = E_{bp}(I\ell)_p$. For $(I\vec{\ell})_a$
and $(I\vec{\ell})_b$ to be the equivalent of each other, they must generate the
same fields $E_{ap} = E_{bp}$ anywhere in space. This requirement results
in the following condition for equivalence $E_{pa}(I\ell)_a = E_{pb}(I\ell)_b$.

In case of dipoles a and b oriented parallel to the boundary, we
have $E_{pa} = E_{pb}$ and consequently $(I\ell)_b = (I\ell)_a$. In case of dipoles
perpendicular to the boundary, we have $n_a^2 E_{pa} = n_b^2 E_{pb}$. With all
these relations we may now take account of any displacement $f(y,x)$,
positive or negative, of the boundary element dydz by the electric
dipole moment

$$\vec{J}(y,z)\, dV = j\omega\varepsilon_o\, (n_b^2-n_a^2)\, \vec{E}_a\, f(y,z)\, dydz \qquad (3.4)$$

on the a-side next to the boundary.

To evaluate the radiation loss from such dipoles, we need to calculate their radiation field in the presence of both film boundaries. The effect of any boundary on the radiation field, and on the
radiated power, will be more pronounced the closer this dipole is
located to the boundary. Any boundary at a sufficient distance
will still modify the radiated field by reflecting and refracting
the incident radiation, but not so much react on the source, and
therefore not change its radiated power very much. For a loss analysis we are primarily interested in this power and will therefore
take account only of the boundary to which the dipole is immediately adjacent.

The assertion that the film boundary to which the dipole is not directly adjacent, does not react on the dipole to change its total
radiated power significantly is corroborated by an accurate com-
putation of the dipole radiation in the presence of both film
boundaries /2/. If the film guides any modes which the dipole
excites, then not all the power is radiated into the space above
and substrate below the film, but part of it is guided by these
film modes which propagate radially away from the dipole with circularly cylindrical phase fronts. The total power, however, as the
sum of radiated power and the power into film modes remains nearly
the same as the radiated dipole power when only the adjacent film
boundary is present.

The far field of electric current elements at the plane boundary
between two different media has the following magnetic-field vector
components in the coordinates of Fig.3.3, with the dipole at their
origin /3/.

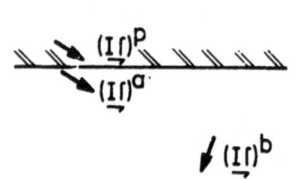

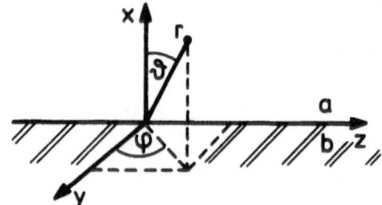

Fig.3.2 Equivalent dipole   Fig.3.3 Coordinates of dipole at
        moments on both sides   the boundary between a
        next to the boundary   and b and for its radia-
                                             tion field

$$
\vec{H} = \begin{bmatrix} H_r \\ H_\vartheta \\ H_\varphi \end{bmatrix} = \begin{bmatrix} 0 & 0 & 0 \\ 0 & g_{\vartheta y} & g_{\vartheta z} \\ g_{\varphi x} & g_{\varphi y} & g_{\varphi z} \end{bmatrix} \begin{bmatrix} (I\ell)_x \\ (I\ell)_y \\ (I\ell)_z \end{bmatrix} \qquad . \qquad (3.5)
$$

For the components of the Green's-function dyadic in this expression, we must distinguish, in which of the two regions the field
is to be determined. With the wave numbers $k_a = n_a \omega \sqrt{\mu_o \epsilon_o}$,
$k_b = n_b \omega \sqrt{\mu_o \epsilon_o}$ in each region and

$$
\delta = (n_b^2 - n_a^2)/ n_a^2 \qquad (3.6)
$$

as the relative difference in permittivities, the dyadic has the
following components in the respective region:

$$g_{\varphi x}^{a} = j\,\frac{k_a}{2\pi r}\,e^{-jk_a r}\,\frac{\sin\vartheta}{1+\frac{1}{1+\delta}\sqrt{1+\delta\,\cos^{-2}\vartheta}}$$

$$g_{\varphi x}^{b} = j\,\frac{k_b}{2\pi r}\,e^{-jk_b r}\,\frac{\sin\vartheta}{1+(1+\delta)\sqrt{1-\frac{\delta}{1+\delta}\cos^{-2}\vartheta}}$$

$$g_{\vartheta y}^{a} = j\,\frac{k_a}{2\pi r}\,e^{-jk_a r}\,\frac{\sin\varphi}{1+\sqrt{1+\delta\,\cos^{-2}\vartheta}} \tag{3.7}$$

$$g_{\vartheta y}^{b} = j\,\frac{k_b}{2\pi r}\,e^{-jk_b r}\,\frac{\sin\varphi}{1+\sqrt{1-\frac{\delta}{1+\delta}\cos^{-2}\vartheta}}$$

$$g_{\varphi y}^{a} = j\,\frac{k_a}{2\pi r}\,e^{-jk_a r}\,\frac{\cos\varphi\,\sqrt{\delta+\cos^2\vartheta}}{1+\delta+\sqrt{1+\delta\cos^{-2}\vartheta}}$$

$$g_{\varphi y}^{b} = j\,\frac{k_b}{2\pi r}\,e^{-jk_b r}\,\frac{\cos\varphi\,\sqrt{(1+\delta)\cdot\cos^2\vartheta-\delta}}{1+\sqrt{1+\delta-\delta\cos^{-2}\vartheta}}$$

$g_{\vartheta z}^{a,b}$ and $g_{\varphi z}^{a,b}$ follow from $g_{\vartheta y}^{a,b}$ and $g_{\varphi v}^{a,b}$, respectively, when φ is replaced by $\varphi+\pi/2$. Note that the square roots of $1-\delta/(1+\delta)\cos^2\delta$ in the expressions for the g^b go to zero at $\vartheta_c = \arcsin(n_a/n_b)$, the limiting angle of total reflection at the boundary from n_b to n_a. Note also that in the boundary plane for $\vartheta=\pi/2$ the $g_{\varphi y}$ as well as the $g_{\varphi z}$ vanish.

The power which the single dipole at the boundary radiates into the space above and below the boundary obtains, when we integrate the far-field Poynting vector

$$S = n^{-1}\sqrt{\mu_o/\varepsilon_o}\;|H|^2 \tag{3.8}$$

over all directions Ω in space

$$P_r = \iint r^2\,S(\Omega)\,d\Omega\;. \tag{3.9}$$

Depending on the region, the refractive index n in eqn.(3.8) is either n_a or n_b.

When we separate the contributions from dipoles which lie horizontal, that is parallel to the boundary from those which are vertical or oriented perpendicular to it, the radiated power may be written as

$$P = P_a \left[h(\delta) + v(\delta) \right] \qquad (3.10)$$

where

$$P_a = (2\pi/3) \sqrt{\mu_o/\varepsilon_o} \; I^2 \; (\ell^2/\lambda^2) \; n_a \; , \qquad (3.11)$$

is the power radiated by a dipole of moment $I\ell$ into an unlimited space of refractive index n_a. The factors

$$h(\delta) = h_a(\delta) + h_b(\delta)$$

$$v(\delta) = v_a(\delta) + v_b(\delta) \qquad (3.12)$$

account for the change in radiated power due to the boundary, when the dipole at the boundary is oriented either parallel or perpendicular to it. They consist of two terms each of which, according to their index a or b, describe the power fractions, which are radiated into the half space a or b, respectively.

Fig.3.4 shows the factor $h(\delta)$ for a horizontal dipole, and Fig.3.5 shows $v(\delta)$ for a vertical dipole, both together with their constituent parts for region a and b. The radiated power from an electric current dipole that lies parallel on an interface changes only slightly when the index differences increases. It increases somewhat when the region b increases its index with respect to region a. It decreases, but at an even smaller rate, when the region below assumes an index n_b smaller than n_a. In this range $h(\delta)$ even goes through a minimum. The constituent parts $h_a(\delta)$ and $h_b(\delta)$ show a more pronounced change. For $n_a < n_b$ most of the power radiates into the lower half space with n_b, while for $n_a > n_b$ it radiates more and more into the upper half space with n_a.

Fig.3.4 Power radiated from an electric dipole which lies in a halfspace with n_a parallel and on the interface to a halfspace with n_b. h_a and h_b are the relative powers radiated into halfspace a and b, respectively.

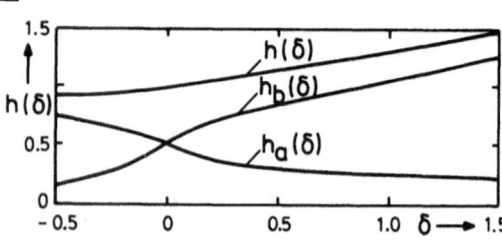

Fig.3.5 shows similar characteristics for the power that a dipole perpendicular to the interface radiates relative to the power radiated into the free space. They are, however, much more pronounced in that they depend strongly on the relative difference δ in permittivities.

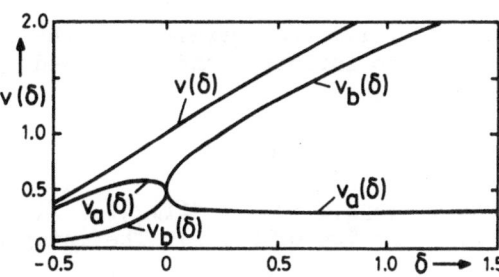

Fig.3.5 Power radiated from
an electric dipole that sits
in halfspace a vertically
on the interface to half-
space b relative to the
dipole radiation in homo-
geneous space with n_a.

With the far field magnetic vector of a single dipole available in
the form of eqn.(3.5), we can now return to the film guide with its
boundary distortions and their effective current distributions. If
these distortions extend from $0 < y < L_y$ in y-direction and from
$0 < z < L_z$ in z-direction, they generate a far field magnetic vector
according to

$$\vec{H} = \sum_{t=x,y,z} \vec{g}_t \int_o^{L_y} \int_o^{L_z} \int J_t(y,z) \, e^{j\psi(y,z)} \, dV \qquad (3.13)$$

Any of the vectors $\vec{g}_t$ in this expression has the spherical components $g_{\vartheta t}$ and $g_{\varphi t}$ and, from eqn(3.5), gives the magnetic field,
which the Cartesian component $(\vec{I\ell})_t = J_t \, dV$, with t=x,y, or z, of
the effective current element $J_t \, dV$ generates. The difference $\psi(y,z)$
in the phase retardation between a source at y=z=0 and one at (y,z)
on the boundary to any far field point (r,ϑ,φ) in space is given by
$$\psi(y,z) = nk \sin\vartheta \, (y \cos\varphi + z \sin\varphi) \qquad (3.14)$$

For the refractive index n in this expression, n_a or n_b must be
substituted, depending on the region in which the far field is to
be evaluated. For the total power, which this field radiates, the
Poynting vector according to eqn.(3.8), must be integrated over all
directions in space, just as in eqn.(3.9).

In case of random distributions for f(y,z), only certain statistical characteristics are known. Here we will evaluate the average
of radiated power and the average mode loss associated with it.
To this end we take the ensemble average of eqn.(3.9)

$$P = \int_\Omega r^2 <S(\Omega)> d\Omega \qquad (3.15)$$

We assume the random distribution f(y,z) to be stationary and
ergodic with a mean sqare

$$\overline{f^2} = \frac{1}{L_y L_z} \int_o^{L_y} \int_o^{L_z} f^2(y,z) \, dz \, dy. \qquad (3.16)$$

Because of limited correlation, the integrals in eqn.(3.13) may be extended to infinity, and expressed by the power spectral distribution

$$\phi(\xi,\zeta) = \overline{f^2} \int\!\!\int_{\infty}^{\infty} \varphi(u,v) \ e^{-j(\xi u+\zeta v)} \ du \ dv \tag{3.17}$$

where $\varphi(y_1-y_2,z_1-z_2) = <f(y_1,z_1) \ f(y_2,z_2)> \ /\overline{f^2}$ is the auto-correlation function of $f(y,z)$. The integral

$$<|F|^2> = \int\limits_{y_1=0}^{L_y} \int\limits_{z_1=0}^{L_z} \int\limits_{y_2=0}^{L_y} \int\limits_{z_2=0}^{L_z} <f(y_1,z_1)f(y_2,z_2)>$$

$$\exp\{-j|\beta(z_1-z_2)-\psi(y_1-y_2,z_1-z_2)|\}dz_2 \ dy_2 \ dz_1 \ dy_1 \tag{3.18}$$

which we need to evaluate in order to determine $< S(\Omega)>$ in eqn.(3.15) may, by way of eqn.(3.17), be expressed by

$$<|F|^2> = L_y L_z \ \phi(\xi,\zeta) \tag{3.19}$$

with the power spectrum at spatial frequency

$$\xi = -nk \ \sin\vartheta \ \cos\varphi \tag{3.20}$$

of the y-distribution in $f(y,z)$ and at spatial frequency

$$\zeta = \beta- nk \ \sin\vartheta \ \sin\varphi \tag{3.21}$$

of the z-distribution in $f(y,z)$. n is equal to n_a or n_b depending into which region, a or b, the radiation is considered. For a specified auto-correlation function, or power spectrum of boundary displacements, the average radiated power $<P>$ may now be evaluated. It substracts from the power P_m of the guided mode

$$\frac{dP_m}{dz} = - \frac{<P>}{L_z L_y} \tag{3.22}$$

and causes P_m to decay exponentially according to $P_m=P_o \exp(- \alpha z)$. The mean attenuation constant $\overline{\alpha}$ follows from these two relations as the power ratio

$$\overline{\alpha} = <P>/(2L_y L_z P_m) \tag{3.23}$$

For lack of better information on the nature of surface roughness and its statistics, we resort to a particularly simple example of auto-correlation in form of the exponential function

$$\varphi(u,v) = \exp(-|u|/B_y -|v|/B_z) \tag{3.24}$$

The quantities B_y and B_z in the exponential give the distances at
which the auto-correlation drops from unity at $u = v = 0$ to $1/e$
in y-direction or in z-direction, respectively. As such characteristic quantities they are called auto-correlation distances of
the random distribution, B_y in y-direction, and B_z in z-direction.

The power spectrum in case of the exponential auto-correlation has
the two-dimensional distribution

$$\phi(\xi,\zeta) = \overline{4f^2} \; B_y \; B_z \; \frac{1 + B_y B_z \xi \zeta}{(1+\xi^2 B_y^2)(1+\zeta^2 B_z^2)} \; . \tag{3.25}$$

If the correlation in surface roughness extends only over distances

$$B_y \ll \frac{1}{nk} \tag{3.26}$$

transverse to the direction of wave propagation, and only over
distances

$$B_z \ll \frac{1}{nk + \beta} \tag{3.27}$$

parallel to the direction of wave propagation, the power spectrum
remains flat up to the highest spatial frequencies of $f(y,z)$,
which contribute to radiation.

Under these condtions, we can set

$$\phi(\xi,\zeta) = \overline{4f^2} \; B_y \; B_z \; . \tag{3.28}$$

For such short correlation distances, the effective sources of all
displaced surface elements radiate independently of each other,
and the total scattered power obtains simply as the sum of contributions from all individual surface scatterers.

Relatively simple expressions for the scattering loss result under
these conditions for H_m-modes in the film. If the surface roughness occurs at the boundary between film with n_1 and upper space
with n_0, the mean attenuation constant may be expressed as

$$\bar{\alpha}_{01} = \frac{2}{3\pi} \; \overline{f^2} \; B_y \; B_z \; \frac{n_0 k^3 v^2}{\beta d_e d^2} \; (1-B)(n_1^2-n_0^2)h \left[\frac{n_1^2-n_0^2}{n_0^2} \right] . \tag{3.29}$$

If the boundery between film and substrate with n_2 shows such
roughness, it causes scattering loss to H_m-waves according to

$$\bar{\alpha}_{12} = \bar{\alpha}_{01} \frac{n_2}{n_0} \frac{n_1^2 - n_2^2}{n_1^2 - n_0^2} \frac{h((n_1^2 - n_2^2)/n_2^2)}{h((n_1^2 - n_0^2)/n_0^2)} \quad . \tag{3.30}$$

d_e in eqn.(3.29) is the effective width of the particular H_m-mode which, according to

$$d_e = d(1 + 1/v + 1/w), \tag{3.31}$$

not only takes account of the film thickness d but also of the penetration of fields into substrate and cover. β is the phase constant of the particular H_m-mode under consideration, while w and v are its transverse attenuation parameters, according to which the evanescent fields decay into cover and substrate, respectively. The function h(δ), as plotted in Fig.3.4 represents the change in radiated power of the horizontally oriented dipole due to the presence of the respective boundary.

Because of the factor $k^3 v^2/\beta$ in eqn.(3.29), the scattered loss depends essentially as $1/\lambda^4$ on wavelength. This corresponds to the Rayleigh-law of scattering and is typical for scatterers, which are small compared to the wavelength. For surface roughness with the same value for the product $f^2 B_y B_z$ at both film boundaries, the ratio of scattering loss is essentially that of the differences $(n_1^2 - n_2^2)/(n_1^2 - n_0^2)$ in permittivities. The boundary with larger index differences therefore scatters more power according to this ratio. For weakly guiding films with a small index difference between film and substrate, scattering from the upper film surface dominates and this surface must be particulary smooth for low scattering loss.

For a particular H_m-mode, the scattering loss depends as $1/d_e$ on the effective mode width. Near cut-off, the fields extend far into the substrate, and suface roughness, which then is relatively small compared to the mode width, causes only little scattering. As we move away from cut-off, d_e goes through a minimum. In this minimum surface roughness becomes most effective in scattering guided mode power. Beyond this minimum, d_e increases monotonically and approaches d while B approaches unity. In this region of thicker and thicker films, a particular film mode consists of rays, which propagate at steadily decreasing angles θ in Fig.2.1 and undergo fewer and fewer reflections at the boundaries. The scattering at rough boundaries will then also decrease and with it the scattering loss.

Fig.3.6 shows the scattering loss of the fundamental H_0-mode in the form $W = \bar{\alpha} d^5/(\bar{f}^2 B_y B_z)$ for three different values of the film parameter V plotted as a function of the asymmetry parameter a_H from eqn.(2.11). For small a_H-values in nearly symmetrical film

Fig.3.6 Scattering loss $\bar{\alpha}_{Q1}$ from
surface roughness of the upper
film boundary and $\bar{\alpha}_{12}$ from roughness of the film substrate
boundary

$$a_H = (n_2^2-n_0^2)/(n_1^2-n_2^2); \quad n_0=1; n_2=1.5$$

$$V = kd \sqrt{n_1^2-n_2^2} = \begin{cases} 1.87 & \cdots\cdots \\ 2.57 & \text{-----} \\ 4.57 & \text{———} \end{cases}$$

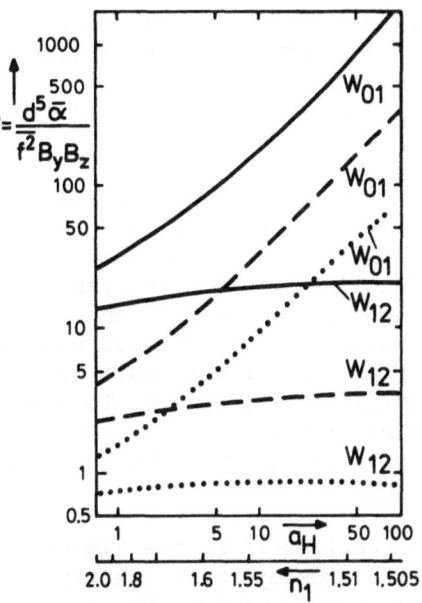

guides, W_{o1} differs only little
from W_{12}. When a_H increases, we
approach the case of weakly guiding films. W_{12} then stays nearly
constant while W_{o1} increases almost linearly with a_H.

The range of correlation distances B_y and B_z for which these results
apply, is limited by the assumptions (3.26) and (3.27). Of these
two conditions, the inequality (3.27) is more restrictive than
(3.26). For boundary perturbations of increasing correlation
distance, this condition will therefore be the first one not to be
satisfied any more.

4. GRADED-INDEX FILMS

The film and slab guides, which we considered in previous sections,
were all assumed to have well defined surface or interfaces with
abrupt transitions in refractive index from one medium to the other
and constant values for the refractive index in each particular region.

Some fabrication methods, however, and especially those, which
involve diffusion processes, lead to smoothly graded transitions
in index. Such grading of index transitions changes the cut-off
and dispersion characteristics of the guides and may also influence the scattering loss. Whether these changes are objectionable
or not, depends on the particular application. We even foresee
the possibility that suitably graded transitions may yield desirable characteristics for guided modes.

For the analysis of such films with graded index profiles, we will
find it useful to distinguish between symmetric profiles, such as
in Fig.4.1a, and strongly asymmetric index distributions with one
abrupt transition such as Fig.4.1b. The first case occurs inten-

Fig.4.1 Refractive index profiles
of graded index films

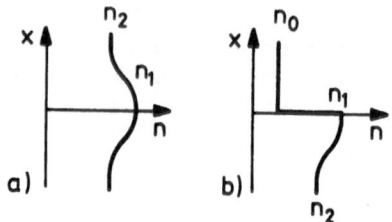

a) symmetric profile of
buried layer

b) asymmetric profile of the
film on substrate

---

tionally, but also unintentionally, when a buried layer is produced by some diffusion process. The second case occurs, when
a guiding film is formed near a substrate surface by in- or out-
diffusion. The very surface of the substrate or film has the abrupt index transition. To a certain approximation the set of
modes of the symmetric profile contains the modes, which are guided by one half of the corresponding profile in the strongly
asymmetric case.

We consider first the symmetric profile of Fig.4.1 and try the
ray optics approach. The index profile is a special case of an inhomogeneous medium. We assume the refractive index to change only
gradually over the distance of a wavelength. A uniform plane
wave will then only be slightly modified and its ray path obeys
the following differential equation

$$\frac{d}{ds} \left(n \frac{d\vec{r}}{ds} \right) = \nabla n \qquad\qquad (4.1)$$

where $\vec{r}$, according to Fig.4.2, is the position vector along the ray
and s the length of this ray from a fixed reference point. According to this equation light rays essentially curve in the direction
of ∇n.

For a homogeneous medium with $\nabla n = 0$, eqn.(4.1)
integrates to $\vec{r} = s\vec{a} + \vec{b}$, with constant vectors $\vec{a}$ and $\vec{b}$, representing straight rays.

Light rays, which are guided by graded-index
films run in planes y=const. of the coordinate system in Fig.4.3. They are also nearly
parallel to the z-axis with the transverse
component dx/ds of their unit direction
vector dr/ds always small compared to the
longitudinal component dz/ds. Under these
condition we have ds ≈ dz. Also for such
films, n does not change in z-direction.
The general ray equation (4.1) then reduces
to the paraxial approximation

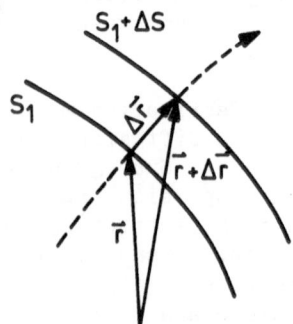

Fig.4.2 Phase fronts
and light ray in an
inhomogeneous medium

Fig.4.3 Symmetric index profile
         and rays of guided and
         radiation modes

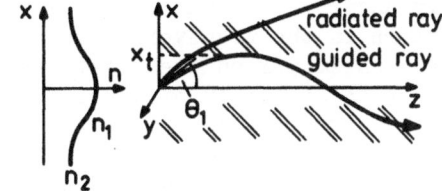

$$\frac{d^2x}{dz^2} = \frac{d}{dx} \log_e n \qquad (4.2)$$

This equation allows to trace the deviation of a ray from the
centre plane along the film.

To identify different modes of propagation in the layer with
symmetric index distribution, we look at rays that radiate from
a line source in the y-axis of Fig.4.3 at an angle θ_1 with respect
to the centre plane. In the ray equation (4.1) the z-component of
∇n vanishes, because n remains constant in this direction. We
therefore have for the ray in Fig.4.3 a z-component of eqn.(4.1)
which integrates along s to $n(d\vec{r}/ds) = n \cos\theta = const$. Snell's
law generalizes for this ray to

$$n \cos\theta = n_1 \cos\theta_1 . \qquad (4.3)$$

Any ray, starting with θ_1 at the centre plane, curves towards
this plane so that the longitudinal component $nk \cos\theta$ of its wave-
vector $n\vec{k}$ remains constant. If only $\theta_1 < \arccos(n_2/n_1)$, with n_2 as
the lower limit of the refractive index beyond the tail of the
profile, this ray will bend back towards the centre plane, and
the profile will guide it. Its turning point x_t occurs within the
profile. Such rays can form guided modes of the profile. Rays with
initial angles $\theta_1 > \arccos(n_2/n_1)$ will experience some bending
within the profile due to the index gradient, but not enough to
remain inside the profile; they will radiate out into the surrounding medium and therefore represent radiation modes.

Not all rays with angles $\theta_1 < \arccos(n_2/n_1)$ form actually guided
modes of propagation; to do this, their locally plane waves need
to satisfy a phase condition for selfconsistent fields within the
profile. This phase condition corresponds to eqn.(2.1) for the
film with abrupt boundaries, which postulates that after two
successive reflections at the upper and lower boundary the wave
repeats ih phase, in order to fit into the film. We formulate this
phase condition for the profile modes by utilizing the transverse
component k_x of the wave vector $n\vec{k}$.

The phase integral between a lower and an upper turning point
must be an integer multiple $m\pi$ of π

$$\int_{-x_t}^{x_t} k_x \, dx = (m + 1/2)\pi . \qquad (4.4)$$

Only under this condition will we obtain the required standing wave pattern in the transverse field distribution with m nodes. The phase shift $\pi/2$ on the right-hand side takes account of the peculiar situation at the turning point. Looking in x-direction transverse to the direction of propagation, the wave impedances are $Z = \omega\mu_o/k_x$ for H-waves and $Z = k_x/(\omega\varepsilon_o n^2)$ for E-waves. At the turning point, we have $k_x = 0$, and the wave impedance change from real values for $|x|<|x_t|$ to purely imaginary values for $|x|>|x_t|$. They have equal magnitude at small but equal distances to both sides from turning point. With $Z^{(-)} = jZ^{(+)}$, therefore, the waves experience a reflection at the turning point with a reflection coefficient $r=(j-1)/(j+1)$ of unit magnitude but $\pi/2$ phase shift.

To further evaluate eqn.(4.4), we express k_x according to the local separation condition by the local wave number $n(x)k$ and the axial phase constant $\beta= n_1 k\cos\theta_1$. This leads to the universal form

$$\int_{-x_t}^{x_t} \sqrt{n^2(x)\ k^2-\beta^2}\ dx = (m + 1/2)\pi \qquad (4.5)$$

of characteristic equation for any symmetric index profile with sufficiently small index gradient. It holds for any ray that has turning points $\pm x_t$, where the radicand of eqn.(4.5) vanishes. Any β-value in the range $n_2< \beta/k<n_1$ will yield such turning points. If, in addition, this β-value satisfies eqn.(4.5) for an integer m, it represents the phase constant β_m of the respective H_m - and E_m - modes. Eqn.(4.5) provides us therefore with a general method to determine the phase constant of guided modes in symmetric index profiles.

For an evaluation of the phase integral in eqn.(4.5), we need to specify a particular index profile $n(x)$. As an example, we take the truncated parabolic profile

$$n = n_1\left[1- \Delta(x/d)^2\right] \qquad \text{for } |x|<d$$
$$n = n_2 \qquad \text{for } |x|>d \qquad (4.6)$$

of Fig.4.4.

$$\Delta = \frac{n_1-n_2}{n_1} \qquad (4.7)$$

represents the maximum relative index difference of this profile. The paraxial ray equation (4.2) for this profile

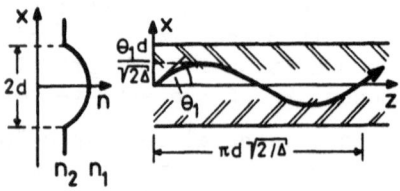

Fig.4.4 Truncated parabolic index profile and ray path of a guided mode

$$\frac{d^2x}{dz^2} + 2(\Delta/d^2)x = 0 \tag{4.8}$$

has the particular solution

$$\frac{x}{d} = (\theta_1/\sqrt{2\Delta}) \sin(\sqrt{2\Delta}\, z/d). \tag{4.9}$$

Rays starting at z=0 with x=0 and θ_1 undulate as sine waves with period length $\pi d\sqrt{2/\Delta}$ and turning points $x_t = \pm d\theta_1/\sqrt{2\Delta}$. The phase integral (4.5) leads to a finite number of modes with phase constants

$$\beta = \sqrt{n_1^2 k^2 - (2m+1) n_1 k\sqrt{2\Delta}/d} \ . \tag{4.10}$$

The highest order mode has $|x_t| = d$ and just grazes the edge of the profile. Its phase constant is $\beta = n_1 k \cos\sqrt{2\Delta}$, which when substituted into eqn.(4.10) yield the highest order m of a guided mode as the next integer smaller than $n_1 kd\sqrt{\Delta/2} - 1/2$. This quantity as well as eqn.(4.10) suggests a finite cut-off even for the lowest order m=0. A more accurate field analysis shows, however, that the H_o - and E_o -modes of the truncated parabolic profile in Fig.4.4 have zero cut-off.

If the index profile has the strong asymmetry which Fig.4.1b indicates, the large index step from n_1 down to n_o will have a relative large limiting angle $\theta_{oc} = \arccos(n_o/n_1)$ for total reflection. Any modes, which are still guided by the comparatively weak index change in the profile will be associated with rays that have angles $|\theta_1| < \arccos(n_2/n_1)$, that are quite small compared to θ_{oc}. Their total reflection leads to a very low field at the boundary between n_1 and n_o. In the limit of infinite asymmetry, there is even a field node at the boundary. In the corresponding symmetric profile all odd order H- and E-modes have this field node in the centre plane. One half of their distributions fits therefore also into the strongly asymmetric guide and propagates with the same phase constant as in the symmetric profile. All odd order modes of any symmetric profile are therefore also modes of propagation in the corresponding asymmetric struture.

5. STRIP GUIDES

The planar film guide provides no confinement of light in the film plane. Some planar devices such as thin-film light deflectors for optical data processing need no such confinement. For other devices and in plane guiding applications transverse confinement may be provided by periodic focussing. Fig 5.1 shows the planar equivalent of a lens as focussing element of a planar beam waveguide. The phase constant of any film mode increases with film thickness slowing the mode down in its phase velocity. Alternatively this effect may also be described by an effective refractive index N = β/k for any

particular film mode; this effective
index increases with film thickness.
The lens-like region of raised film
thickness then acts as a focussing
lens for a film mode. Launching a
film through a prism or grating
coupler from a laser beam with
Gaussian distribution, periodic focussing compensates for transverse
spreading by transforming the phase
fronts from a dispersing into a
converging film wave.

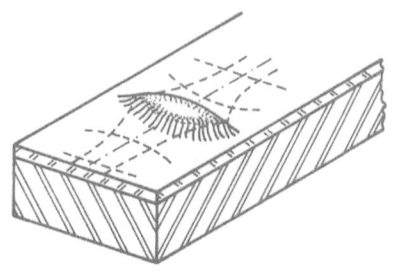

Fig.5.1 Planar film lens re-
focussing a film-mode beam

Limiting the film to a finite width leads to the strip guides of
Fig.5.1 which provide transverse confinement in a continuous fashion.
The film with **refractive index** n_1
extends now in transverse direction
only over the limited width b. This
strip may be deposited on a substrate surface in form of the
raised strip in Fig.5.2a; it may also be embedded in the substrate with
its cover boundary flush with the
substrate surface as in Fig.5.2b. As
a third version, it may be buried
completely inside the substrate medium with the same refractive index n_2
surrounding it on all **four** sides,
as in Fig5.2c. The general strip
guide struture of Fig. 5.3 includes
all three cases of Fig.5.2 and serves
us a sufficiently general model to
analyze guided modes in all three
structures. It represents the raised
strip for $n_3 = n_o$, the embedded strip
for $n_3 = n_2$, and the buried strip for
$n_3 = n_2 = n_o$.

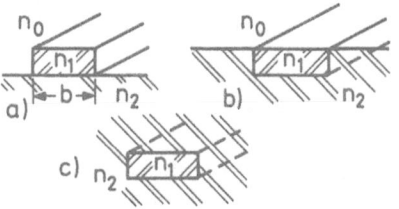

Fig.5.2 Basic strip guide
        structures
        a) raised strip
        b) embedded strip
        c) buried strip

Fig.5.3 General model for
        raised, embedded
        and buried strips

The confinement may be regarded as total reflection of a film wave
inside the strip at the side walls of the strip. For any of the
guided modes of the strip, we start with the corresponding H_m - or
E_m-wave of the film with index n_1 between the substrate of index
n_2 and the cover of index n_o. We let this film wave with its phase
constant $\beta = N_s k$ propagate at an angle θ with respect to the strip
axis as indicated by Fig.5.4. For this film wave, the region has
the effective index of refraction N_s. When incident on the side
walls of the strip, this film wave excites a field distribution

in the region beyond, which is similar to
the field of the corresponding film wave
in a film with index n_3 between the same
substrate and cover as the strip region.
In our situation, however, we have either
$n_3 = n_o$, or $n_3 = n_2$, or $n_3 = n_2 = n_o$ and hence
no film waves exist in the regions with
n_3. The film wave which is incident from
the strip, will therefore excite a radiation field and will then not be guided by
the strip. If, however,

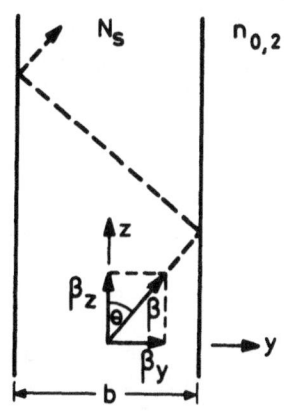

$$\beta_z = \beta \cos\theta > n_{o,2} k \quad , \qquad (5.1)$$

the incident film wave from the strip
excites only evanescent fields in the
region beyond the strip. It will then be
totally reflected by the side walls and
travel on a zig-zag path inside the strip.
If now, in addition, it satisfies a phase
condition corresponding to eqn.(2.2) for

Fig.5.4 Film mode on
zig-zag-path
in strip guide
forms strip mode

film waves, it repeats in phase after two successive reflections.
Its field distribution then fits into the film and forms guided
modes of propagation of the particular strip.

If the particular strip mode grows out of an H_m-mode of the film
in the strip region, we designate this strip mode as HE_{ml}-mode,
if it grows out of an E_m-mode, we designate it as EH_{ml}-mode. This
designation with both letters H and E accounts for the hybrid
character of strip modes. The first index m in the HE_{ml}- and EH_{ml}-
designation of strip modes continues to count the order of the film
mode out of which it grows. The second index l is also an integer
number and designates the transverse order of the respective strip
mode parallel to the bottom and cover of the strip. It is the number
of periods by which the direct shift of a phase front differs from
the detour via two successive reflections at the side walls with
their additional phase shifts. In zig-zaging along the strip, the
film mode fields superimpose to standing waves in transverse direction also parallel to the bottom and cover of the strip. The index
l counts the number of nodes of their transverse standing wave pattern.

To actually formulate the phase condition for strip modes that would
correspond to eqn.(2.2) for film modes, we neglect any direct
interaction between total reflection at opposite side walls of the
strip. In addition, we use eqns. (1.1) and (1.2) to evaluate the
phase shift for total reflection. These phase shift expressions
were derived for plane uniform waves and total reflection at a
plane interface. To apply them for the total reflection of a film
wave at the side wall of the strip, we replace k_{1x} by $\beta_y = \beta \sin\theta$
and k_{2x} by $(j\beta_z^2 + (u/d)^2 - n_3^2 k^2)^{1/2}$. The first of these substitutions

introduces the transverse y-component of the relevant wave vector
in the strip region, while the second substitution introduces the
corresponding quantity in the evanescent field region beyond the
side walls. This latter substitution follows with the wave vector
components parallel to the strip region. To facilitate formulation
of the characteristic equations for strip modes, we abbreviate
these substitution by

$$u_s/b = \beta_y = (n_1^2 k^2 - (u/d)^2 - \beta_z^2)^{1/2} \tag{5.2}$$

$$v_s/b \quad = (u/d)^2 + \beta_z^2 - n_3^2 k^2)^{1/2}; \tag{5.3}$$

the quantities u_s and v_s then represent the transverse phase parameter inside the strip and the transverse attenuation parameter
beyond the strip, respectively, both in y-direction. They correspond
to the transverse parameters u and v in x-direction of the constituent film modes.

The phase angles of total reflection follows from eqns.(1.1) and
(1.2) and with the above substitutions appear as

$$\phi_m = 2\arctan(n_1^2 \, v_s/n_3^2 \, u_s) \tag{5.4}$$

for an H_m-film mode and

$$\phi_e = 2\arctan(v_s/u_s) \tag{5.5}$$

for an E_m-film mode. Note that an H_m-film has a magnetic field H_x
as dominant component parallel to the side wall, while for an E_m
film mode it is the electric field component E_x.

The characteristic equations of the various strip modes can now be
formulated from the general equation (2.2) and its special forms
for symmetric films. We obtain the characteristic equation for HE_{ml}-
modes with even order l:

$$u_s \tan(u_s/2) = (n_1/n_3)^2 v_s \tag{5.6}$$

and for HE_{ml}-modes with odd orders l

$$-u_s \cot(u_s/2) = (n_1/n_3)^2 v_s \tag{5.7}$$

In case of EH_{ml}-modes with even order l, we obtain

$$u_s \tan(u_s/2) = v_s \tag{5.8}$$

and for EH_{ml}-modes with odd order l

$$-u_s \cot(u_s/2) = v_s. \tag{5.9}$$

To evaluate these characteristic equations for any of the strip modes, we would start from the film parameter $V = kd \sqrt{n_1^2 - n_2^2}$ and determine first the transverse phase parameter u of the particular film mode, that will form the strip mode. We will next take the corresponding parameter $V_s = kb\sqrt{n_1^2 - n_3^2}$ of the strip and, by solving the respective characteristic equation, determine the transverse phase parameter u_s in the strip. The phase constant of the strip mode will then follow from eqn.(5.2)

$$\beta_z = \sqrt{n_1^2 k^2 - (u/d)^2 - (u_s/b)^2} \qquad (5.10)$$

Practical strip guides have quite often only a slightly higher refractive index n_1 than the index n_2 of the sustrate. Both, n_1 and n_2 differ, however, substantially from $n_0 = 1$, if the cover is free space. Under these weak guiding conditions, the above approximate evaluation of characteristic equation simplifies still further and the results may be represented in terms of the universal phase and film parameters B and V, that were also applied to represent film mode characteristics.

For the raised strip, we have $n_3 = n_0$ and in case of weak guidance with $n_1 \approx n_2$, but a pronounced difference between n_1 and $n_0 = 1$, the large index step from n_1 to n_0 renders the fields nearly zero at the side walls of the strip, so that

$$u_s = (\ell+1)\pi. \qquad (5.11)$$

The other transverse phase parameter u follows from

$$u = \pm V \sin u \qquad (5.12)$$

Fig.5.5 shows the phase parameter B plotted versus the film parameter V, from this approximation for a few of the low order modes of the weakly guiding raised strip.

For the embedded strip, we have $n_3 = n_2$. If, in addition, we assume again weak guidance with $n_1 \approx n_2$, but a pronounced difference between n_1 and n_3, the transverse phase parameter u follows again from eqn.(5.12), while u_s as the transverse phase parameter of the weakly guiding symmetric film follows from eqn.(5.8) for even orders ℓ and from eqn.(5.9) for odd orders ℓ. The transverse attenuation parameter v_s in these equations is related to u_s and the strip parameter V_s by $u_s^2 + v_s^2 = V_s^2$, so that instead of eqns.(5.8) and (5.9), the following forms of the characteristic equation

$$u_s = \pm V_s \cos(u_s/2) \qquad (5.13)$$

$$u_s = \pm V_s \sin(u_s/2) \qquad (5.14)$$

give u_s directly, if not explicitly, in terms V_s. Fig.5.6 shows

the phase parameter B as
calculated according to
this approximation for embedded strips of weak guidance. Any particular mode
of theembedded strip has a
lower cut-off and a larger
phase constant than the same
mode in the raised strip.

An accurate, but computer generated solution for strip
modes is based on an expansion
of the electromagnetic fields
in terms of circular harmonics
/4/. For the buried strip in
Fig.5.2c, the longitudinal
components of the electric
and magnetic fields inside
the strip are set up in the
form:

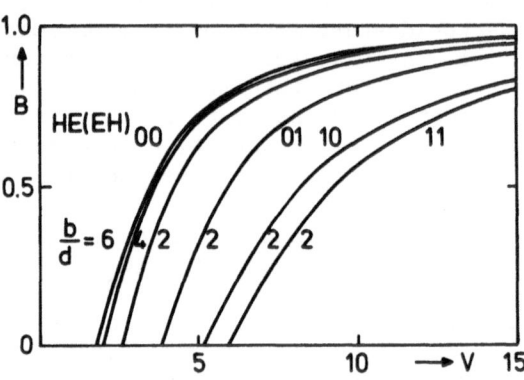

Fig.5.5 Phase parameter $B=(\beta/k-n_2)/$ (n_1-n_2) of low order modes in a raised strip with $(n_1-n_2)<<(n_1-n_o)$ versus the film parameter $V=kd\sqrt{n_1^2-n_2^2}$

$$E_{z1} = \sum_{p=0}^{N} a_p J_p(k_r r) \left[\sin(p\varphi+\varphi_p)\right] e^{-j\beta z}$$

$$(5.15)$$

$$H_{z1} = \sum_{p=0}^{N} b_p J_p(k_r r) \left[\sin(p\varphi+\psi_p)\right] e^{-j\beta z}$$

and outside the strip as:

$$E_{z2} = \sum_{p=0}^{N} c_p K_p(\alpha_r r) \left[\sin(p\varphi+\varphi_p)\right] e^{-j\beta z}$$

$$(5.16)$$

$$H_{z2} = \sum_{p=0}^{N} d_p K_p(\alpha_r r) \left[\sin(p\varphi+\psi_p)\right] e^{-j\beta z}$$

with $k_r = (n_1^2 k^2-\beta^2)^{1/2}$ as well as $\alpha_r = (\beta^2-n_2^2 k^2)^{1/2}$ and J_p and K_p as Besselfunctions and modified Hankelfunctions, respectively, of the integer order p. The transverse fields for these longitudinal components follow readily from Maxwell's equations. The circular harmonic expansions of eqns. (5.15) and (5.16), when extended to infinitely many terms form complete sets, which allow to represent any source free field distribution with exp(-jβz) as its z-depen-dence. They therfore lend themselves also for any strip mode with phase constant β. To determine the expansion coefficient a_p, b_p, c_p, and d_p, we choose N points along the boundary of the strip and match all four tangential field components inside and outside the strip at these points. This provides us with just as many equations as there are unknown quantities in the expansions (5.15) and (5.16). They form a homogeneous and linear system for the expansion coefficients. For nontrivial solutions, its coefficient determinant must vanish, which condition is the characteristic equation for strip modes. Values for β which solve this equation are phase constants of strip modes.

Numerical computations show that for moderate aspect

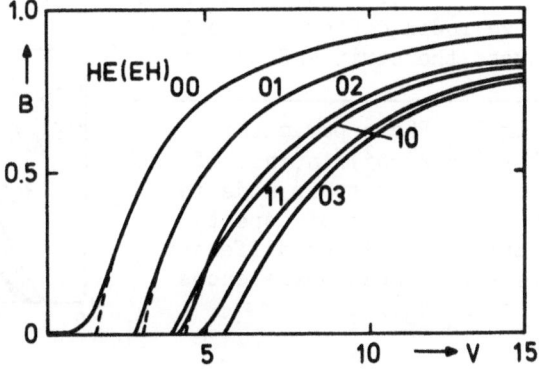

Fig.5.6 Phase parameter $B=(\beta/k-n_2)/(n_1-n_2)$ of low order modes in an embedded strip with $(n_1-n_2)\ll(n_1-n_0)$ versus the film parameter $V = kd\sqrt{n_1^2-n_2^2}$

ratios b/d of the strip only a limited number of circular harmonics are needed to achieve sufficient accuracy for the phase constant. For larger aspect ratios, however, the convergence is not very good. Fig. 5.7 shows accurate results for the phase parameter B of a buried strip as a function of the film parameter V. The broken lines in Fig.5.7 represent results from eqn.(5.12), which are based on the film mode approximation for the strip mode. These approximations agree so well with the accurate numerical results, that the two types of curves cannot be distinguished from each other over most of their range. A noticeable deviation occurs only close to their respective cut-offs. This discrepancy between the accurate numerical solution and the film mode approximations appears most pronounced for the lowest order modes in their two polarizations HE_{oo} and EH_{oo}. While the accurate numerical solution display the

Fig.5.7 Phase parameter $B = (\beta/k-n_2)/(n_1-n_2)$ versus the film parameter $V = kd\sqrt{n_1^2-n_2^2}$ for low order modes in a buried strip with $(n_1-n_2)\ll n_1$ and b/d = 2. Broken lines represent the approximation from eqn.(5.12) with film modes in the strip.

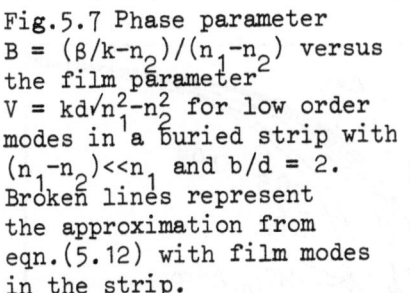

dispersion curve down to their zero cut-off, the film mode approximation ends with B = 0 at a finite cut-off value for V.

The raised strip and the embedded strip have fundamental modes with
finite cut-off. For them, the film mode approximation will not show
this relatively large discrepancy at cut-off. We can therefore pro-
nounce the film mode approximation to be sufficiently accurate for
most practical purposes in all three types of strip guides of Fig.
5.2, except near cut-off, in particular for the lowest order modes
of the buried strip with their zero-cut-off.

Fig. 5.7 gives the phase parameter of the buried strip only for the
aspect ratio b/d = 2. For other aspect ratios, strip modes have different phase constants. The total range over which the phase constant
varies is illustrated by Fig.5.8 for the two lowest order modes of
the buried strip. It compares the phase constant for the limiting
cases of a square strip and an infinitely wide strip to the phase
constant of a strip with aspect ratio
b/d = 2. The curves for b/d = 1 and 2 are
from an exact numerical evaluation of
Goell's computer analysis while the curve
for b/d → ∞ reproduces the phase constant
for the H_o - and E_o -modes of a weakly guiding symmetrical film. The latter case actually represents a symmetrical film.

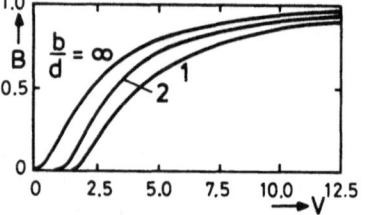

Fig.5.8 Phase parameter
B = $(\beta/k-n_2)/(n_1-n_2)$
of the lowest order
strip modes HE_{oo} and
EH_{oo} in a buried strip
of different aspect
ratios /4/;$(n_1-n_2)<<n_1$

All the dispersion curves, which have been
presented so far, are for weakly guiding
strips with $(n_1-n_2)<<(n_1-n_o)$. In this
limiting case any HE_{m1} -mode of a particular
strip has the same phase constant as the
EH_{m1} -mode of the same transverse orders m
and 1. As in case of film modes, any substantial difference in refractive index
of strip and substrate breaks this degen-
eracy between HE_{m1} - and EH_{m1} -modes. Fig. 5.9 shows the dispersion
curves of the lowest order modes in their two polarisation HE_{oo}
and EH_{oo} for a buried strip of aspect ratio b/d=2 and various relative index differences Δ be-
tween the strip and the em-

Fig.5.9 Phase parameter
B = $(\beta^2/k^2-n_2^2)/(n_1^2-n_2^2)$ of the
lowest order strip mode in
a buried strip of aspect
ratio b/d = 2 for different
relative index differences
Δ = $(n_1-n_2)/n_2$

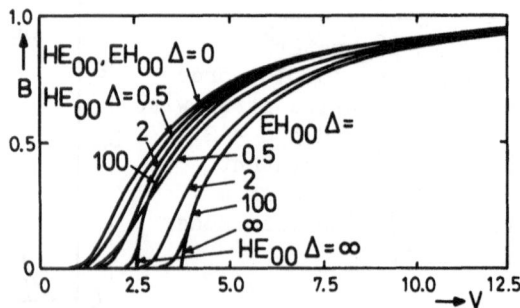

bedding material. The phase parameter in this general case follows
from the exact definition $B = (\beta^2/k^2 - n_2^2)/(n_1^2 - n_2^2)$, which holds for any
value of $(n_1 - n_2)$. The curves represent accurate results, again from
Goell's computer analysis. A significant difference between the phase
parameter of the HE_{oo} -mode and the EH_{oo} -mode appears already at
$n_1 - n_2 = 0.5$. The EH_{oo} -phase constant decreases more from its limiting value for the weakly guiding strip than does the HE_{oo} -phase
constant. The HE_{oo} -mode with the highest of all phase constants is
at all wavelengths and for any aspect ratio $b/d > 1$ best confined
to the strip. It therefore forms the fundamental or dominant mode
of propagation of the strip.

The dispersion characteristics of strip modes are similar to those
of film modes. Near cut-off the strip modes propagate with the wave
number of the substrate. Far above cut-off, for wide enough strips
of relativly thick films or large index-differences, they propagate
with the wave number of the strip material. The raised strip of
Fig. 5.2a can be fabricated from a film by masking the strip and
removing the surrounding film, by reverse sputtering, ion beam
etching, or chemical etching. The embedded strip of Fig. 5.2b is
fabricated by embossing or ion implantation.

6. STRIP-LOADED FILM GUIDE

When films are limited to a finite width in form of strips, they
provide transverse confinement of film waves by total reflection
at the side walls of the strip. The side walls of the raised strip
with their large index step reflect film waves more effectively,
and fields concentrate more inside the strip than in case of the
embedded strip with its weakly guiding side walls. This characteristic speaks in favour of the raised strip. The side walls of such
a raised strip must, however, be very smooth in order not to cause
too much scattering loss. The embedded strip may be allowed to have
more side wall imperfections without excessive scattering loss.
From eqns.(3.29) and (3.30) the scattering loss at rough interface
with different index steps increases with the difference in relative
permittivities; it is therefore much larger for the raised strip
than for the embedded strip.

Another advantage of the embedded strip as compared to the raised
strip of the same size may be its wider extension of transverse
fields. The evanescent fields beyond both side walls of the strip
for any particular mode are much stronger in case of the embedded
strip and extend further sideways than in case of the raised strip.
The larger mode size in the embedded strip facilitates launching of
these modes and jointing of strip guides.

To gain still more in mode size, and further reduces the scattering
at rough edges of the structure, the transverse confinement should
be provided by an even smaller effective difference in refractive

index at the side walls of the structure.
The strip-loaded film of Fig.6.1 offers
this possibility. It allows to control
the effective change in refractive index
quite accurately to very small differences. According to Fig.6.1 the strip-
loaded film consists of a thin film of

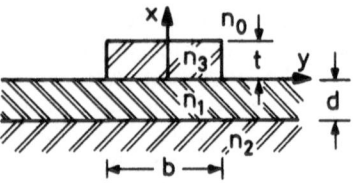

index n_1, deposited on a sustrate of
index n_2, loaded with a raised strip of   Fig.6.1 Strip-loaded
index n_3. The index of the cover region   film guide
n_o is normally that of free space with
$n_o = 1$. The strip can have the same index as the substrate; it
could also have an index between substrate and film index. However,
to maintain the favourable characteristics of the strip loaded
film guide, the strip index should neither be much lower than the
substrate index nor much higher than the film index.

We learn to understand wave guidance and field confinement in the
strip-loaded film by considering the strip region as a four-layer
film guide. A particular film mode of this four-layer structure has
a phase constant β_s and an effective index $N_s = \beta_s/k$ which are always
larger than the phase constant β_f and effective index N_f, respectively, of the corresponding film mode in the unloaded side regions of
the film. If the four-layer film mode propagates at an angle θ with
respect to the strip axis, as shown in Fig. 6.2, it will experience
total reflection on the edges of the four-layer region, when

$$N_s \cos\theta > N_f \qquad\qquad (6.1)$$

If in addition, this film mode satisfies a phase
condition corresponding to eqn.(2.1), it repeats
in phase after two successive reflections at the
opposite strip edges.It then fits into the strip
loaded film region and forms selfconsistent field
solutions for guided modes of propagation in the
strip-loaded film.

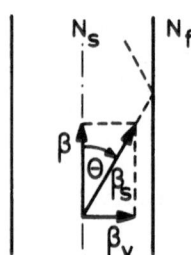

To evaluate the phase condition for these guided
modes, we need to know the phase shift ϕ,
which the film mode of the strip-loaded
region experiences upon total reflection
at the edges of the four-layer region. An
analytical solution to this problem is not   Fig.6.2 Ray at angle
available. For the general structure of θ representing a film
Fig.6.1, we will therefore resort to the   mode in the four
same approximate procedure as we applied   layer strip-loaded
it to strip guides.   region

For our present approximation of film mode reflection at the edge
of the four-layer struture, we consider the impedances, which film

modes have in their transverse direction parallel to the film. Let
an H_m-film mode of the strip-loaded region propagate with phase
constant

$$\beta_s = \sqrt{n_1^2 k^2 - (u_s/d)^2} \tag{6.2}$$

at an angle θ with respect to the strip axis. In the expression
(6.2), we have related β_s to the wave number $n_1 k$ and the transverse
phase parameter u_s of the film mode inside the film. This film mode
then has the axial phase constant

$$\beta = \beta_s \cos\theta \tag{6.3}$$

and the transverse phase constant

$$\beta_y = \beta_s \sin\theta \tag{6.4}$$

From Maxwell's equation its transverse magnetic field component H_x
relates to the electric field component as

$$-j\omega\mu_o H_x = -j\beta_y E_z + j\beta E_y. \tag{6.5}$$

From $\nabla \vec{E} = 0$ follows $E_y = -(\beta/\beta_y)E_z$, so that the transverse wave
admittance $Y = H_x/E_z$ in the strip loaded region obtains as

$$Y_s = \beta_s^2/(\omega\mu_o\beta_y). \tag{6.6}$$

The same considerations hold for an H_m-film mode in the unloaded
side-region, except that in case of total reflection at the edge
of the four-layer region its fields decay exponentially in y-direction with an attenuation constant

$$\alpha_y = \sqrt{\beta^2 - \beta_f^2}. \tag{6.7}$$

The phase constant β_f in this expression is that of the respective
H_m-film mode in the unloaded region. It relates by

$$\beta_f^2 = n_1^2 k^2 - (u_f/d)^2 \tag{6.8}$$

to the wave number $n_1 k$ of the film and the transverse phase parameter u_f of this film mode. With β_f and α_y of the transversely
evanescent film mode in the unloaded film region, its transverse
wave admittance obtains as

$$Y_f = j \beta_f^2/(\omega\mu_o\alpha_y). \tag{6.9}$$

Before we proceed to evaluate total reflection at the wave admittance
step from Y_s to Y_f, we will first determine the corresponding wave
impedances for E_m-film modes. Noting the duality between H_m- and

E_m-waves in dual structures, the wave impedances of E_m-waves constitute dual quantities to the admittances of corresponding H_m-waves.
They therefore follow from eqns.(6.6) and (6.9) with the permeability
μ replaced by the permittivity $\varepsilon = n^2 \varepsilon_o$. The transverse E_m-wave impedance in the strip-loaded region is

$$Z_s = \beta_s^2 / (\omega n^2 \varepsilon_o \beta_y) ,$$ (6.10)

while for the evanescent fields of an E_m-wave in the unloaded region,
it is

$$Z_f = j \, \beta_f^2 / (\omega n^2 \varepsilon_o \alpha_y) .$$ (6.11)

For the reflection of an E_m-wave at the edge of the four-layer region
we have the E_x-component perpendicular to the plane of incidence
and therefore take the reflection coefficient

$$r_e = \frac{Z_f - Z_s}{Z_f + Z_s} .$$ (6.12)

For the reflection of an H_m-wave on the other hand, we have the
H_x-component perpendicular to the plane of incidence, and the reflection coefficient is

$$r_m = \frac{Y_f - Y_s}{Y_f + Y_s} .$$ (6.13)

In determining the reflection from transverse wave impedance, we
achieve a fairly good match of field components inside the film,
and almost as good also inside the substrate. The fields above the
film, in the loading strip on one side of the edge, and in the cover
space on the other side, are, however not accounted for in this
impedance match. The reflection coefficients will therefore be
accurate onlyas long as the fields are well confined to the film.

When we substitute the transverse wave impedances and admittances
into the respective reflection coefficient, we obtain one and the
same expression for the phase shift of total reflection for E_m-film
waves as well as H_m-film waves. This phase shift is

$$\phi = 2\arctan\left[\alpha_y \beta_s^2 / (\beta_y \beta_f^2)\right] .$$ (6.14)

It shows similarity with the phase angles in eqns.(1.1) and (1.2)
for total reflection of the plane uniform wave at a plane interface.
If we replace the transverse attenuation constant jk_{2x} in eqn.(1.2)
by our present transverse attenuation constant α_y and the transverse
phase constant k_{1x} by our present transverse phase constant β_y, and
if furthermore the ratio n_1/n_2 of refractive indices in eqn.(1.2)
is replaced by the ratio $N_s'/N_f^2 = \beta_s/\beta_f$ of our present effective indices of refraction, we obtain eqn.(6.14) for the phase angle of

total reflection at the edge of the four-layer region. With these
substitutions, therefore, the reflection corresponds to that of a
plane uniform E-wave at the plane interface.

This correspondence leads us directly to the characteristic equations
of guided modes in the strip loaded film. The structure is symmetric
with   respect to the longitudinal centre-section at y=0. With
eqn.(6.14) corresponding to the reflection of E-waves, we therefore
take the characteristic equations of even and odd order E_m-waves of
the symmetrical slab, and obtain

$$\beta_y \tan(\beta_y b/2) = \alpha_y (\beta_s/\beta_f)^2 \qquad\qquad (6.15)$$

for HE_{ml}- and EH_{ml}-modes of even order l and

$$-\beta_y \cot(\beta_y b/2) = \alpha_y (\beta_s/\beta_f)^2 \qquad\qquad (6.16)$$

for HE_{ml}- and EH_{ml}-modes of odd order l. The designation HE_{ml} and
EH_{ml} corresponds in all details to that of strip modes: HE_{ml}-modes
obtain from H_m-film modes and EH_{ml}-from E_m-film modes. The transverse order l represents again the integer multiple of 2π by which
the phase shift along the direct path of a film mode differs from
the detour via the two successive reflections at the opposite edges
of the strip-loaded region. l also equals the number of nodes of
the standing wave pattern which the transverse field distribution
forms under the loading strip in y-direction.

For a universal representation of dispersion characteristics of
film modes, we have used the film parameter $V = \sqrt{u^2+v^2}$ and the
phase parameter $B = (\beta^2/k^2-n_2^2)/(n_1^2-n_2^2)$. We have then obtained B as
a function of V for each particular film mode with n_1/n_2 as an additional parameter. In the present case of the strip-loaded film,
the corresponding substitutions lead to the new waveguide parameter

$$V_s = b\sqrt{\beta_y^2 + \alpha_y^2} . \qquad\qquad (6.17)$$

This parameter can also be written in terms of the effective indices
of refraction for the film mode in the strip-loaded region

$$N_s = \beta_s/k = \sqrt{\beta^2+\beta_y} /k \qquad\qquad (6.18)$$

and in the unloaded region

$$N_f = \beta_f/k = \sqrt{\beta^2-\alpha_y^2}/k , \qquad\qquad (6.19)$$

so that

$$V_s = kb \sqrt{N_s^2-N_f^2} . \qquad\qquad (6.20)$$

Substituting N_s for n_1 and N_f for n_2 in B leads to the new phase parameter

$$B_s = (\beta^2/k^2 - N_f^2)/(N_s^2 - N_f^2) . \qquad (6.21)$$

For any HE_{ml}- or EH_{ml}-mode, now in the strip-loaded film, B_s depends on V_s and on N_s/N_f, just as B depends on V and on n_1/n_2 for the E_1-mode of the same order l as the strip-loaded film mode has it in y-direction. Weakly guiding films with $n_1 \approx n_2$ show the additional simplification that with $n_1/n_2 \approx 1$, B becomes a unique function of V. By the same token, for the strip-loaded film, if the loading is weak enough to let $\beta_s \approx \beta_f$ and consequently $N_s \approx N_f$, the phase parameter B_s is the same unique function of V_s as $B(V)$ for the weakly guiding symmetrical film.

To actually evaluate the dispersion characteristic B_s as a function of V_s and N_s/N_f for a particular mode in a specific strip loaded film, we first need to know the effective indices N_s and N_f of the constituent film mode in the loaded and unloaded regions. The effective N_f of the film mode in the unloaded region may be read from Fig.2.2 or computed more accurately as a solution of the characteristic equation (2.7) or (2.8), respectively, of the asymmetrical film.

The effective index N_s, however, belongs to the film mode in the four-layer structure. Under present conditions, we have evanescent fields in the cover region and possibly also in the strip region of the four-layer structure. The interface at x=0 between film and strip region will then show total reflection for any plane uniform wave which forms film waves in its zig-zag propagation down the film. For an H-wave incident on this interface at x=0, the reflection coefficient is determined by the impedance which the strip layer of thickness t and transverse wave impedance $Z_3 = \omega\mu_o/k_{3x}$ presents at the interface, when it is backed up by the cover region of transverse wave impedance $Z_o = \omega\mu_o/k_{ox}$. The strip layer acts as a transmission line of wave impedance Z_3 that transforms its load impedance Z_o into

$$Z_{13} = Z_3 \frac{Z_o + Z_3 \tanh jk_{3x}t}{Z_3 + Z_o \tanh jk_{3x}t} . \qquad (6.22)$$

If, for total reflection, both the cover region and the strip region are excited with evanescent fields, we have $k_{ox} = -j\alpha_o$ and $k_{3x} = -j\alpha_3$ as well as $Z_o = j\omega\mu_o/\alpha_o$ and $Z_3 = j\omega\mu_o/\alpha_3$. The reflection coefficient

$$r_e = (Z_{13} - Z_1)/(Z_{13} + Z_1) \qquad (6.23)$$

at the film strip interface has unit amplitude under these conditions, and with $Z_1 = \omega\mu_o/k_{1x}$ its phase angle amounts to

$$\phi_e = 2 \, \arctan\left[\frac{\alpha_3(\alpha_o + \alpha_3 \tanh\alpha_3 t)}{k_{1x}(\alpha_3 + \alpha_o \tanh\alpha_3 t)}\right] \qquad (6.24)$$

with the transverse phase constant

$$k_{1x} = \sqrt{n_1^2 k^2 - \beta_s^2} \qquad (6.25)$$

inside the film, and transverse attenuation constants

$$\alpha_3 = \sqrt{\beta_s^2 - n_3^2 k^2}$$
$$\alpha_o = \sqrt{\beta_s^2 - n_o^2 k^2} \qquad (6.26)$$

in strip and cover region, respectively. At the film-substrate interface the total reflection has the phase

$$\phi_e = 2 \, \arctan(\alpha_2/k_{1x}) \qquad (6.27)$$

with

$$\alpha_2 = \sqrt{\beta_s^2 - n_2^2 k^2} \, . \qquad (6.28)$$

Substituting these phase shifts for ϕ_{10} and ϕ_{12} into eqn.(2.2) leads to the characteristic equation for H-waves

$$\tan k_{1x} d = \frac{k_{1x}\left[\alpha_2 + \alpha_3 \, \dfrac{\alpha_o + \alpha_3 \tanh\alpha_3 t}{\alpha_3 + \alpha_o \tanh\alpha_3 t}\right]}{k_{1x}^2 - \alpha_2 \alpha_3 \, \dfrac{\alpha_o + \alpha_3 \tanh\alpha_3 t}{\alpha_3 + \alpha_o \tanh\alpha_3 t}} \qquad (6.29)$$

Its solutions for a given four-layer structure at a specified wave number k yield the effective index $N_s = \beta_s/k$ of H_m-modes in the four-layer region.

For a loading strip of sufficient thickness t, so that the argument $\alpha_3 t$ is large enough for $\tanh\alpha_3 t \approx 1$, eqn.(6.29) reduces to the characteristic equation

$$\tan k_{1x} d = \frac{k_{1x}\left[\alpha_2 + \alpha_3\right]}{k_{1x}^2 - \alpha_2 \alpha_3} \qquad (6.30)$$

of the asymmetric film, for which the strip material with index n_3 forms the cover. Under these conditions, the effective index may again be read directly from Fig. 2.2 or obtained from a more accurate numerical solution of eqn. (2.7).

The special case of $n_3 = n_2$ deserves some attention because with the limited choice of suitable materials, we might want to use the same material for substrate and loading strip. If for $n_3 = n_2$, the film is weakly guiding with $n_1 \approx n_2$, but the strip thick enough for eqn. (6.30) to apply in the loaded region, we have a symmetric slab for the N_s of the film modes in the loaded region, and a strongly asymmetric film for the N_f of film modes in the unloaded side regions. Under these conditions, the phase parameter $B \approx (\beta/k - n_2)/(n_1 - n_2)$ follows for each mode of the strip-loaded film as a unique function of the film parameter $V = kd\sqrt{n_1^2 - n_2^2}$ with the aspect ratio b/d as the only additional parameter. Fig. 6.3 shows this phase parameter for the two polarizations of the fundamental mode of this weakly guiding structure with substrate and strip of the same index. Under these conditions of Fig.6.3 the strip-loaded film guides only the fundamental mode. The next higher order mode in its two polarizations HE_{01} and EH_{01} has its cut-off at $b/d = 7.2$.

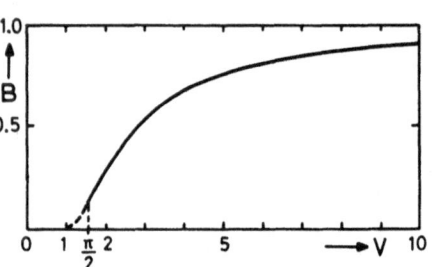

Fig. 6.3 Phase parameter $B = (\beta/k - n_2)/(n_1 - n_2)$ versus frequency parameter $V = kd\sqrt{n_1^2 - n_2^2}$ in a strip-loaded film with $n_1 - n_2 = n_1 - n_3 \ll n_1 - n_0$ and $\tanh \alpha_3 t \approx 1, b/d = 2$.

7. RIB GUIDE

Another special guiding structure obtains from the strip-loaded film in Fig. 6.1, when film and loading strip have the same index $n_3 = n_1 > n_2 > n_0$. Such a structure results when we first deposit a film onto a substrate, and then remove part of this film down to a lower thickness, except for the strip region. The strip like region of raised film thickness remains as a ridge or rib on top of the film of reduced thickness; it lends the name ridge or rib guide to this type of planar optical waveguide. The rib guide offers the same advantages as the strip loaded guide to nearly the same extent. It relaxes the stringent requirements on the smoothness of side walls, which low scattering losses impose on strip guides, particularly on the raised strip.

Fig. 7.1 shows a rib guide with a rib of height h and width b and the film of thickness d extending symmetrically to both sides of the rib. The modes which are guided by the rib may again be considered to consist of film modes of the rib region which experience total reflection at the opposite edges of the rib and propagate on a zigzag path down the rib. The rib itself forms a

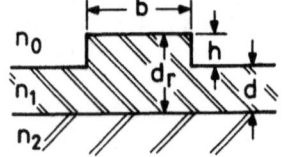

Fig.7.1 Rib guide of index n_1 on a substrate of index n_2.

simple film guide which is asymmetrical if the cover index n_o differs
from the substrate index n_2. At the edge of the rib, the film steps
down to a thinner layer. For any of the film modes, this step means
also that its effective index steps down from its value $N_r = \beta_r/k$
inside the rib to $N_f = \beta_f/k$ in the thinner film to both sides of the
rib. The phase shift for total reflection of a film mode in the rib
at its edges obtains approximately from the same considerations, that
led to eqn. (6.14) for the phase shift of total reflection at the
edges of the strip-loaded film. We again match transverse impedances
of a particular film mode inside the film at the rib's edge. By
matching these impedances we ensure a fairly good, but not complete,
match also for tangential field components at this interface. With
this impedance match, we account for field continuity inside the film,
but neglect the evanescent fields in cover and substrate. We therefore approximate the actual situation the better the more the fields
of constituent film modes are confined to the film. Under these
conditions, and with β_s replaced by β_r, the same expression (6.14)
applies also for the phase shift of total reflection at the rib's
edges. As a further consequence, the condition for film modes to repeat in phase after two successive reflections leads to the same
characteristic equations (6.15) and (6.16) for rib modes of even and
odd symmetry with respect to the rib's longitudinal plane of symmetry.
As the only change in eqns. (6.15) and (6.16), we replace β_s by β_r
and interpret β_y as the transverse phase constant, $\beta_y = \beta_r \sin\theta$
of the film mode in the rib region on its zig-zag path down the rib.
We thus obtain the characteristic equation

$$\beta_y \tan(\beta_y b/2) = \alpha_y \ (\beta_r/\beta_f)^2 \qquad\qquad (7.1)$$

for HE_{ml}- and EH_{ml}-modes of even order l, and the characteristic
equation

$$-\beta_y \cot(\beta_y b/2) = \alpha_y \ (\beta_r/\beta_f)^2 \qquad\qquad (7.2)$$

for HE_{ml}- and EH_{ml}-modes of odd order l.

To solve these equations, we proceed very much as in case of the
strip-loaded film. We first determine $N_r = \beta_r/k$ and $N_f = \beta_f/k$ of
the constituent H- or E-mode in the films of thickness d_r and d,
respectively. We then obtain the phase parameter

$$B_{rf} = (\beta^2/k^2 - N_f^2)/(N_r^2 - N_f^2) \qquad\qquad (7.3)$$

as a function of the rib parameter

$$V_{rf} = kb\sqrt{N_r^2 - N_f^2} \qquad\qquad (7.4)$$

from the solution of the equivalent symmetrical film problem.

For a rib guide which consists of weakly guiding or strongly asymmetric films with $(n_1-n_2)<<(n_1-n_0)$, the phase parameter

$$B = (\beta^2/k^2-n_2^2)/(n_1^2-n_2^2) \tag{7.5}$$

may be presented as a function of the film parameter

$$V_r = kd_r\sqrt{n_1^2-n_2^2} \tag{7.6}$$

in the rib region of the film, with the aspect ratio b/d_r and the film thickness ratio d/d_r as the only additional parameters. To show that such a representation is possible, we note that for any constituent H_m- or E_m-mode in weakly guiding films, we have always $N_r \approx N_f$ for their effective indices, so that $\beta_r/\beta_f \approx 1$ in the characteristic equations (7.1) and (7.2). The phase parameter B_{rf} depends under these conditions only on the rib parameter V_{rf}. Furthermore for such weakly guiding films, the phase parameters

$$B_r = (N_r^2-n_2^2)/(n_1^2-n_2^2) \tag{7.7}$$

and

$$B_f = (N_f^2-n_2^2)/(n_1^2-n_2^2) \tag{7.8}$$

for any prticular film modes depend only on the respective film parameters $V_r = kd_r (n_1^2-n_2^2)^{1/2}$ and $V_f = kd (n_1^2-n_2^2)^{1/2}$. With $B_r(V_r)$, we therefore have $B_f(V_f) = B_r(V_r d/d_r)$. Expressing B_{rf} and V_{rf} by these parameters, we obtain

$$B_{rf} = (B-B_f)/(B_r-B_f)$$

and   (7.9)

$$V_{rf} = (b/d_r)\sqrt{(B_r-B_f)}\ V_r$$

We solve eqn.(7.9) for B and obtain

$$B = B_{rf} (B_r-B_f) + B_f .$$

Hence with V_{rf} as well as B_r and B_f as unique functions of V_r, b/d_r, and d/d_r, the phase parameter B depends likewise only on these three parameters.

By way of example, Fig.7.2 shows this phase parameter B for a number of low order modes in a rib guide of aspect ratio $b/d_r = 4$ and two different ratios of film thickness. We note, that in this weakly-guiding film approximation, HE_{ml}-modes are degenerate with EH_{ml}-modes of equal transverse orders m and l. This degeneracy exists already for the constituent H_m- and E_m-modes of the weakly guiding film and transfers to the corresponding rib guide modes. Any rib

guide mode in Fig. 7.2 with a
transverse order 1 different
from zero start with a finite
value B at its respective
cut-off point.

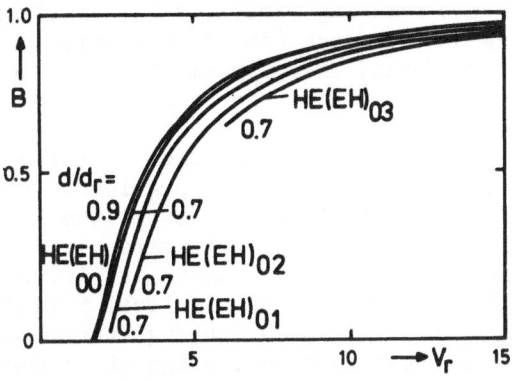

Fig.7.2 Phase parameter B =
$(\beta^2/k^2-n_2^2)/(n_1^2-n_2^2)$ of rib
guide modes versus rib para-
meter $V_r = kd_r\sqrt{n_1^2-n_2^2}$ for the
aspect ratio $b/d_r=4$ and
$(n_1-n_2)<<(n_1-n_0)$.

---

8. BULGE GUIDES

The rib waveguide on a substrate in Fig. 7.1 steps down abruptly in
its film thickness at the junction between rib region and film.
Such abrupt transitions are difficult to manufacture; in actual waveguides, these transitions will be more or less gradual, and the
structure in Fig. 7.1 would then only be a mathematical model for
the actual waveguide, with gradual transverse changes in film
thickness in the rib region.

A more realistic model for this structure is shown in Fig. 8.1. It
takes account of these gradual tran-
sitions and , instead of a rectangular rib, has a longitudinal centre
region of the film that bulges gradually to a larger film thickness.
The maximum bulge thickness in Fig.
8.1 is d_0, $d_b(y)$ designates the
varying film thickness in the bulge
region, which tapers out to the
uniform thickness d of the film at
both sides of the bulge. The total
width of the bulge is 2b. Fields

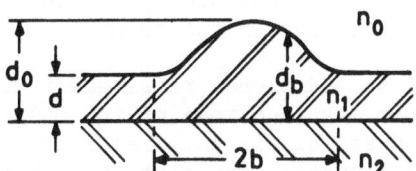

Fig. 8.1 Bulge guide on a
substrate

and energy of waves that are guided by such a structure, are confined
to or near the bulge. Hence, the structure may be called a bulge
guide. The gradual transverse transition in film thickness in rib
or bulge guides does not only occur due to manufacturing deficien-
cies. Such gradual transitions are also introduced intentionally
because they may offer certain advantages. As one desirable characteristic, the scattering at rough edges of a rectangular rib
guide reduces substantially when. instead of such an edge with its
inherent roughness, the rib or bulge thickness changes gradually
in transverse direction. Any longitudinal imperfections, remaining
in an otherwise smooth bulge will not nearly cause as much scattering

as the same imperfections in a rectangular rib. Such smooth bulges
may also be much easier to fabricate and, when tailored to specific
bulge forms, they may offer certain desirable dispersion characteristics for their guided modes.

To analyse wave propagation in the bulge guide, we start, as in
case of the rib and strip guides, with film modes in the bulged
region of the film. We assume the local bulge thickness to change
gradually enough in transverse direction so that this slight
change in thickness causes but little modification of local film
modes. We represent this local film mode by the ray in Fig 8.2
and let this ray run through the longitudinal
centre section of the bulge at an angle θ_1
with respect to its axis. This ray then pene-
trates into bulge regions with decreasing bulge
thickness; here it propagates with a decreasing
phase constant β_b or, equivalently, decreasing
effective index of refraction $N_b = \beta_b/k$, according to the decreasing bulge thickness. The
change in effective index has the same effect
on the ray of the film mode as the index change
in a graded-index medium has on a locally plane
wave. According to the ray equation (4.1), it
causes the ray to curve into the direction of
increasing effective index N_b. As a result, the
ray bends towards the axis and turns back to
it, as long as a turning point $y = y_t$ occurs
within the bulge. In this case, it runs
along an undulating path and can form a
guided mode of the bulge.

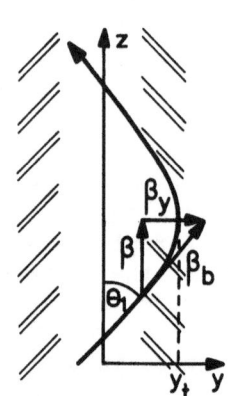

Fig. 8.2 Undulating
path of film mode
ray in a bulge guide

As an additional requirement for the local film mode to actually
form a guided mode of the bulge, it must satisfy a phase condition
for selfconsistent fields in the bulge. For the analogous case of
the graded-index slab, the phase condition was formulated in eqn.
(4.4) with the phase integral in transverse direction. The same
procedure may be adopted here.

$$\beta_y = \sqrt{\beta_b^2 - \beta^2} \qquad\qquad (8.1)$$

is the transverse phase constant, where β designates the phase constant of the bulge-guide mode, which is yet to be determined. The
phase constant β_b of the constituent film or slab mode depends on
the transverse coordinate y and follows from

$$\beta_b = n_1^2 k^2 - u_b^2/d_b^2(y) \qquad\qquad (8.2)$$

with u_b as the transverse phase parameter of the constituent film
mode according to eqn. (2.9).

To formulate the characteristic equation for bulge guide modes
from a phase condition, such as eqn. (4.4), we replace k_x in eqn.
(4.4) by β_y, integrate over y as the relevant transverse coordinate,
instead of over x, and replace m by the integer 1, as the transverse
order in y-direction of the particular HE_{ml}- or EH_{ml}-mode of the
bulge. We thus obtain

$$\int_{-y_t}^{y_t} \sqrt{n_1^2 k^2 - \beta^2 - u_b^2/d_b^2} \, dy = (1 + 1/2)\pi \qquad (8.3)$$

The limits of integration $\pm y_t$ in the phase integral of this characteristic equation represent the turning points of the undulating
ray path. At these points, we have $\beta_y = 0$, so that they follow from

$$u_b^2/d_b^2 (y_t) = n_1^2 k^2 - \beta^2 \qquad (8.4)$$

The phase integral in the characteristic equation (8.3) accounts
only approximately for the phase shift in total reflection at the
turning points $\pm y_t$. It therefore gives accurate results for the
dispersion characteristic of a particular guided mode only sufficiently far above its cut-off point. It is hence well suited to
analyze bulge guides, which have many guided modes, most of which
then are high above cut-off. For a bulge guide with only a few,
possibly only the fundamental mode, a more accurate method of
analysis must be used /5/.

References

/1/ Kogelnik, H. and Ramaswany, V., Scaling rules for thin-film
    optical waveguides, Applied Optic 13 |1974|, 1857-1862

/2/ Hinken, J.H., Film mode exitation by Hertzian dipoles,
    Ar. Electr. & Übertrag.techn. 31 |1977|, 315-320

/3/ Hinken, J.H., Film mode attenuation due to random surface
    irregularities. Proc. 6th European Microwave Conference,
    Rome 1976, pp. 513-517

/4/ Goell, J.E., A circular-harmonic computer analysis of rectan-
    gular dielectric waveguides, Bell Syst. tech. J.
    48 |1969|, 2133-2160

/5/ Petermann, K., Theory of single-mode single-material fibres,
    Ar. Electr. & Übertrag.techn. 30 |1976|, 147-153

THE COUPLED-MODE FORMALISM IN GUIDED-WAVE OPTICS

Amnon Yariv

California Institute of Technology

Pasadena, California 91125

ABSTRACT: The problem of propagation and interaction of optical radiation in dielectric waveguides is cast in the coupled-mode formalism. This approach is useful for treating problems involving energy exchange between modes. A derivation of the general theory is followed by application to the specific cases of electrooptic modulation, photoelastic and magnetooptic modulation, and optical filtering. Also treated are nonlinear optical applications such as second-harmonic generation in thin films and phase matching.

I. INTRODUCTION

A growing body of theoretical and experimental work has been recently building up in the area of guided-wave optics, which may be defined as the study and utilization of optical phenomena in thin dielectric waveguides [1,2]. Some of this activity is due to the hopes for integrated optical circuits in which a number of optical functions will be performed on small solid substrates with the interconnections provided by thin-film dielectric waveguides [3,4]. Another reason for this interest is the possibility of new nonlinear optical devices and efficient optical modulators which are promised by this approach [5-7].

A variety of theoretical *ad hoc* formalisms have been utilized to date in treating the various phenomena of guided-wave optics. In this paper we present a unified theory cast in the coupled-mode form to describe a large number of seemingly diverse phenomena. These include: 1) nonlinear optical interactions; 2) phase matching

by periodic perturbations; 3) electrooptic switching and modulation; 4) photoelastic switching and modulation; and 5) optical
filtering and reflection by a periodic perturbation.

II.   THE COUPLED-MODE FORMALISM

We will employ, in what follows, the coupled-mode formalism
[8] to treat the various phenomena listed in Section I. Before em-
barking on a detailed analysis it will prove beneficial to consider
some of the common features of this theory. Consider two electromagnetic modes with, in general, different frequencies whose complex
amplitudes are A and B. These are taken as the eigenmodes of the
unperturbed medium so that they represent propagating disturbances

$$a(z,x,t) = Ae^{i(\omega_n t \pm \beta_n z)} f_a(x)$$

$$b(z,x,t) = Be^{i(\omega_h t \pm \beta_h z)} f_b(x)$$

(1)

with A and B constant.

In all the cases to be considered in this article and in most
of the cases of practical interest, the perturbation can be viewed
as a change of the dielectric constant $\varepsilon(\vec{r})$ from its initial value.
When this happens the modes a and b are no longer true eigenmodes,
i.e., they no longer satisfy Maxwell's equation and/or the boundary
condition. If the perturbation is sufficiently mild we can often
satisfy Maxwell's equation by a solution

$$E = A(z) e^{i(\omega_a t \pm \beta_a z)} + B(z)e^{i(\omega_b t \pm \beta)}$$

where the amplitudes A and B are now functions of z. They will be
shown to obey equations of the type

$$\frac{dA}{dz} = \kappa_{ab} Be^{-i\Delta z}$$

$$\frac{dB}{dz} = \kappa_{ba} Ae^{+i\Delta z}$$

(2)

where the phase-mismatch constant Δ will be shown to depend on the
propagation constants β_a and β_b as well as on the spatial variation
of the coupling perturbation. The coupling coefficients κ_{ab} and
κ_{ba} are determined by the physical situation under consideration
and their derivation constitutes a major part of solving a given
problem. Before proceeding, however, with the specific experimental situations, let us consider some general features of the solutions of the coupled-mode equations.

A. Codirectional Coupling

We take up, first, the case where modes a and b carry (Poyn-ting) power in the *same* direction. It is extremely convenient to define A and B in such a way that $|A(z)|^2$ and $|B(z)|^2$ correspond to the power carried by modes a and b, respectively. The conser-vation of total power is thus expressed as

$$\frac{d}{dz} (|A|^2 + |B|^2) = 0 \tag{3}$$

which, using (2), is satisfied when [9]

$$\kappa_{ab} = -\kappa_{ba}^* \quad . \tag{4}$$

If the boundary conditions are such that a single mode, say b, is incident at z = 0 on the perturbed region occupying the half space z > 0, we have

$$b(0) \equiv B_o , \quad a(0) = 0 \tag{5}$$

Subject to these conditions the solutions of (2) become

$$A(z) = B_o \frac{2\kappa_{ab}}{(4\kappa^2 + \Delta^2)^{1/2}} e^{-i\Delta z/2} \sin[\tfrac{1}{2}(4\kappa^2 + \Delta^2)^{1/2} z]$$

$$B(z) = B_o e^{i\Delta z/2} \{\cos[\tfrac{1}{2}(4\kappa^2 + \Delta^2)^{1/2} z]$$

$$\quad - i \frac{\Delta}{(4\kappa^2 + \Delta^2)^{1/2}} \sin[\tfrac{1}{2}(4\kappa^2 + \Delta^2)^{1/2} z]\} \tag{6}$$

where $\kappa^2 \equiv |\kappa_{ab}|^2$. Under phase-matched condition $\Delta = 0$, a complete spatially periodic power transfer between modes a and b takes place with a period $\pi/2\kappa$

$$a(z,t) = B_o \frac{\kappa_{ab}}{\kappa} e^{i(\omega_a t - \beta_a)} \sin(\kappa z)$$

$$b(z,t) = B_o e^{i(\omega_b t - \beta_b z)} \cos(\kappa z) \tag{7}$$

A plot of the mode intensities $|a|^2$ and $|b|^2$ is shown in Fig. 1. This figure demonstrates the fact that for phase mismatch $\Delta \gg |\kappa_{ab}|$ the power exchange between the modes is negligible. Specific physical situations which are describable in terms of this picture will be discussed further below.

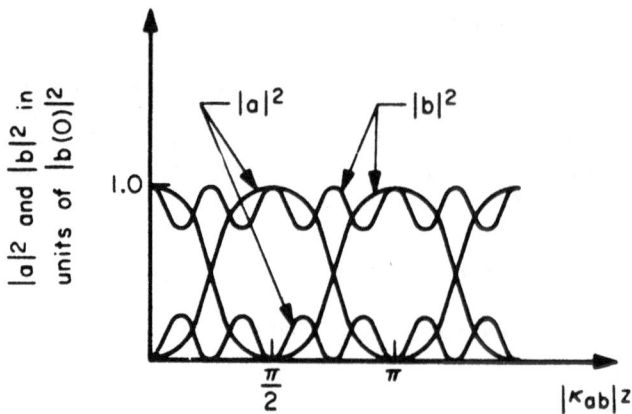

Fig. 1. The variation of the mode power in the case of codirec-
         tional coupling for phase-matched and unmatched operation

B.   Contradirectional Coupling

In this case the propagation in the unperturbed medium is
described by

$$a = Ae^{i(\omega_a t + \beta_a z)}$$

$$b = Be^{i(\omega_b t - \beta_a z)} \qquad\qquad (8)$$

where A and B are constant. Mode a corresponds to a left (−z)
traveling wave while b travels to the right. A time-space periodic perturbation can lead to power exchange between the modes.
Conservation of total power can be expressed as

$$\frac{d}{dz}\left(|A|^2 - |B|^2\right) = 0 \qquad\qquad (9)$$

which is satisfied by (2) if we take

$$\kappa_{ab} = \kappa^*_{ba} \qquad\qquad (10)$$

so that

$$\frac{dA}{dz} = \kappa_{ab} B e^{-i\Delta z} \qquad \frac{dB}{dz} = \kappa^*_{ab} A e^{i\Delta z} \qquad\qquad (11)$$

In this case we take the mode b with an amplitude B(0) to be incident at z = 0 on the perturbation region which occupies the space
between z = 0 and z = L. Since mode a is generated by the perturbation we have a(L) = 0. With these boundary conditions the solution of (11) is given by

$$A(z) = B(0) \frac{2i\kappa_{ab} e^{-i(\Delta z/2)}}{-\Delta \sinh \frac{SL}{2} + iS \cosh \frac{SL}{2}} \sinh[\frac{S}{2}(z-L)]$$

$$B(z) = B(0) \frac{2i\kappa_{ab} e^{-i(\Delta z/2)}}{-\Delta \sinh \frac{SL}{2} + iS \cosh \frac{SL}{2}}$$

$$\cdot \{\Delta \sinh[\frac{S}{2}(z-L)] + iS \cosh[\frac{S}{2}(z-L)]\} \qquad (12)$$

$$S \equiv \sqrt{4\kappa^2 - \Delta^2}, \qquad \kappa \equiv |\kappa_{ab}| \qquad (13)$$

Under phase-matching conditions $\Delta = 0$ we have

$$A(z) = B(0)(\frac{\kappa_{ab}}{\kappa}) \frac{\sinh[\kappa(z-L)]}{\cosh(\kappa L)}$$

$$\qquad (14)$$

$$B(z) = B(0) \frac{\cosh[\kappa(z-L)]}{\cosh(\kappa L)}$$

A plot of the mode powers $|B(z)|^2$ and $|A(z)|^2$ for this case is shown in Fig. 2. For sufficiently large arguments of the cosh and sinh functions in (14), the incident-mode power decays exponentially along the perturbation region. This decay, however, is due not to absorption but to reflection of power into the backward traveling mode a. This case will be considered in detail in the following sections, where acoustooptic, electrooptic, and spatial index perturbation will be treated. The exponential-decay behavior of Fig. 2 will be shown in Section VIII to correspond to the stopband region of periodic optical media.

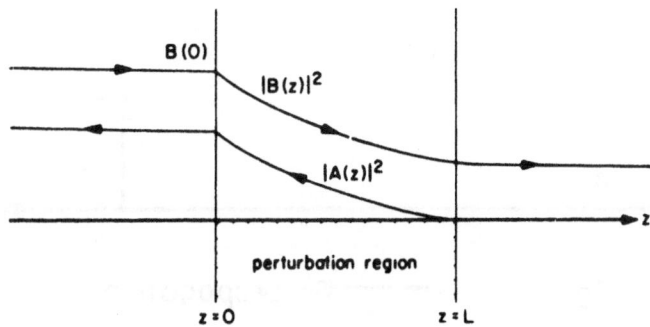

Fig. 2.   The transfer of power from an incident forward wave B(z) to a reflected wave A(z) in the case of contradirectional coupling

III.   ELECTROMAGNETIC DERIVATIONS OF THE
COUPLED-MODE EQUATIONS

A.   TE Modes

Consider the dielectric waveguide sketched in Fig. 3. It consists of a film of thickness t and index of refraction n_2 sandwiched
between media with indices n_1 and n_3. Taking $(\partial/\partial y) = 0$, this guide
can, in the general case, support a finite number of confined TE
modes with field components E_y, H_x, and H_z, and TM modes with components H_y, E_x, and E_z. The "radiation" modes of this structure
which are not confined to the inner layer are not considered in this
paper and will be ignored. The field component E_y of the TE modes,
as an example, obeys the wave equation

$$\nabla^2 E_y = \frac{n_i^2}{c^2} \frac{\partial^2 E_y}{\partial t^2} \quad , \qquad i = 1,2,3 \tag{15}$$

We take $E_y(x,z,t)$ in the form

$$E_y(x,z,t) = \mathcal{E}_y(x) \, e^{i(\omega t - \beta z)} \tag{16}$$

The transverse function $\mathcal{E}_y(x)$ is taken as

$$\mathcal{E}_y(x) = \begin{cases} C \exp(-qx) \quad , & 0 \le x < \infty \\ C[\cos(hx) - (q/h)\sin(hx)] \quad , & -t \le x \le 0 \\ C[\cos(ht) + (q/h)\sin(ht)]\exp[p(x+t)] \quad , & \\ & -\infty < x \le -t \end{cases} \tag{17}$$

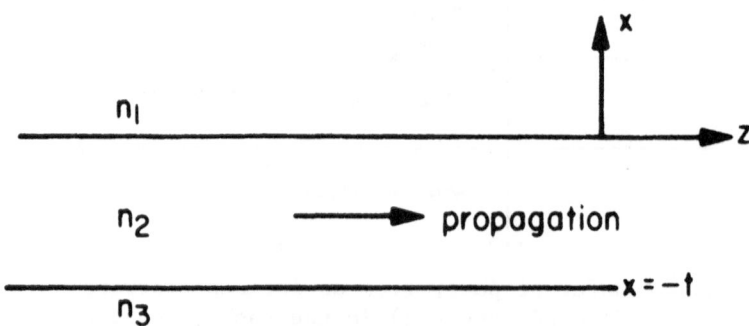

Fig. 3.   The basic configuration of a slab dielectric waveguide

which, applying (15) to regions 1,2,3, yields

$$h = (n_2^2 k^2 - \beta^2)^{1/2} \qquad\qquad p = (\beta^2 - n_3^2 k^2)^{1/2}$$

$$q = (\beta^2 - n_1^2 k^2)^{1/2} \qquad\qquad k \equiv \omega/c \qquad\qquad (18)$$

From the requirement that E_y and H_z be continuous at $x = 0$ and $x = -t$, we obtain[1]

$$\tan(ht) = \frac{q + p}{h(1 - \frac{pq}{h^2})} \qquad\qquad (19)$$

This equation in conjunction with (18) is used to obtain the eigen-values β of the confined TE modes.

The constant C appearing in (17) is arbitrary. We choose it in such a way that the field $\mathcal{E}_y(x)$ in (17) corresponds to a power flow of 1 W (per unit width in the y direction) in the mode. A mode whose $E_y = A\mathcal{E}_y(x)$ will thus correspond to a power flow of $|A|^2$ W/m. The normalization condition is thus

$$-\frac{1}{2} \int_{-\infty}^{\infty} E_y H_z^* \, dx = \frac{\beta_m}{2\omega\mu} \int_{-\infty}^{\infty} [\mathcal{E}_y^{(m)}(x)]^2 \, dx = 1 \qquad\qquad (20)$$

where the symbol m denotes the mth confined TE mode corresponding to mth eigenvalue of (19).

Using (17) in (20) we determine

$$C_m = 2h_m \left[\frac{\omega\mu}{|\beta_m|(t + \frac{1}{q_m} + \frac{1}{p_m})(h_m^2 + q_m^2)}\right]^{1/2} \qquad\qquad (21)$$

Since the modes $\mathcal{E}_y^{(m)}$ are orthogonal we have

$$\int_{-\infty}^{\infty} \mathcal{E}_y^{(\ell)} \mathcal{E}_y^{(m)} \, dx = \frac{2\omega\mu}{\beta_m} \delta_{\ell,m} \qquad\qquad (22)$$

B.   TM Modes

The field components are

---

[1] The assumed form of E_y in (17) is such that $\mathcal{E}_y$ and $\mathcal{H}_z = (i/\omega\mu)$, $\partial\mathcal{E}_y/\partial x$ are continuous at $x = 0$, and that $\mathcal{E}_y$ is continuous at $x=-t$. All that is left is to require continuity of $\partial\mathcal{E}_y/\partial x$ at $x=-t$. This leads to (19).

$$H_y(x,z,t) = \mathscr{H}_y(x) \; e^{i(\omega t - i\beta z)}$$

$$E_x(x,z,t) = \frac{i}{\omega \varepsilon} \frac{\partial H_y}{\partial z} = \frac{\beta}{\omega \varepsilon} \mathscr{H}_y(x) \; e^{i(\omega t - \beta z)}$$

$$E_z(x,z,t) = -\frac{i}{\omega \varepsilon} \frac{\partial H_y}{\partial x} \qquad\qquad (23)$$

The transverse function $\mathscr{H}_y(x)$ is taken as

$$\mathscr{H}_y(x) = \begin{cases} -C[\dfrac{h}{q}\cos(ht) + \sin(ht)]e^{p(x+t)}, & x < -t \\[3mm] C[-\dfrac{h}{q}\cos(hx) + \sin(hx)] & , & -t < x < 0 \\[3mm] -\dfrac{h}{q}\,Ce^{-qx} & , & x > 0 \end{cases} \qquad (24)$$

The continuity of H_y and E_z at the interfaces requires that the various propagation constants obey the eigenvalue equation

$$\tan(ht) = \frac{h(\bar{p} + \bar{q})}{h^2 - \bar{p}\bar{q}} \qquad\qquad (25)$$

where

$$\bar{p} \equiv \frac{n_2^2}{n_3^2} p \quad , \quad \bar{q} \equiv \frac{n_2^2}{n_1^2} q$$

The normalization constant C is chosen so that the field represented by (23) and (24) carries 1 W per unit width in the y direction.

$$\frac{1}{2} \int_{-\infty}^{\infty} H_y E_x^* \; dx = \frac{\beta}{2\omega} \int_{-\infty}^{\infty} \frac{\mathscr{H}_y^2(x)}{\varepsilon} \; dx = 1$$

or using $n_1^2 \equiv \varepsilon_1/\varepsilon_o$

$$\int_{-\infty}^{\infty} \frac{[\mathscr{H}_y^{(m)}(x)]^2}{n^2(x)} \; dx = \frac{2\omega\varepsilon_o}{\beta_m} \qquad\qquad (26)$$

This condition determines the value of C_m as [10]

$$C_m = 2 \sqrt{\frac{\omega\varepsilon_o}{\beta_m t_{eff}}}$$

$$t_{eff} \equiv \frac{\bar{q}^2 + h^2}{\bar{q}^2} [\frac{1}{n_2^2} + \frac{q^2 + h^2}{\bar{q}^2 + n^2} \frac{1}{h_1^2 q} + \frac{p^2 + h^2}{\bar{p}^2 + h^2} \frac{1}{n_3^2 p}]$$ (27)

C. The Coupling Equation

The wave equation obeyed by the unperturbed modes is

$$\nabla^2 \vec{E}(\vec{r},t) = \mu\varepsilon \frac{\partial^2 \vec{E}}{\partial t^2}$$ (28)

We will show below that in most of the experiments of interest to us we can represent the perturbation as a distributed polarization source $\vec{P}_{pert}(\vec{r},t)$, which accounts for the deviation of the *medium polarization from that which accompanies the unperturbed mode.* The wave equation for the perturbed case follows directly from Maxwell's equations if we take $\vec{D} = \varepsilon_o \vec{E} + \vec{P}$.

$$\nabla^2 E_y(\vec{r},t) = \mu\varepsilon \frac{\partial^2 E_y}{\partial t^2} + \mu \frac{\partial^2}{\partial t^2} [P_{pert}(\vec{r},t)]_y$$ (29)

with similar equations for the remaining Cartesian components of $\vec{E}$.

We may take the eigenmodes of (28) as an orthonormal set in which to expand E_y and write

$$E_y = \sum_\ell \frac{A_\ell(z)}{2} \mathcal{E}_y^{(\ell)}(x) e^{i(\omega t - \beta_\ell z)} + c.c.$$

$$+ \int_{-n_3 k}^{n_3 k} \frac{A(\beta)}{2} e^{i(\omega t - \beta z)} \mathcal{E}_y^{(\beta)}(x) d\beta \qquad n_3 < n_2$$ (30)

where ℓ extends over the discrete set of confined modes and includes both positive and negative traveling waves. The integration over β takes in the continuum of radiation modes, and c.c. denotes complex conjugation. Our chief interest lies in perturbations which couple only discrete modes so that, in what follows, we will neglect the second term on the right side of (30). Problems of coupling to the radiation modes arise in connection with waveguide losses [11] and grating couplers [12].

Substituting (30) into (29), assuming "slow" variation so that $d^2 A_m/dz^2 \ll \beta_m \, dA_m/dz$, and recalling that $\mathcal{E}_y^{(m)}(x) \, e^{i(\omega t - \beta_m z)}$ obeys the unperturbed wave equation (28), gives

$$\sum_{\ell} [-i\beta_\ell \frac{dA_\ell}{dz} \mathcal{E}_y^{(\ell)}(x) \, e^{i(\omega t - \beta_\ell z)}] + c.c. = \mu \frac{\partial^2}{\partial t^2} (P_{pert})_y \quad (31)$$

Multiplying (31) by $\mathcal{E}_y^{(m)}(x)$, and integrating and making use of the orthogonality relation (22) yields

$$\frac{dA_m^{(-)}}{dz} \, e^{i(\omega t + \beta_m z)} - \frac{dA_m^{(+)}}{dz} \, e^{i(\omega t - \beta_m z)} + c.c.$$

$$= \frac{-i}{2\omega} \frac{\partial^2}{\partial t^2} \int_{-\infty}^{\infty} [P_{pert}(\vec{r},t)]_y \, \mathcal{E}_y^{(m)}(x) \, dx \quad (32)$$

where $A_m^{(-)}$ is the complex normal mode amplitude of the negative traveling TE mode, while $A_m^{(+)}$ is that of the positive one. Equation (32) is the main starting point for the following discussion in which we will consider a number of special cases.

IV. NONLINEAR INTERACTIONS

In this section we consider the exchange of power between three modes of different frequencies brought about through the nonlinear optical properties of the guiding or bounding layers. The relevant experimental situations involve second-harmonic generation, frequency up-conversion, and optical parametric oscillation. To be specific we consider first the case of second-harmonic generation from an input mode at $\omega/2$ to an output mode at ω. The perturbation polarization is taken as

$$P_i^{(\omega)}(\vec{r},t) = \frac{1}{2}[P_i^{(\omega)} e^{i(\omega t - \beta z)} + c.c.] \quad (33)$$

The complex amplitude of the polarization is

$$P_i^{(\omega)} = d_{ijk}^{(\omega)}(\vec{r}) \, E_j^{\omega/2} \, E_k^{\omega/2} \quad (34)$$

where $d_{ijk}^{(\omega)}$ is an element of the nonlinear optical tensor and summation over repeated indices is understood. We have allowed, in (34), for a possible dependence of d_{ijk} on the position $\vec{r}$.

A. Case I: $TE_{input}-TE_{output}$. Without going, at this point, into considerations involving crystalline orientation, let us assume that an optical field parallel to the waveguide y direction will generate a second-harmonic polarization along the same direction

$$P_y^{(\omega)} = dE_y^{(\omega/2)} \, E_y^{(\omega/2)} \tag{35}$$

where P and E represent complex amplitudes, and d corresponds to a linear combination of d_{ijk} which depends on the crystal orientation. In this special case an input TE mode at $\omega/2$ will generate an output TE mode at ω. Using (30) in (35) gives

$$P_y(\vec{r},t) = \frac{1}{2} d(\vec{r}) \sum_n \sum_p A_n^{(\omega/2)} A_p^{(\omega/2)} \mathcal{E}_y^{(n,\omega/2)} \mathcal{E}_y^{(p,\omega/2)}$$

$$\times e^{i[\omega t - (\beta_n \omega/2 + \beta_p \omega/2)z]} + c.c. \tag{36}$$

We consider a case of a single mode input, say n. In that case the double summation of (36) collapses to a single term $n = p$. If we then use $P_y(\vec{r},t)$ as $[P_{pert}(\vec{r},t)]_y$ in (32) we get

$$\frac{dA_m^{(\omega)}}{dz} = - \frac{i\omega d(z)}{4} A_n^{(\omega/2)\,2} e^{-i(2\beta_n \omega/2 - \beta_m \omega)z} S^{(n,n,m)} \tag{37}$$

with

$$S^{(n,n,m)} \equiv \int_{-\infty}^{\infty} \mathcal{E}_y^{(n,\omega/2)} \mathcal{E}_y^{(n,\omega/2)} \mathcal{E}_y^{(m,\omega)} f(x) \, dx \tag{38}$$

where we took $d(\vec{r}) \equiv d(z) \, f(x)$.

In the interest of conciseness let us consider the case where the inner layer 2 is nonlinear and where both the input and output modes are well confined. We thus have $q_m, p_m \gg h_m$ and $h_m d \simeq \pi$. From (17) and (21) we get

$$\mathcal{E}_y^{(m,\omega)} \rightarrow 2\sqrt{\frac{\omega\mu}{\beta_m t}} \sin \frac{m\pi x}{t} , \qquad -t \le x \le 0$$

The overlap integral $S^{(n,n,m)}$ is maximum for $n = m = 1$, i.e., fundamental mode operation both at ω and $\omega/2$. For this case the overlap integral becomes

$$S^{(1,1,1)} = \int_{-\infty}^{\infty} \mathcal{E}_y^{(1,\omega/2)} \mathcal{E}_y^{(1,\omega/2)} \mathcal{E}_y^{(1,\omega)} f(x) \, dx$$

$$= \int_{-t}^{0} \mathcal{E}_y^{(1,\omega/2)} \mathcal{E}_y^{(1,\omega/2)} \mathcal{E}_y^{(1,\omega)} \, dx \tag{39}$$

$$= \frac{1.2\sqrt{2}}{\sqrt{t}} \frac{(\omega\mu)^{3/2}}{(\beta^\omega)^{1/2}\beta^{\omega/2}}$$

and (37) can be written as

$$\frac{dA^{(\omega)}}{dz} = -i \ \frac{2^{1/2} \times 1.2}{4} \ \frac{d}{\sqrt{t}} \ \frac{\mu^{3/2}\omega^{5/2}}{(\beta^{\omega})^{1/2}\beta^{\omega/2}} \ (A^{(\omega/2)})^2 \ e^{-i\Delta z} \qquad (40)$$

with

$$\Delta \equiv \beta^{\omega} - 2\beta^{\omega/2} \qquad (41)$$

and where the now-superfluous mode-number subscripts have been
dropped. Integrating (40) over the interaction distance ℓ gives

$$\left|A^{(\omega)}(\ell)\right|^2 = \frac{\omega^5\mu^3(1.2)^2 d^2}{8\beta^{\omega}(\beta^{\omega/2})^2 t} \ \left|A^{\omega/2}\right|^4 \ \ell^2 \frac{\sin^2(\Delta\ell/2)}{(\Delta\ell/2)^2} \qquad (42)$$

The normalization condition (20) was chosen so that $\left|A\right|^2$ is the
power per unit width in the mode. We can thus rewrite (42) as

$$\frac{P^{\omega}}{P^{\omega/2}} = 0.72(\frac{\mu}{\varepsilon_o})^{3/2} \frac{\omega^2 d^2 \ell^2}{n^3} (\frac{P^{\omega/2}}{wt}) \frac{\sin^2(\Delta\ell/2)}{(\Delta\ell/2)^2} \qquad (43)$$

where we used $\beta^{\omega} \simeq \omega\sqrt{\mu\varepsilon}$, $\varepsilon/\varepsilon_o = n^2$. Note that $(P^{\omega/2}/wt)$ is the intensity (watts/square meter) of the input mode. Except for a
numerical factor of 1.44, this expression is similar to that derived for the bulk-crystal case [13]. Efficient conversion results
when the phase-matching condition

$$\Delta \equiv \beta^{\omega} - 2\beta^{\omega/2} \ = \ 0 \qquad (44)$$

is satisfied. In this case the factor $\sin^2(\Delta\ell/2)(\Delta\ell/2)^2$ is unity.
Phase-matching techniques will be discussed later.

     B. Case II: TM_{input}-TE_{output}. The anisotropy of the non-
linear optical properties can be used in such a way that the output
at ω is polarized orthogonally to the field of the input mode at
$\omega/2$. To be specific, we consider the case of an input TM mode and
an output TE mode. If, as an example, the guiding layer (or one of
the bounding layers) belongs to the $\bar{4}$3m crystal class (GaAs, CdTe,
InAs), it is possible to have a guide geometry as shown in Fig. 4.
x,y,z is the waveguide coordinate system as defined in Fig. 4,
while 1, 2, and 3 are the conventional crystalline axes. For in-
put TM mode with $E||x$ we have

$$E_3 = E_1 = \frac{E_x}{\sqrt{2}}$$

The nonlinear optical properties of $\bar{4}$3m crystals are described by
[13]

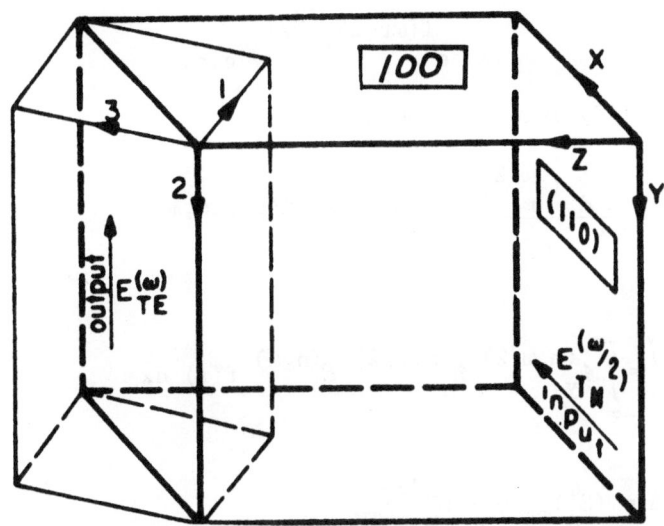

Fig. 4.   The orientation of a 43m crystal for converting a TM input
          at $\omega/2$ to a TE wave at ω.   x,y,z are the dielectric wave-
          guide coordinates, while 1, 2, and 3 are the crystalline
          axes.   Top surface is (100).

$$P_1 = 2d_{123}\, E_2 E_3$$

$$P_2 = 2d_{123}\, E_1 E_3$$

$$P_3 = 2d_{123}\, E_1 E_2$$

so that

$$P_y = P_2 = d_{123} E_x^2 \tag{45}$$

Taking

$$H_y = \frac{1}{2} \sum_{\ell} B_\ell \, \mathcal{H}_y^{(\ell)}(x)\, e^{i((\omega t/2) - \beta^{\ell/2} z)} + c.c.$$

and using $(\partial H_y/\partial z) = -i\omega\varepsilon\, E_x$ gives

$$E_x^{(\omega/2)}(\vec{r},t) = \frac{1}{2}\left(\frac{1}{\omega\varepsilon/2}\right) \sum_{\ell} \beta_\ell B \cdot \mathcal{H}_y^{(\ell)}\, e^{i((\omega t/2) - \beta^{\omega/2} z)} + c.c. \tag{46}$$

Using (45) and assuming a single, say m, mode input at $\omega/2$ results
in

$$P_y(\vec{r},t) = \frac{d}{2}\left[\left(\frac{\beta_m^{\omega/2}}{\omega\varepsilon/2}\right) B_m^{(\omega/2)}\, \mathcal{H}_y^{(m,\omega/2)}(x)\right]^2$$

$$\times \ e^{i(\omega t - 2\beta_m^{\omega/2} z)} \ + \ c.c. \tag{47}$$

Substituting (47) into (32) we obtain

$$\frac{dA_n^{(\omega)}}{dz} = - \ \frac{i\omega d}{4} \ \frac{(\beta_m^{\omega/2})^2}{\omega^2 \varepsilon^2} \ (B_m^{(\omega/2)})^2 \ e^{i\Delta z} \ S^{(m,m,n)} \ + \ c.c. \tag{48}$$

where

$$S^{(m,m,n)} = \int_{-\infty}^{\infty} \mathcal{H}_y^{(m,\omega/2)} \ \mathcal{H}_y^{(m,\omega/2)} \ \mathcal{E}_y^{(n,\omega)} \ f(x) \ dx \tag{49}$$

and

$$\Delta = (\beta_n^{\omega})_{TE} - 2(\beta_m^{\omega/2})_{TM}$$

For the special case $m = n = 1$ and for well-confined modes we have, using (17) and (22),

$$S^{(1,1,1)} = (\frac{\omega \varepsilon}{\beta_{TM}^{\omega/2}}) (\frac{2\omega \mu}{\beta_{TE}^{\omega}})^{1/2} \frac{1.2}{\sqrt{t}} \tag{50}$$

Proceeding as in the previous section leads finally to

$$\frac{(P^{\omega})_{TE}}{(P^{\omega/2})_{TM}} = 0.72 (\frac{\mu}{\varepsilon_o})^{3/2} \frac{\omega^2 d^2 \ell^2}{n^3} (\frac{P^{\omega/2}}{wt}) \frac{\sin^2 \Delta \ell/2}{(\Delta \ell/2)^2} \tag{51}$$

an expression identical to that obtained in (43) for TE-TE conversion. We must recall, however, that the nonlinear coefficient d in (51) is not necessarily the same as that appearing in (43), reflecting the differences in crystalline orientation needed to achieve coupling in each case.

C.   Phase Matching

It follows from (43) or (51) that a necessary condition for second-harmonic generation is $\Delta \ell/2 \ll \pi$ so that the factor $\sin^2 (\Delta \ell/2)/(\Delta \ell/2)^2$ is near unity. In this case the conversion efficiency is proportional to ℓ^2. This phase-matching condition can be satisfied by using the dependence of the propagation constants β of the various modes on the waveguide dimensions [7]. An alternate approach is to introduce a space-periodic perturbation into the waveguide with a period Λ satisfying

$$\Lambda = \frac{2\pi}{\Delta} q \quad , \qquad q = 1, 2, 3, \cdots \tag{52}$$

Schemes based on waveguide corrugation and on modulating the non-linear coefficient d have been proposed [14]. In this section we will consider the case of d modulation. We go back to (37), but allow explicitly for a spatial modulation of d by taking $d(z)$ as

$$d(z) = \frac{d}{2} + \sum_{\substack{q \text{ odd integer}}} \frac{2d}{q\pi} \sin \frac{2\pi q}{\Lambda} z \tag{53}$$

corresponding to a square-wave alternation between 0 and d with a period Λ. Instead of (37) we now have

$$\frac{dA_m^{(\omega)}}{dz} = -\frac{i\omega}{4} \left[\frac{d}{2} - \sum_{q \text{ odd}} \frac{id}{q\pi} (e^{i2\pi q z/\Lambda} - e^{-i2\pi q z/\Lambda}) \right]$$

$$\times [A_n^{(\omega/2)}]^2 e^{-i(2\beta_n \omega/2 - \beta_m \omega)z} S^{(n,n,m)} \tag{54}$$

We can choose the period Λ such that for some value of q

$$-\frac{2\pi q}{\Lambda} + \beta_m^\omega - 2\beta_n^{\omega/2} = 0 \tag{55}$$

This results in a synchronous term (i.e., one with a zero exponent) on the right side of (54) so that

$$\frac{dA_m^{(\omega)}}{dz} = \frac{\omega d}{4q\pi} [A_n^{(\omega/2)}]^2 S^{(n,n,m)} \tag{56}$$

where the nonsynchronous terms have been neglected. A comparison to (37) shows that the effective nonlinear coefficient is now re-duced to

$$d_{eff} = \frac{d}{q\pi}$$

and that instead of (43)

$$\frac{P^\omega}{P^{\omega/2}} = 0.72 \left(\frac{\mu}{\varepsilon_o}\right)^{3/2} \frac{\omega^2 d_{eff}^2 \ell^2}{n^3} \left(\frac{P^{\omega/2}}{wt}\right) \tag{57}$$

operation based on $q = 1$ is thus most efficient, leading to a re-duction by a factor of π^2 in the conversion efficiency. We note, however, that the factor $\sin^2(\Delta\ell/2)/(\Delta\ell/2)^2$ is now unity, which makes it possible to take advantage of the ℓ^2 dependence of the conversion efficiency.

V.   ELECTROOPTIC MODE COUPLING

The electrooptic effect in thin-film configurations can be used in a variety of switching applications. Its use as a polarization switch in a GaAs waveguide at 1.15μ has been demonstrated [6].   In contrast to the conventional bulk [15] treatment of the electrooptic effect which relies heavily on the concept of induced retardation, we view the process as that of coupling between TE and TM modes brought about by the applied low-frequency electric field.

The linear-electrooptic effect is conventionally defined [16] in terms of a third-rank tensor r_{ijk} which relates the changes in the constants of the index ellipsoid to the applied field according to

$$\Delta(\frac{1}{n^2})_{ij} = r_{ijk} E_k \qquad (58)$$

It follows from (58) that an alternative and equivalent definition would be to specify the changes of the dielectric tensor ε_{ij} as

$$\Delta\varepsilon_{ij} = \frac{\varepsilon_i \varepsilon_j}{\varepsilon_o} r_{ijk} E_k^{(0)} \qquad (59)$$

where the (0) superscript denotes a "low" frequency, i.e., a frequency well below the crystal's Reststrahl band. Using the relations

$$\vec{D} = \varepsilon_o \vec{E} + \vec{P}$$

$$D_i = \varepsilon_{ij} E_j$$

and choosing a principal coordinate system so that

$$\varepsilon_{ij}(E^{(0)} \neq 0) = \varepsilon_{ij}\delta_{ij} + \Delta\varepsilon_{ij}$$

leads to

$$P_1^{(\omega)} = [\frac{\varepsilon_i \varepsilon_j}{\varepsilon_o} r_{ijk} E_k^{(0)} + (\varepsilon_{ij} - \varepsilon_o)\delta_{ij}] E_j^{(\omega)} \qquad (60)$$

where we used the convention $\varepsilon_i \equiv \varepsilon_{ii}$. The perturbation polarization to be used in (32) is that part of $P_i^{(\omega)}$ which is proportional to the "low"-frequency electric field, i.e.,

$$[P_{pert}(\vec{r},t)]_i = \frac{\varepsilon_i \varepsilon_j r_{ijk}}{\varepsilon_o} E_k^{(0)} [\frac{E_j^{(\omega)}}{2} e^{i(\omega t - \beta z)} + c.c.] \qquad (61)$$

To be specific, we assume that the input is a TM mode with $\vec{E}^{(\omega)} \parallel \vec{a}_x$ which is coupled by the electrooptic properties of the bounding media or the guiding layer to the TE mode with $\vec{E}^{(\omega)} \parallel \vec{a}_y$.[2] The starting point is again (32) where the mode m corresponds to the output TE, and $[P_{pert}]_y$ is the y component of the polarization (61) induced by the x (and z*) electric-field components of the input TM mode. Using (61) we get

$$P_y^{(\omega)} = \frac{\varepsilon_i \varepsilon_j r_{ijk} \ell_{iy} \ell_{jx} E_k^{(0)}}{\varepsilon_o} E_x^{(\omega)} \tag{62}$$

where the ℓ's are direction cosines. Defining

$$\varepsilon_i \varepsilon_j r_{ijk} \ell_{iy} \ell_{jx} E_k^{(0)} \equiv \varepsilon^2 r E^{(0)} \tag{63}$$

(62) becomes

$$P_y^{(\omega)} = \frac{\varepsilon^2 r E^{(0)}}{\varepsilon_o} E_x^{(\omega)} \tag{64}$$

where $P_y^{(\omega)}$ is the complex amplitude of the polarization. In most cases of practical interest the choice of crystal orientation and the field $E_k^{(0)}$ is such as to simplify (63) to a simple form resembling (64); an example is provided at the end of Section VI. In any case, the definition of (63) applies to the most general case.

Using (22) the E_x component of a single forward-traveling TM mode is given by

$$E_x^{(\ell)}(\vec{r},t) = \frac{\beta_\ell}{2\omega\varepsilon(x)} B_\ell \mathcal{H}_y^{(\ell)}(x) e^{i(\omega t - \beta_\ell z)} \tag{65}$$

where the normalization (26) is such that $|B_\ell|^2$ is the power per unit width in the mode. From (64) and (65) we obtain

$$P_y(\vec{r},t) = \frac{\varepsilon^2 r(x,z) E^{(0)}}{2\omega\varepsilon_o \varepsilon(x)} \beta_\ell B_\ell \mathcal{H}_y^{(\ell)}(x) e^{i(\omega t - \beta_\ell z)} + c.c. \tag{66}$$

Substitution of (66) into the wave equation (32) leads to

$$\frac{dA_m^+}{dz} \exp(-i\beta_m^{TE}z) - \frac{dA_m^-}{dz} \exp(i\beta_m^{TE}z)$$

---

[2] The E_z component of a TM mode can also cause coupling, but this will typically be a smaller effect, since $E_z \ll E_x$.

$$= -\frac{i}{4} \int_{-\infty}^{\infty} \frac{\varepsilon^2 r(x,z) E^{(0)}(x,z)}{\varepsilon(x)\varepsilon_o} \beta_\ell B_\ell \mathcal{H}_y^{(\ell)}(x) \mathcal{E}_y^{(m)}(x) dx$$

$$\times \exp(-i\beta_\ell^{TM} z) + \text{c.c.} \qquad (67)$$

Equation (67) is general enough to apply to a large variety of cases. The dependence of $E^{(0)}$ and $r(x,z)$ on x allows for coupling by electrooptic material in the guiding or in the bounding layers. The z dependence allows for situations where $E^{(0)}$ or r depend on position. To be specific, we consider first the case where the guiding layer $-t < x < 0$ is uniformly electrooptic and where $E^{(0)}$ is uniform over the same region, so that the integration in (67) is from $-t$ to 0. In that case, the overlap integral of (67) is maximum when the TE(m) and TM(ℓ) modes are well confined and of the *same* order so that $\ell = m$. Under well-confined conditions p,q $\gg$ h and the expressions (17) for $\mathcal{E}_y^{(m)}(x)$ and (24) for $\mathcal{H}_y^{(m)}(x)$ in the guiding layer become

$$\mathcal{E}_y^{(m)}(x) \rightarrow \left(\frac{4\omega\mu}{t\beta_m^{TE}}\right)^{1/2} \sin\frac{m\pi x}{t}$$

$$\mathcal{H}_y^{(m)}(x) \rightarrow \left(\frac{4\omega\varepsilon_o n_2^2}{t\beta_m^{TM}}\right)^{1/2} \sin\frac{m\pi x}{t} \qquad (68)$$

where for well-confined mode $\beta_\ell^{TM} \simeq \beta_m^{TE} \equiv \beta = kn_2$. In this case the overlap integral becomes

$$\int_{-1}^{0} \mathcal{H}_y^{(m)}(x) \mathcal{E}_y^{(m)}(x) dx = \frac{4\omega\sqrt{\mu\varepsilon_2}}{t\beta} \int_{-t}^{0} \sin^2\frac{m\pi x}{t} dx = 2 \qquad (69)$$

Having chosen the case of a uniform $E^{(0)}$ and r, the only z dependence on the right side of (67) is that of the $\exp(-i\beta_\ell^{TM} z)$ factor. Since $\beta_m^{TM} \simeq \beta_m^{TE}$ ($\ell = m$) we may neglect the term involving A_m^-. The coupling thus involves only the forward TE and TM modes. Using (69), (67) becomes

$$\frac{dA_m}{dz} = -i\kappa B_m \exp[-i(\beta_m^{TM} - \beta_m^{TE})z] \qquad (70)$$

while from (4)

$$\frac{dB_m}{dz} = -i\kappa A_m \exp[i(\beta_m^{TM} - \beta_m^{TE})z]$$

$$\kappa = \frac{n_2^3 kr E^{(0)}}{2} \qquad (71)$$

The form of (70) will apply to the general case involving arbitrary spatial dependence of r and $E^{(0)}$. In that case we need to perform the integration in (67) to evaluate the coupling coefficient κ.

The form of (70) is identical to that of (2). The solution of (70) is thus given by (6) with

$$\Delta = \beta_m^{TM} - \beta_m^{TE} \qquad\qquad (72)$$

The transfer of power between the modes for the phase-matched ($\Delta = 0$) and $\Delta \neq 0$ case are as shown in Fig. 1. A complete transfer of power between the modes thus requires that $\Delta = 0$, i.e., phase matching. Means for phase matching will be discussed in Section VI. For the meantime let us assume that $\kappa \gg \Delta$ so that, according to (6), the effects of phase mismatch can be neglected. A complete power transfer in this case occurs in a distance ℓ such that

$$\kappa\ell = \pi/2$$

or using (71)

$$\ell E = \lambda_o / 2n_2^3 r \qquad\qquad (73)$$

where $\lambda_o = 2\pi/k$. The product ℓE is identical to the "half-wave" voltage of bulk electrooptic modulators [15]. The "half-voltage" in the bulk case, we recall, is the field-length product which causes a 90° rotation in the plane of polarization of a wave incident on an electrooptic crystal.

Unlike the bulk case, the coupling between the two guided modes can take place even when the electrooptic perturbation is limited to an arbitrarily small portion of the transverse dimensions [6] or when the two modes are of different order ($\ell \neq m$).

To appreciate the order of magnitude of the coupling, consider a case where the guiding layer is GaAs and $\lambda_o = 1$ μm. In this case [15]

$$n_2 \simeq 3.5 \quad , \qquad\qquad n_2^3 r = 59 \times 10^{-12} \frac{m}{V}$$

Taking an applied field $E^{(0)} = 10^6$ V/m we obtain from (71)

$$\kappa = 1.85 \text{ cm}^{-1}$$

$$\ell \equiv \pi/2\kappa = 0.85 \text{ cm}$$

for the coupling constant and the power-exchange distance, respectively.

VI.   PHASE MATCHING IN ELECTROOPTIC COUPLING

In general, $\beta^{TM} \neq \beta^{TE}$ even for the same order mode so that the fraction of the power exchanged in the electrooptic coupling case described previously does not exceed, according to (6), $\kappa^2/(\kappa^2 + \Delta^2)$. If $\Delta \gg \kappa$, the coupling is negligible. To appreciate the importance of this fact, let us use the numerical data of the example considered at the end of Section V. We have $\kappa = 1.85 \text{ cm}^{-1}$ and $\beta \simeq n_2 k \simeq 2.2 \times 10^5 \text{ cm}^{-1}$. The exchange factor $\kappa^2/(\kappa^2 + \Delta^2)$ is thus reduced to 0.5 when $\Delta/\beta \simeq [(B_{TE} - B_{TM})/B_{TE}] \sim 10^{-5}$. The critical importance of phase matching is thus manifest. Since the dispersion due to the waveguide will in general be such as to make $\Delta \gg \kappa$, some means for phase matching are necessary. We start by considering again the coupled-mode equations (70), reintroducing the possible z dependence of κ

$$\frac{dA_m}{dz} = - i\kappa(z) \, B_m e^{-i\Delta z}$$

$$\frac{dB_m}{dz} = - i\kappa(z) \, A_m e^{-i\Delta z}$$

$$\Delta \equiv \beta_m^{TE} - \beta_m^{TM} \qquad\qquad\qquad (74)$$

with

$$\kappa(z) = n_2^3 \, kr(z) \, E^{(0)}(z)$$

As in the case of second-harmonic generation, we can use a spatial modulation of r or the field $E^{(0)}$ for phase matching. Consider, for example, the case where the field $E^{(0)}(z)$ reverses its direction periodically as with the electrode arrangement of Fig. 5. Approximating the electric field in the guiding layer by

$$E^{(0)}(z) = \sum_{q \text{ odd}} \frac{4E_o}{q\pi} \sin \frac{2q\pi}{\Lambda} z \qquad\qquad (75)$$

corresponding to a field reversal between E_o and $-E_o$ every Λ meters, we can take $\kappa(z)$ in (74) as

$$\kappa(z) = -i\kappa_o \sum_q \frac{2}{q\pi} (e^{(i2\pi q/\Lambda)z} - e^{-(i2\pi q/\Lambda)z})$$

$$\kappa_o = n_2^3 \, kr \, E_o \qquad\qquad\qquad (76)$$

If we substitute (76) in (74) we obtain on the right-side terms with exponential dependence of the type

$$\exp\ i(\pm\Delta\ \pm\ \frac{2q\pi}{\Lambda})z$$

One can choose Λ such that, for some q, $(2\pi q/\Lambda) = \Delta$. This results in a synchronous driving term (i.e., one with a zero exponent). To be specific, let us choose

$$2\pi/\Lambda = \Delta \tag{77}$$

and keeping only the synchronous term, obtain from (74)

$$\frac{dA_m}{dz} = \frac{\kappa_o}{(\pi/2)}\ B_m$$

$$\frac{dB_m}{dz} = -\ \frac{\kappa_o}{(\pi/2)}\ A_m \tag{78}$$

This corresponds to phase-matched operation with an effective coupling coefficient reduced by $\pi/2$ relative to phase-matched operation with a uniform field $E^{(0)}(z) = E_o$. The solution of (78) is given by (7).

We close this section by considering again the use of $\overline{4}3m$ crystals for the phase-matching scheme just discussed. The nonvanishing elements of the r_{ijk} tensor are [15] $r_{321} = r_{312} = r_{123}$. From (61) it follows directly that a $\overline{4}3m$ crystal, oriented as in Fig. 5 so

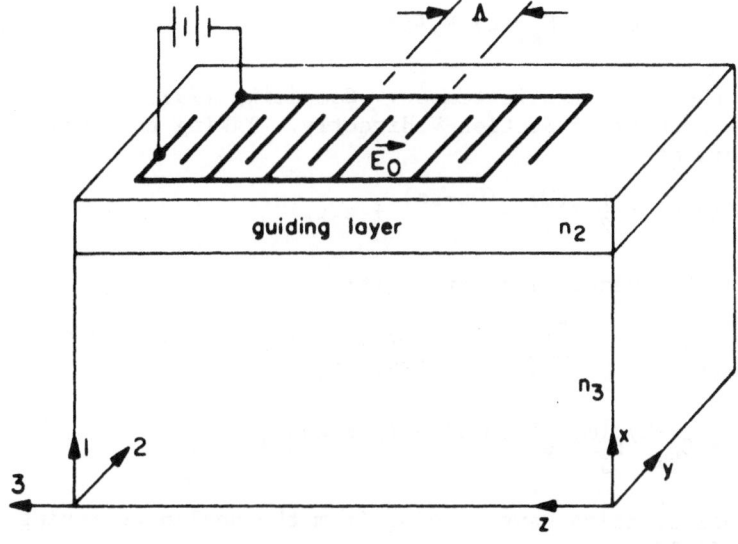

Fig. 5.   An interdigital-electrode structure for applying a spatially modulated electric field in electrooptic phase matching.   x, y, and z are the waveguide coordinates, while 1, 2, and 3 refer to the cubic [100] axes of a $\overline{4}3m$ crystal.

that its cubic 1,2,3 axes coincide, respectively, with the x,y,z directions of the waveguide, is optimal since in this case

$$P_x^{(\omega)} = \frac{\varepsilon^2}{\varepsilon_o} \, r_{123} \, E_z^{(0)} \, E_y^{(\omega)} \quad , \quad P_y^{(\omega)} = \frac{\varepsilon^2}{\varepsilon_o} \, r_{123} \, E_z^{(0)} \, E_x^{(\omega)}$$

thus coupling the TE mode $(E_y^{(\omega)})$ to the TM $(E_x^{(\omega)})$ and vice versa, in the presence of a longitudinal d.c. field $E_z^{(0)}$.

VII.   PHOTOELASTIC COUPLING

The possibility of coupling dielectric waveguide optical modes through the intermediary of sound waves has been demonstrated [17]. In this section we will treat this class of interactions using the coupled mode formalism.

The photoelastic effect is defined by relating the effect of strain $S_{k\ell}$ on the constants of the index ellipsoid through [18]

$$\Delta(\frac{1}{n^2})_{ij} = P_{ijk\ell} \, S_{k\ell} \tag{79}$$

so that the perturbation polarization is

$$P_i = \frac{\varepsilon_i \varepsilon_j}{\varepsilon_o} \, P_{ijk\ell} \, S_{k\ell}(t) \, E_j(t)$$

$P_{ijk\ell}$ is the photoelastic tensor. Comparing (79) to (58) we can apply the results of Section V directly. Taking the strain field in the form of

$$S_{k\ell}^{(\Omega)}(\vec{r},t) = \frac{1}{2} \, S_{k\ell}^{(\Omega)} \, e^{(\Omega t - Kz)} + c.c. \tag{80}$$

we obtain in a manner similar to (61)

$$[P_{pert}(\vec{r},t)]_i = \frac{\varepsilon_i \varepsilon_j P_{ijk\ell}}{4\varepsilon_o} \, [S_{k\ell}^{(\Omega)} \, E_j^{(\omega)} \, e^{i[(\omega+\Omega)t - (\beta+K)z]}$$

$$+ \, S_{k\ell}^{(-\Omega)} E_j^{(\omega)} \, e^{i \, (\omega-\Omega)t - (\beta-K)z]} + c.c. \tag{81}$$

for the polarization wave arising from the nonlinear mixing of an electric field

$$\frac{1}{2} \, E_j^{(\omega)} \, e^{i(\omega t - \beta z)} + c.c. \tag{82}$$

and a sound strain wave (80).

To be specific, we will assume again that the input optical field is a TM mode and will derive the equation governing the evolution of the TE mode due to the coupling. In a manner similar to (63) we abbreviate the information relating to crystal symmetry and orientation by defining

$$\varepsilon^2 pS \equiv \varepsilon_i \varepsilon_j P_{ijk\ell} \, S_{k\ell} \, \ell_{jk} \, \ell_{iy} \tag{83}$$

and instead of (81) use

$$[P_{pert}(\vec{r},t)]_y = \frac{\varepsilon^2}{4\varepsilon_o} \, p[S^{(\Omega)} E_x^{(\omega)} \, \exp\{i[(\omega+\Omega)t - (\beta_{TM}+K)z]\}$$

$$+ S^{(-\Omega)} E_x^{(\omega)} \, \exp\{i[(\omega-\Omega)t - (\beta_{TM}-K)z]\}] \quad + c.c.$$

In a manner identical to that leading to (67) we obtain

$$\frac{dA_m^{(+)}}{dz} \, \exp[i(\omega_{TE}t - \beta_{TE}z)]$$

$$- \frac{dA_m^{(+)}}{dz} \, \exp[i(\omega_{TE}t + \beta_{TE}z)]$$

$$= - \frac{i}{8} \int_{-\infty}^{\infty} \frac{\varepsilon^2 p(x,z) \, S^{\Omega}(x)}{\varepsilon(x)\varepsilon_o} \, \beta_\ell B_\ell \, \mathcal{H}_y^{(\ell)}(x) \, \mathcal{E}_y^{(m)}(x) \, dx$$

$$\cdot \, [\exp\{i[(\omega+\Omega)t - (\beta_{TM}+K)z]\}$$

$$+ \, \exp\{i[(\omega-\Omega)t - (\beta_{TM}-K)z]\}] \tag{84}$$

A few comments may be in order here. Each of the two terms on the right-hand side of (84) represents a traveling polarization wave. Both input waves, i.e., $S^{(\Omega)}$ and $E_x^{(\omega)}$, we recall, are taken as traveling in the +z direction. Ordinarily, β_{TE} is close to, but slightly larger than, β_{TM}. In this case the coupling is via the first term on the right side of (84) and the wavelength of the sound wave is adjusted so that

$$\beta_{TE} = \beta_{TM} + K \tag{85}$$

and the resulting TE mode is shifted up in frequency to[3]

---

[3] A quantum mechanical analysis of this phenomenon [19] shows that in the section of the waveguide in which the TE mode grows, phonons combine with TM photons on a one-to-one basis to generate TE photons so that $\omega_{TE} = \omega_{TM} + \Omega$.

$$\omega_{TE} = \omega + \Omega$$

Since the sign of β_{TE} and β_{TM} is the same, the coupling is codirectional. This is the case which we consider in detail below. Since $K/\beta = (c/v_s)(\Omega/\omega)$, where v_s is the sound velocity, it is possible for reasonable values of the sound frequency Ω to have $K \simeq 2\beta$. In this case the second term on the right side of (84) represents a polarization wave traveling in the $-z$ direction with a phase velocity $-\omega/(K-\beta) \simeq (-\omega/\beta)$. This wave is capable of coupling to the backward TE (or TM) mode. In this case we have

$$\beta_{TE} = B_{TM} - K < 0$$

$$\omega_{TE} = \omega + \Omega \tag{86}$$

Another possibility exists when the sound wave travels oppositely to the input TM mode. In this case we merely reverse the sign of K in (84). Codirectional coupling is now provided by the second term on the right side of (84) with

$$\beta_{TE} = \beta_{TM} + K$$

$$\omega_{TE} = \omega - \Omega \tag{87}$$

where the fact that now $\omega_{TE} < \omega$ can be understood by noting that for each photon removed by the interaction from the input TM mode one new (negative traveling) phonon and one new TE photon are generated. Contradirectional coupling can take place due to the first term when

$$\beta_{TE} = \beta_{TM} - K < 0$$

$$\omega_{TE} = \omega + \Omega \tag{88}$$

Returning to the codirectional-coupling case represented by (85), we obtain, following the same steps leading to (70),

$$dA_m^{(\omega+\Omega)}/dz = i\kappa B_m^{(\omega)} e^{-i\Delta z}$$

$$dB_m^{(\omega)}/dz = -i\kappa A_m^{(\omega+\Omega)} e^{i\Delta z} \tag{89}$$

$$\Delta \equiv K - (\beta_{TE} - \beta_{TM})$$

where we assumed $\omega \gg \Omega$. In the case of well-confined modes and of a photoelastic medium filling uniformly the guiding region 2, the coupling constant, following the procedure leading to (71), is found to be

$$\kappa = \frac{\pi p S^{(\Omega)} n_2^3}{2\lambda_o} \tag{90}$$

which is similar to κ of (71) except that the photoelastic constant
p replaces r, the electrooptic constant, and a factor of 2 appears
in the denominator. The latter is due to the fact that the sound
strain was taken as a time-harmonic field while, in the electrooptic case, the modulation field $E^{(0)}$ was taken as a d.c. field. The
solution of (89) is given by (6) and illustrated by Fig. 1. Complete power transfer can take place only when $\Delta = 0$, i.e., when

$$K = \beta_{TE} - \beta_{TM} \tag{91}$$

Since $K = \Omega/v_s$, this condition can be fulfilled by adjusting the
sound frequency Ω. Under phase-matched conditions we have, according to (6)

$$|A_m^{(\omega+\Omega)}|^2 = |B_m(0)|^2 \sin^2 \kappa z$$

$$|B_m^{(\omega)}|^2 = |B_m(0)|^2 \cos^2 \kappa z \tag{92}$$

with complete power exchange in a distance

$$\ell = \pi/2\kappa \tag{93}$$

It is of interest to estimate the acoustic power needed to satisfy
the switching condition (93). Solving (93) for the strain using
(90) gives

$$s^2 = \frac{\lambda_o^2}{\ell^2 p^2 n^6}$$

The corresponding acoustic intensity $I(W/m^2)$ can be obtained using
the relation $I = [(\rho v_s^3 s^2)/2]$ where ρ is the mass density. The result is

$$I_{switching} = \frac{\lambda_o^2 \rho v_s^3}{2\ell^2 p^2 n^6} = \frac{\lambda_o^2}{2\ell^2 M} \tag{94}$$

where $M \equiv n^6 p^2/\rho v_s^3$ is the acoustic figure of merit [18].

In a GaAs crystal, as an example, using the following data:
$M \simeq 10^{-13}$, $\ell = 5$ mm, and an optical wavelength $\lambda_o = 1$ μm, we get

$$I_{switching} = 20 \ W/cm^2$$

The corresponding strain amplitude is

$$s^{(\Omega)} \simeq 2.3 \times 10^{-5}$$

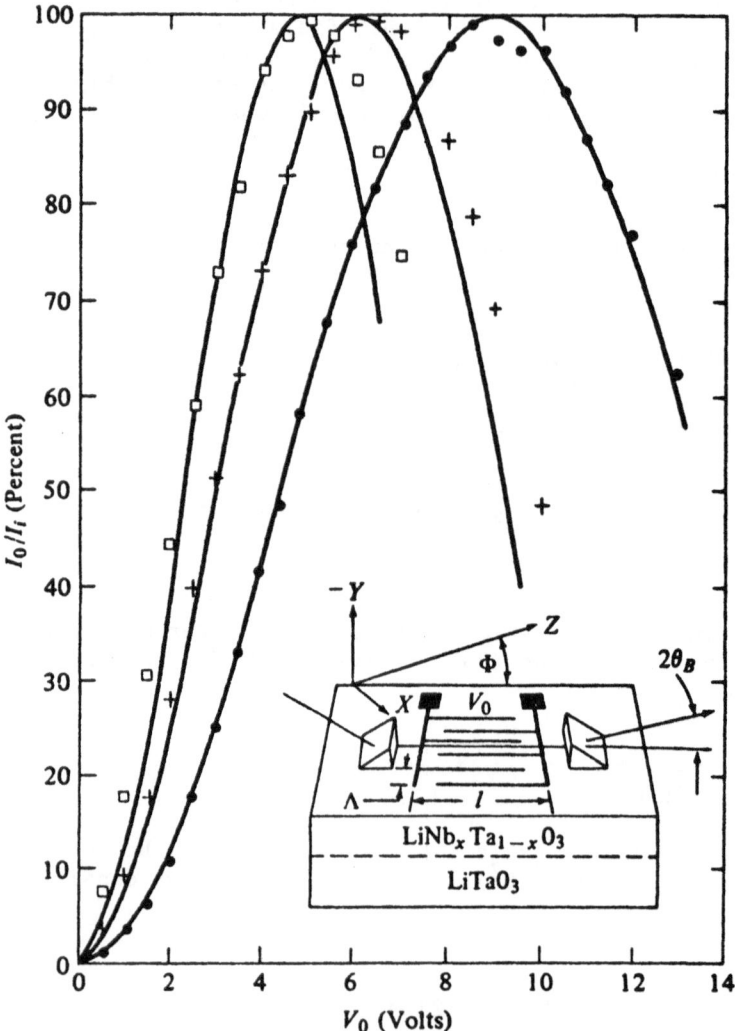

Fig. 6.   Schematic diagram of an electrooptic grating modulator in
          a $LiNb_xTa_{1-x}O_3$-$LiTaO_3$ waveguide. The input mode, which is
          coupled into the waveguide via a prism, is deflected
          through an angle $2\theta_B$ when a voltage is applied to the in-
          terdigital electrode structure. Λ = 7.6 µm and ℓ = 0.3 cm.
          The curves show percentages of light diffracted as a func-
          tion of voltage. Open squares 4976Å (He-Se laser) crosses
          5598Å (He-Se laser), and solid circles 6328Å (He-Ne laser).
          The solid curves are plots of $sin^2(BV_o)$ normalized to the
          data at 75 percent. (After J. Hammer and W. Phillips,
          Appl. Phys. Lett. 24, 545 (1974).

where we used

$$\rho = 5.34 \text{ g/cm}^3 \qquad \text{and} \qquad v_s = 5.15 \times 10^3 \text{ m/s}$$

An example of photoelastic coupling is shown in Fig. 6.

VIII.   COUPLING BY A SURFACE CORRUGATION

Consider an idealized dielectric waveguide such as that in
Fig. 3. Let us next perturb the spatial distribution of n^2
slightly from that shown in the figure. If the perturbation is
small, it is useful to consider its effect in terms of coupling
of the modes of the unperturbed system [11]. In this section we
will consider a perturbation due to a mechanical corrugation of the
interface as shown in Fig. 7. Using the relation

$$\vec{P} = [\varepsilon(\vec{r}) - \varepsilon_o] \vec{E}$$

we get

$$\vec{P}_{pert} = \Delta\varepsilon(\vec{r}) \vec{E}(r,t) = \Delta n^2(\vec{r}) \varepsilon_o \vec{E}(r,t) \qquad (95)$$

where $\varepsilon(\vec{r}) = n^2(\vec{r})\varepsilon_o$ so that coupling is only between TE or TM
modes but not from TE to TM. To be specific, consider the case of
a TE mode of order m propagating in the +z direction in a smooth
waveguide. At z = 0 it encounters a corrugated region, as shown in
Fig. 7, extending to z = L.

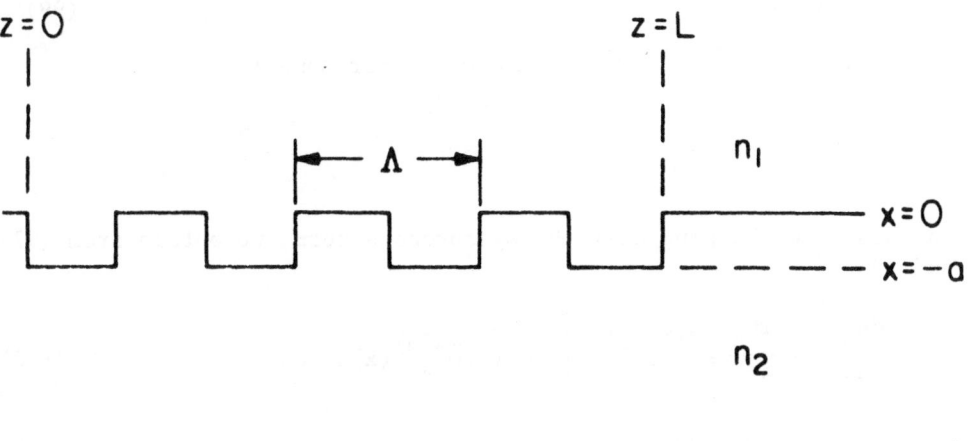

Fig. 7. A corrugated dielectric waveguide

Using (30) for E_y and limiting the summation to a single term $\ell = m$ gives upon substitution in (95)

$$[P_{pert}(\vec{r},t)]_y = \frac{\Delta n^2(\vec{r})\epsilon_o}{2} [A_m^{(+)}(z) \, \mathcal{E}_y^{(m)}(x)$$

$$\cdot \, e^{i(\omega t - \beta_m z)} + c.c.] \tag{96}$$

We anticipate that the period Λ will be chosen so that $2\pi/\Lambda \simeq 2\beta_m$ and the coupling will thus take place predominantly between the forward and backward modes of order m. Substituting (96) in (32) and limiting the left side of the latter to the backward $A_m^{(-)}$ term gives

$$\frac{dA_m^{(-)}}{dz} = \frac{i\omega\epsilon_o}{4} A_m^{(+)} e^{-2i\beta_m z} \int_{-\infty}^{\infty} \Delta n^2(x,z)[\mathcal{E}_y^{(m)}(x)]^2 \, dx \tag{97}$$

From Fig. 7 we have

$$\Delta n^2(x,z) = \Delta n^2(x)[\frac{1}{2} + \frac{2}{\pi}(\sin \eta z + \frac{1}{3}\sin 3\eta z + \cdots)]$$

$$\equiv \sum_{\ell} \Delta n_{\ell}^2(x,z) \tag{98}$$

where

$$\Delta n^2(x) = \begin{cases} n_2^2 - n_1^2 & , \quad -a \le x \le 0 \\ \\ 0 & , \quad \text{elsewhere} \end{cases}$$

$$\eta \equiv 2\pi/\Lambda \tag{99}$$

Coupling from $A_m^{(+)}$ to $A_m^{(-)}$ occurs when, for some ℓ,

$$\ell\eta \equiv \ell\frac{2\pi}{\Lambda} \simeq 2\beta_m$$

In this case, keeping only the synchronous term, we obtain from (97) and (98)

$$\frac{dA_m^{(-)}}{dz} = \frac{\omega\epsilon_o}{4\pi\ell} A_m^{(+)} e^{i\Delta z} \int_{-\infty}^{\infty} \Delta n^2(x)[\mathcal{E}_y^{(m)}(x)]^2 \, dx \tag{100}$$

$$\Delta \equiv \ell\eta - 2\beta_m$$

The next task is to evaluate the integral in (100). Using (99), the integral can be written as

$$\int_{-\infty}^{\infty} \Delta n^2(x) [\mathcal{E}_y^{(m)}(x)]^2 \, dx = (n_2^2 - n_1^2)$$

$$\int_{-a}^{0} [\mathcal{E}_y^{(m)}(x)]^2 \, dx = (n_2^2 - n_1^2) \, c_m^2$$

$$\int_{-a}^{0} [\cos(h_m x) - \frac{q_m}{h_m} \sin(h_m x)]^2 \, dx \qquad (102)$$

Although the integral in (102) can be calculated exactly using (19) and (21), an especially simple result ensues if we consider that operation is sufficiently above the propagation cutoff, so that $q_m \gg h_m$.[4]   Performing the integration and assuming $ha \ll 1$ results in

$$(n_2^2 - n_1^2) \int_{-a}^{0} [\mathcal{E}_y^{(m)}(x)]^2 \, dx$$

$$= \frac{c_m^2 \, q_m^2 \, a^3}{3} \, (1 + \frac{3}{q_m a} + \frac{3}{qm^2 a^2}) \qquad (103)$$

In the well-confined regime, $q_m, P_m \gg h_m$ so that (21) becomes

$$c_m^2 \rightarrow \frac{4 h_m^2 \omega \mu}{\beta_m t q_m^2} \qquad (104)$$

Using $\beta_m \rightarrow n_2 k$, $h_m \rightarrow \pi/t$, and (104) in (103) leads to

$$(n_2^2 - n_1^2) \int_{-a}^{0} [\mathcal{E}_y^{(m)}(x)]^2 \, dx = (n_2^2 - n_1^2) \frac{4\pi^2 \omega \mu}{3 n_2 k} (\frac{a}{t})^3 \, \cdot$$

$$\cdot \; [1 + \frac{3}{q_m a} + \frac{3}{q_m^2 a^2}]$$

which upon substitution into (100) gives

$$\frac{dA_m^{(-)}}{dz} = \kappa_\ell \, A_m^{(+)} \, e^{i\Delta z} \qquad (105)$$

and using (10)

---

[4] Well above threshold $q_m/h_m \rightarrow (n_2^2 - n_1^2)^{1/2} (2t/\lambda_o)$.

$$\frac{dA_m^{(+)}}{dz} = \kappa_\ell \, A_m^{(-)} \, e^{-i\Delta z}$$

$$\kappa_\ell = \frac{\pi k}{3\ell} \frac{(n_2^2 - n_1^2)}{n_2} \left(\frac{a}{t}\right)^3 \left[1 + \frac{3(\lambda_o/a)}{2\pi(n_2^2 - n_1^2)^{1/2}}\right.$$

$$\left. + \frac{3(\lambda_o/a)^2}{4\pi^2(n_2^2 - n_1^2)}\right] \tag{106}$$

where ℓ, we recall, is the order of the (corrugation function) harmonic responsible for the coupling, and $\lambda_o \equiv 2\pi/k$.

The behavior of the incident and reflected waves $A_m^{(+)}$ and $A_m^{(-)}$ is given by (14) and is illustrated by Fig. 2 (in which the reflected wave is A(z) and the incident wave B(z)). The exponential decay behavior occurs only for a narrow range of frequencies which satisfy, according to (13), the condition

$$\Delta \equiv \eta - 2\beta(\omega) \lesssim 2\kappa \tag{107}$$

where $\eta = 2\pi/\Lambda$ and we consider the case $\ell = 1$ only. This behavior is formally analogous to Bragg scattering of Bloch electron waves in a crystal from one edge of the Brillouin zone to the other by the crystal periodicity [20]. The latter phenomenon is responsible for the appearance of forbidden energy gaps. The behavior of the corresponding optical gap can be elucidated by considering the total propagation constant of, say, the incident wave B(z) of (12) [$A^+(z)$ in the notation of this section]. From (8) and (12) we can write it as

$$\beta'(\omega) = \beta(\omega) - \frac{\Delta}{2} \pm i\frac{S}{2} = \frac{\eta}{2} \pm i\frac{S}{2}$$

$$S = \sqrt{4\kappa^2 - \Delta^2} \tag{108}$$

The imaginary part of β' is then given by

$$\operatorname{Im}\beta'(\omega) = \sqrt{\kappa^2 - (\beta - \eta/2)^2}$$

$$\approx \sqrt{\kappa^2 - \frac{n_{eff}^2}{c^2}(\omega - \omega_o)^2} \tag{109}$$

where ω_o, the midgap frequency, is defined by $\beta(\omega_o) = \eta/2$; and to get the second equality we approximated the unperturbed behavior of β by $\beta(\omega) \approx (\omega/c)n_{eff}$. The height of the energy gap is thus the frequency region over which β' is complex. Using (109) it is given by

$$(\Delta\omega)_{gap} \equiv \omega_u - \omega_\ell = \frac{2\kappa c}{n_{eff}} \qquad (110)$$

where ω_u and ω_ℓ are the upper and lower gap frequencies, respectively.

The behavior of β' at $\omega > \omega_u$ and $\omega < \omega_\ell$ is likewise derivable from (108). It is given for $\omega > \omega_u$, for example, by

$$\beta' = \frac{\eta}{2} \pm \sqrt{\frac{n_{eff}^2}{c^2}(\omega - \omega_u)^2 + \frac{2n_{eff}\kappa}{c}(\omega - \omega_u)} \qquad (111)$$

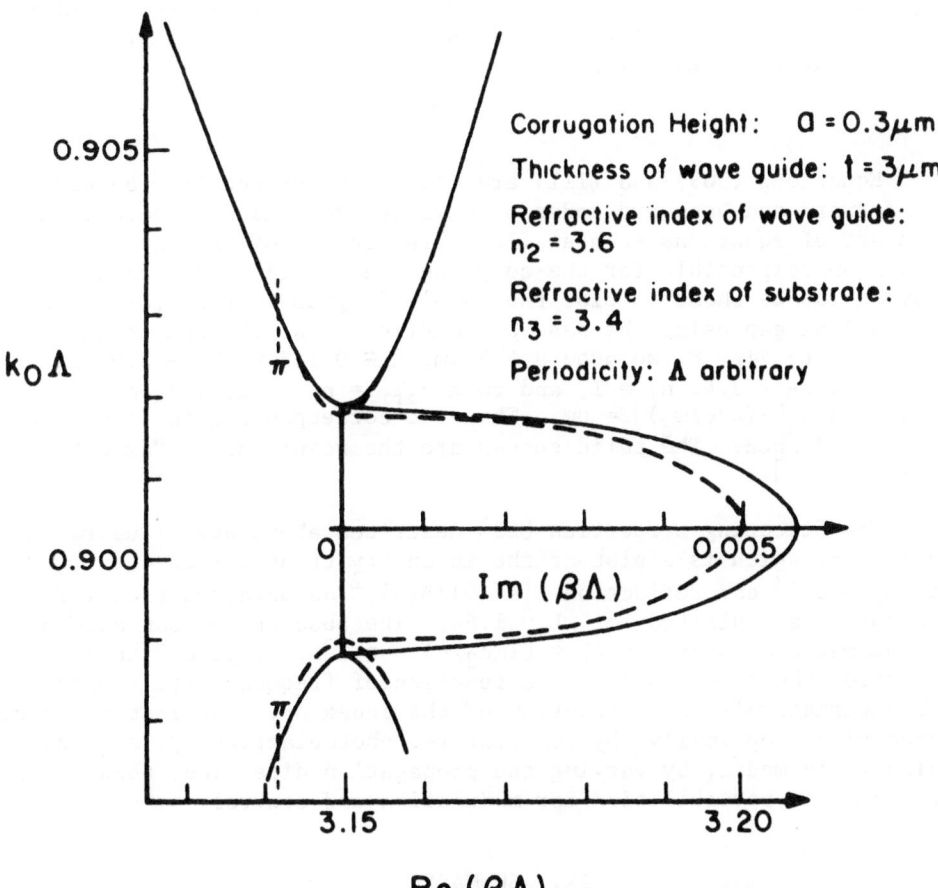

$k_0 \Lambda$

0.905

0.900

π

π

O

0.005

Im $(\beta\Lambda)$

3.15

3.20

Re $(\beta\Lambda)$

Corrugation Height: $a = 0.3\mu m$

Thickness of wave guide: $t = 3\mu m$

Refractive index of wave guide:
$n_2 = 3.6$

Refractive index of substrate:
$n_3 = 3.4$

Periodicity: Λ arbitrary

Fig. 8. A plot of the dispersion ($k\Lambda$ versus $\beta\Lambda$) diagram in the vicinity of the optical gap

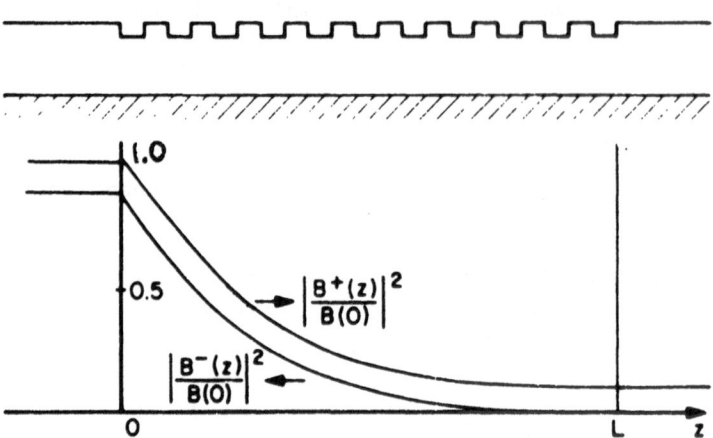

Fig. 9. A theoretical plot of the transmitted and reflected inten-
         sities (12) of a periodic waveguide near the Bragg
         (optical-gap) regime

Equations (109) and (111) are valid for any coupling between
the forward and backward modes of a waveguide which is describable
by a set of equations such as (105), regardless of the physical
mechanism responsible for the coupling. As an illustration of the
above ideas we chose to plot the $\omega - \beta'$ diagram in the vicinity of
the optical gap using the case of coupling by a surface corrugation.
Referring to Fig. 8, we used d = 3 µm, Λ = 0.143 µm, a = 0.3 µm,
n_2 = 3.6, n_3 = 3.4, n_1 = 1, and took n_{eff} = n_2. The midgap wavelength is $\lambda_o [= (2\pi c/\omega_o)]$ = µm. The plot corresponding to (109) and
(111) is dashed. The solid curves are the result of an "exact"
analysis [21].

The filtering properties [22] described above are illustrated
in Fig. 9, which is a plot of the intensity transmission
$|B^+(L)/B(0)|^2$ and reflection $|A^-(0)/B(0)|^2$ as obtained from (12).
The curves are plotted for κL = 1.84. The abscissa ΔL can also be
approximated as above by $\Delta L \simeq [(\omega n_2/c) - \eta/2]L$. Figure 9 thus
describes the transmission as a function of frequency (filtering)
or, alternatively, as a function of the index n_2. The latter can be
tuned electrooptically, by temperature, photoelastically, or, in
anisotropic media, by varying the propagation direction, thus offer-
ing some new possibilities for modulation and control.

                        IX.   SUMMARY

We have applied the formalism of coupled modes to describe a
wide range of experimental situations encountered in guided-wave

optics. Explicit expressions for the coupling coefficients, which play a central role in this theory, are given. The formalism treats the case of slab dielectric waveguides, thus assuming no variation in one (y) direction. The extension to guides where the confinement is in both the x and y directions principally involves replacing the integration over all x in the expressions for the coupling coefficients by an integration over both x and y. For cases where the modes are very well confined in the y direction, the numerical correction is small.

We have not discussed the applications of the coupled-mode formalism to the distributed feedback laser [23,24] and to directional coupling [25], since the original treatments are already cast in this form.

REFERENCES

1. A. Yariv and R.C.C. Leite, "Dielectric waveguide mode of light propagation in p-n junctions," Appl. Phys. Lett. 2, 55 (1963).

2. H. Osterberg and L. W. Smith, "Transmission of optical energy along surfaces," J. Opt. Soc. Amer. 54, 1073 (1964).

3. R. Shubert and J. H. Harris, "Optical surface waves on thin films and their application to integrated data processors," IEEE Trans. Microwave Theory Tech. (1968 Symposium Issue), MTT-16, 1048 (1968).

4. S. E. Miller, "Integrated optics, an introduction," Bell Sys. Tech. J. 48, 2059 (1969).

5. D. F. Nelson and R. K. Rinehart, "Light modulation by the electrooptic effect in reversed-biased GaP functions," Appl. Phys. Lett. 5, 148 (1964).

6. D. Hall, A. Yariv, and E. Garmire, "Optical guiding and electrooptic modulation in GaAs epitaxial layers," Opt. Commun. 1, 403 (1970).

7. D. B. Anderson and J. T. Boyd, "Phase matched second harmonic generation in GaAs thin film waveguides," Appl. Phys. Lett. 19, 266 (1971).

8. J. R. Pierce, "Coupling of modes of propagation," J. Appl. Phys. 25, 179 (1954).

9. A. Yariv, "On the coupling coefficients in the coupled-mode theory," Proc. IRE (Corresp.) 46, 1956 (1958).

10. A. Gover, private communication.

11. D. Marcuse, "Mode conversion caused by surface imperfections of a dielectric slab waveguide," 48, 3187 (1969).

12. M. L. Dakss, L. Kuhn, P. F. Heidrich, and B. A. Scott, "Grating coupler for efficient excitation of optical guided waves in

thin films," Appl. Phys. Lett. <u>16</u>, 523 (1970).

13. A. Yariv, <u>Introduction to Optical Electronics</u> (Holt, Rinehart and Winston, New York, 1971), p. 190.

14. S. Somekh and A. Yariv, "Phase matching by periodic modulation of the nonlinear optical properties," Opt. Commun. <u>6</u>, 301 (1972).

15. See [13] Ch. 9.

16. J. F. Nye, <u>Physical Properties of Crystals</u> (Oxford University Press, New York, 1957).

17. L. Kuhn, M. L. Dakss, P. F. Heidrich, and B. A. Scott, "Deflection of optical guided waves by a surface acoustic wave," Appl. Phys. Lett. <u>17</u>, 265 (1970).

18. R. W. Dixon, "The photoelastic properties of selected materials and their relevance to acoustic light modulators and scanners," J. Appl. Phys. <u>38</u>, 5149 (1967).

19. A. Yariv, "Quantum theory for parametric interactions of light and hypersound," IEEE J. Quant. Electron. <u>QE-1</u>, 28 (1965).

20. C. Kittel, <u>Introduction to Solid State Physics</u> (Wiley, New York, 1971, 4th ed.), Ch. 9.

21. K. Sakuda and A. Yariv, "Analysis of optical propagation in a corrugated dielectric waveguide," Opt. Commun. <u>8</u>, 1 (1973).

22. F. W. Dabby, A. Kestenbaum, and U. C. Paek, "Periodic dielectric waveguides," Opt. Commun. <u>6</u>, 125 (1972); also, F. W. Dabby, M. A. Saifi, and A. Kestenbaum, "High frequency cutoff periodic dielectric waveguides," Appl. Phys. Lett. <u>22</u>, 190 (1973).

23. H. Kogelnik and C. V. Shank, "Coupled wave theory of distributed feedback lasers," J. Appl. Phys. <u>43</u>, 2328 (1972).

24. P. Zory, "Laser oscillation in corrugated leaky optical waveguides," Appl. Phys. Lett. <u>22</u>, 125 (1973).

25. S. Somekh, E. Garmire, A. Yariv, H. L. Garvin, and R. G. Hunsperger, "Channel optical waveguide directional couplers," Appl. Phys. Lett. <u>22</u>, 46 (1972).

OPTICAL WAVEGUIDE MODULATION TECHNIQUES

Jacob M. Hammer

RCA Laboratories
David Sarnoff Research Center
Princeton, NJ 08540

In the following chapter we will briefly review some physical
effects useful in the construction of optical modulators. A description of how these effects are put to use in constructing "bulk"
modulators will lay the groundwork for a description of the principles of optical waveguide modulators. It will be seen that because the available physical effects are small it is necessary to
have a relatively long interaction length to accumulate a useful
effect. In "bulk" modulators the diffraction properties of light
limit the available interaction length for a given optical beam
cross-sectional area. This, in turn, causes the bulk devices to
require very high drive powers and voltages and/or currents. Optical waveguide modulators are able to avoid this limitation and
thus exhibit much lower power-voltage-current requirements than the
bulk devices.

Much of the general background material given in this chapter
has been covered in detail in Chapter IV of the book Integrated
Optics, T. Tamir ed. Thus, we will make general reference to that
work.

I. DEFINITIONS

Modulators are devices that change measurable properties of
light in response to electrical signals. The measurable properties
of light are intensity (amplitude), phase, polarization and wavelength. For intensity modulation, the amplitude, E, of a modulated
wave can be expressed as $E = G(t) \exp(jwx-\beta Z)$, then $I = G*G$. For
phase modulation, $E = \exp(jwt-\beta Z+\phi(t))$, and for polarization modulation $E = E_x\exp(jwt-\beta Z+\phi_x(t) + E_y \exp(jwt-\beta_x+\phi_y(t))$.

Switches are devices that either translate the spacial position
or rotate the angle of propagation of a light beam in response to an
electrical signal. Most modulators and switches use effects that
result in phase rather than amplitude changes. The most widely
available detectors are, however, sensitive to amplitude changes.
It is thus useful to note that phase changes are formally equivalent
to amplitude or intensity changes as follows: Denote a phase change
by $\Delta\phi$. Then for interference modulators such as polarization ro-
tators, grating modulators, interferometers, the amplitude change
goes as $\sin^2(\Delta\phi/2)$. [1] For coupled mode modulators the amplitude
varies as $1-[A+(\beta\Delta\phi)^2]^{-1}$. [2] Light from simple phase modulators
when detected by a heterodyne system gives signals which vary as the
Bessel functions $J_n(\Delta\phi)$. [3]

We now define measures of merit. Referring to Fig. 1, the insertion loss L is given by:

$$
L = \begin{cases}
1 - I_m/I_{in} & I_o < I_m \\
\\
1 - I_o/I_{in} & I_m < I_o
\end{cases}
$$

The modulation depth, η, and the extinction ratio, η_m, are given by:

$$
\eta = \begin{cases}
|I - I_o|/I & I_1 < I \\
|I - I_o|/I_o & I < I_o
\end{cases}
\qquad
\eta_m = \begin{cases}
|I_m - I_o|/I_m & I_o < I_m \\
|I_m - I_o|/I_o & I_m < I_o
\end{cases}
$$

A very important measure of merit is the power per unit bandwidth. To facilitate comparison of a variety of modulators the
power per unit bandwidth at $\Delta\phi = 2$ rad, which has been called the
Specific Energy, $((P/\Delta f)_2)$ is often used. We note that at $\Delta\phi = 2$
rad, $\sin^2\Delta\phi/2 = 0.703$. Thus, the Specific Energy is measured at an

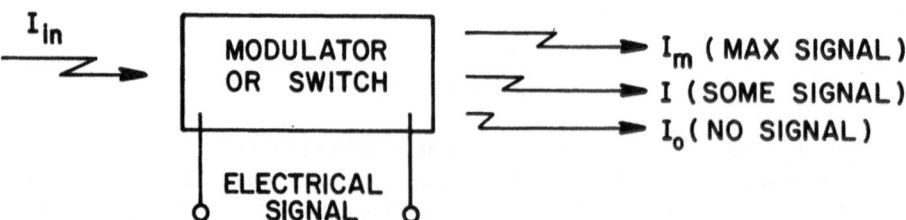

Fig. 1. When the input light intensity is I_{in}, the output intensity
         is I_m for maximum electrical signal, I at some arbitrary
         level of electrical signal and I_o when there is no elec-
         trical signal.

equivalent intensity modulation of 70%. Other important measures of merit are the drive current or voltage, and bandwidth,Δf, and the response time T: T = $2\pi/\Delta f$. The holding power is an important measure of merit for switches. [4]

II. SOME PHYSICAL EFFECTS USED IN OPTICAL MODULATORS AND SWITCHES

A. Electro-Optic Effects

For completeness, we mention electro-absorption or the "Franz-Keldish" [5] effect. We will be mainly concerned, however, with electrically induced changes in refractive index. The Kerr effect is a change in refractive index proportional to the square of the electric field. The principal useful effect is, however, the linear electro-optic or Pockels [6] effect in which the refractive index change varies as the first power of the electric field. The Pockels effect is found only in crystals which lack a center of symmetry. These crystals also show piezoelectricity. Thus, one must distinguish the clamped coefficient used for high frequency (above mechanical resonance) from the unclamped coefficient or low frequency coefficient (below mechanical resonance).

The refractive index change Δn is given by

$$\Delta n = -1/2 n'^3 r' E \tag{1}$$

r' is the effective electro-optic coefficient and is a linear combination of the elements of the electro-optic tensor [6]. The resulting phase change in light traveling a distance ℓ is

$$\Delta\phi = 2\pi\ell\Delta n/\lambda_o \quad . \tag{2}$$

Referring to Fig. 2, we may take E = V/a. When a = ℓ and using Eqs. (1) and (2), the half wave voltage is $V_{\lambda/2} = \lambda_o/n'^3 r'$. n' is the effective refractive index, r' the effective electro-optic coefficient for the chosen crystal orientation and light polarization direction. Some typical values of n', r' and $V_{\lambda/2}$ for a few useful materials are given in Table 1. As can be seen, the effect is small ($\Delta n < 10^{-4}$). It is thus necessary for light to travel a relatively long distance (thousands of wavelengths) in the material containing the electric field to accumulate a sufficiently large phase shift for practical purposes. This requirement leads to the basic difficulty of making good optical modulators and switches.

B. Acousto-Optic Effect

The acousto-optic effect occurs in all materials and phases and

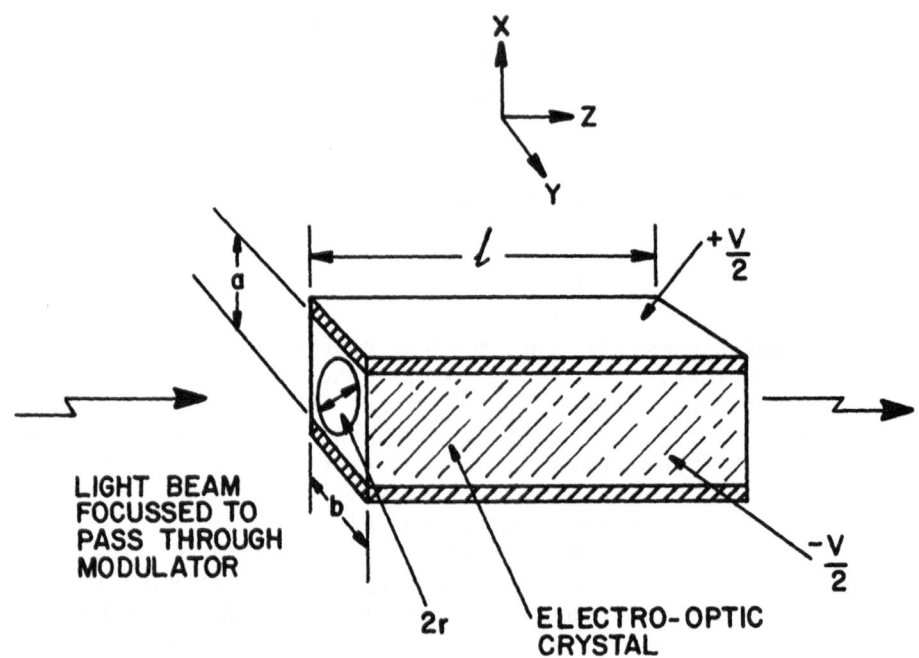

Fig. 2.   A voltage applied to the electrodes induces a refractive
          index change in the electro-optic crystal.   The index
          change depends linearly on the field E = V/a.   The magni-
          tude of the change is described by elements of a third
          rank tensor.   Thus, the crystal orientation and light
          polarization must be taken into account.

TABLE 1.   Some Typical Electro-Optic Coefficients and
           Refractive Index Changes for Applied Fields
           of 10,000 V/cm.   Values of n, r, and ε from
           Kaminow [29].

Relative Dielectric Constant ε	Material	λ (μm)	n'	r' (10^{-12}m/V)	Δn E=10^4V/cm	V$_{\lambda/2}$ (V)
28	LiNbO₃	0.6328	2.203(n_ε)	30 (r_{33})	1.6×10^{-4}	1,970
12.3	GaAs	0.9	3.6 (n)	1.2 (r_{14})	2.8×10^{-5}	16,100
8.2	ZnO	0.6328	2.015(n_ε)	2.6 (r_{33})	1.1×10^{-5}	29,700

is due to change in refractive index caused by strain. The relation between Δn and strain is described by a fourth rank tensor [7]. The acoustic wave is described by the strain-stress tensor. If simple materials and axes are chosen, however, we can think in terms of simple shear or compressional waves. We can write the index change as

$$\Delta n = \sqrt{10^7} \ M_2 P_a /2A \tag{3}$$

where the acousto-optic figure of merit, M_2, is given by Eq. (4)

$$M_2 = n^6 p^2 /\rho v_a^3 \tag{4}$$

v_a = acoustic velocity, p = element of strain optic tensor, ρ = density, P_a = acoustic power, A = area, and n = refractive index.

Some typical acousto-optic parameters for a few useful materials are given in Table 2. As can be seen here, even at the relatively high acoustic power density of 100 W/cm^2 only small changes in refractive index occur. Thus, as with the electro-optic effect it is necessary to have a long interaction length to accumulate a useful phase change.

III.   BULK MODULATORS

A.   Bulk Acousto-Optic Modulators

We can express the phase shift due to the flow of an acoustic wave as $\Delta\phi = (2\pi\ell\Delta n)/\lambda \ \sin(2\pi y/\Lambda)$. (See Fig. 3.) This space periodic variation acts as a phase grating and diffracts the light traveling in the material containing the wave. Such a grating can be characterized by a grating vector $\vec{k}_g$

$$|\vec{k}_g| = 2\pi/\Lambda$$

The direction of $\vec{k}_g$ is well characterized if

$$\ell \gg \Lambda^2/\lambda$$

Here, Bragg diffraction occurs, otherwise thin grating behavior is observed.

For *thin gratings* orders occur at angles

$$\sin\theta_m = m\lambda/\Lambda, \ m = o \pm 1, 2, \ldots$$

TABLE 2. Acousto-Optic Properties of Some Materials Used in Acousto-Optic Modulators. Values Adapted from PINNOW [8]. M_2 for Fused Quartz = $1.51 \times 10^{-18} \text{sec}^3/\text{cm}$. All values are for $\lambda_0 = 0.633$ μm.

Material	Wave Type	Refractive Index	Acoustic Attenuation at 500 MHz (dB/cm)	Acoustic Velocity $v_a (10^5 \text{cm/sec})$	M_2 Relative to Fused Quartz	Δn at P_a/A 100 W/cm^2
TaO$_2$	Shear	2.27	4.9	0.617	525	1.3×10^{-4}
PbMnO$_4$	Longitudinal	2.39	3.3	3.66	23.7	6.2×10^{-4}
LiNbO3	Longitudinal	2.2	.05	6.57	4.6	5.8×10^{-5}
Fused Quartz	Longitudinal	1.46	3.0	5.96	1.0	2.7×10^{-5}

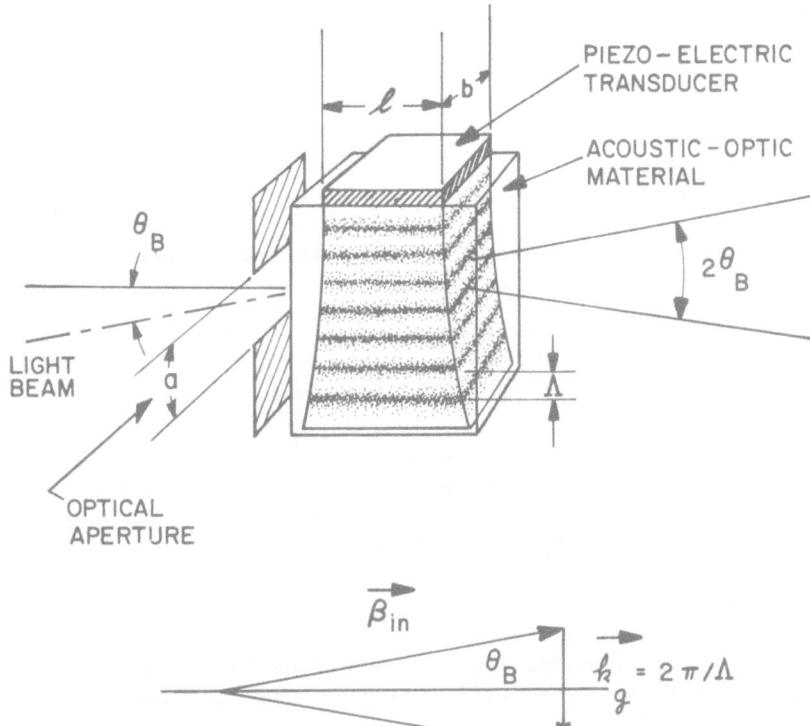

Fig. 3.   Acoustic waves are launched by the piezoelectric trans-
          ducer and travel downward in the acousto-optic crystal.
          Light entering at the Bragg angle θ_B is diffracted. The
          Bragg angle condition is found from the simple vector
          diagram in the lower part of the figure.

The intensities are given by

$$I/I_o = \begin{cases} [J_m(\Delta\phi)]^2/2 & |m| > 0 \\ [J_o(\Delta\phi)]^2 & m = 0 \end{cases}$$

The extinction ratio of the zeroth order is

$$\eta = (I-I_o)/I = 1 - [J_o(\Delta\phi)]^2$$

For *Bragg gratings* (see Fig. 3) only one order exists defined by

$$2\beta \sin\theta_B = k_g \tag{5a}$$

or

$$\sin\phi_B = \lambda/2\Lambda \tag{5b}$$

Here the intensities vary as $\sin^2\Delta\phi/2$, thus

$$\eta = 1-\sin^2(\Delta\phi/2).$$

The magnitudes of β_{in} and β are given by $|\beta| = 2\pi/\lambda$ and the magnitude of k_g is $2\pi/\Lambda$. Using Eqs. (2) and (3)

$$I/I_o = \sin^2 \frac{\pi}{\lambda_o} \sqrt{\frac{10^7 M_2 \ell P_a}{2b}}. \tag{6}$$

Referring to Fig. 3, the dimensions a and b cannot be freely chosen. Briefly, to change from one intensity to another in some time T it is necessary that the acoustic wave transit the optical aperture in a time T or less. Thus, we have the restriction

$$1/T = \Delta f = v_a/a \tag{7}$$

Further, if the Bragg condition is exactly satisfied at one acoustic frequency it will not be satisfied at another. The limit then as found by Gordon [9] is given as

$$\Lambda/\ell = \lambda_o/na \tag{8}$$

Equation (8) implies that light diffraction angle equals the sound diffraction angle. Using Eqs. (2), (3), (7) and (8) and setting $\Delta\phi = 2$ rad, we obtain the specific energy $(P_a/\Delta f_2) = 45\Delta f_a\lambda_o^3 a/nv_a^2M_2$. Units are mW/MHz when cgs units are used for v_a, M_2.

In the bulk case b cannot be freely chosen. Referring to Fig. 4, assume a cylindrical light beam so that $a = b = 2r$. Then it can be shown that r is limited by

$$r \geq \sqrt{\lambda_o\ell/\pi n} . \tag{9a}$$

Using Eqs. (6) and (8)

$$(P_a/\Delta f)_2 = 50.8\lambda_o^3/nv_aM_2 . \tag{9b}$$

As can be seen from Eq. (9b), the lower limit for the specific energy of a bulk acousto-optic modulator is independent of the chosen dimensions and depends only on the wavelengths and material properties. Good design can only result in achieving this lower limit. As will be seen later, a waveguide modulator can avoid this limitation.

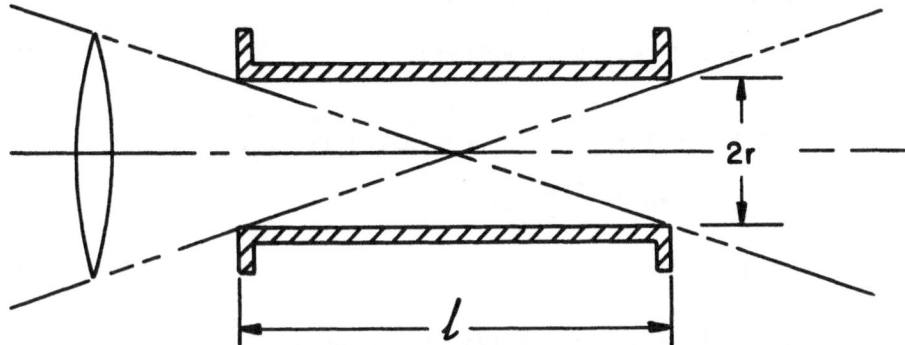

Fig. 4.   In diffraction limited optics the radius r of a cylinder
          through which a light beam may be focussed depends on the
          cylinder's length ℓ and may not be freely chosen.

B. Bulk Electro-Optic Modulators

A typical arrangement of a bulk electro-optic modulator is
shown in Fig. 2. Generally, linearly polarized light with the
polarization direction at 45° to the y axis in the x-y plane is used
at the input. The electro-optic effect gives a different phase shift
for the polarization component parallel to the y axis than the phase
shift undergone by the component perpendicular to the y axis. The
result is that the emergent light is elliptically polarized with the
direction of the major axis of the ellipse a function of the applied
voltage.

For a LiNbO$_3$ type crystal with field applied parallel to "C",
the intensity emerging from the analyzer will be

$$\eta = 1 - I/I_0 = 1 - \sin^2 \Delta\phi/2 = 1 - \sin^2(\pi n_e^3 r_{33} \ell V/2a\lambda_0) \qquad (10)$$

n_ε = extraordinary refractive index.

1. Drive Power Required for Bulk Electro-Optic Modulators.
The actual drive power required will be equal to the *reactive power*
plus any real power. For most crystals used in bulk modulators the
loss tangent is small and may be ignored so that only the reactive
power need be considered.

The reactive power may be calculated from the stored energy in
a very general way. The differential stored energy is given by

$$dU = \frac{\vec{E} \cdot \vec{D}}{2} \, dx \, dy \, dz \quad . \qquad (11)$$

By formal integration and using $p/\Delta f = U$

$$p/\Delta f = U = \pi/2 \ (\varepsilon_o \varepsilon \ E^2) \ (\text{Volume}).$$ (12)

From Eqs. (1) and (2), the phase change associated with E is $\Delta\phi = \pi \ell n'^3 r' E/\lambda$. Solving for E and taking (Volume) = a b ℓ

$$p/\Delta f = \varepsilon_o \varepsilon \ \lambda_o^2/\pi n'^6 r'^2 \ \frac{ab}{\ell} \ \Delta\phi^2$$ (13)

For $\Delta\phi = 2$ rad

$$P/\Delta f = 4 \times 10^9 \ (\varepsilon_o \varepsilon \ \lambda_o^2/\pi n'^6 r'^2) \ \frac{ab}{\ell} \ (\text{mW/MHz})$$ (14)

Again, as in the bulk acousto-optic modulator a, b and ℓ are not independent. For a square aperture it can be shown that

$$a = b = \sqrt{2\ell\lambda_o/n'}$$ (15)

or $ab/\ell = 2\lambda_o/n'$. Hence, from Eqs. (12), (13) and (14) it is seen that $P/\Delta f$ is independent of dimension. If a similar substitution is made in the expression for the modulation depth [Eq. (10)],

$$\eta = 1 - \sin^2 \left(\pi n_\varepsilon^{7/2} r_{33} \sqrt{\frac{\ell}{\lambda_o}} \left(\frac{1}{2\sqrt{2}} \right) V \right)$$ (16)

Thus, increasing ℓ will reduce drive voltage but not the specific energy in a bulk electro-optic modulator.

IV.   OPTICAL WAVEGUIDE MODULATORS

As contrasted to bulk modulators, the cross-section of a guided beam can have dimensions on the order of the size of a wavelength of light for "unrestricted" interaction length. Thus, waveguide modulators require much less power, voltage and/or current than bulk modulators made using the same materials. On the other hand, as compared to bulk modulators, waveguide modulators need a complex material system to achieve both guiding and an active effect. This latter requirement leads to problems in fabrication while at the same time opening opportunities for new types of devices.

A.   Acousto-Optic Waveguide Modulators

As illustrated in Fig. 5, the acousto-optic waveguide modulator is based on the interaction of a surface acoustic wave (SAW) and a guided light beam. The propagation directions are generally

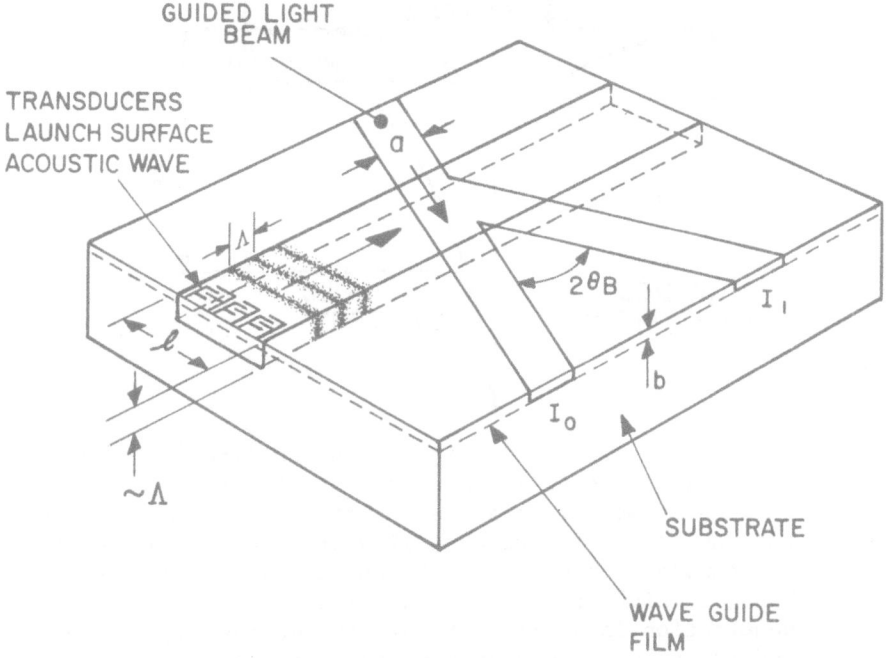

Fig. 5.   A surface acoustic wave (SAW) is launched by the trans-
          ducers. The SAW interacts with light guided in the wave-
          guide film. If the guided light is adjusted to intersect
          the SAW at the Bragg angle, θ_B, strong diffraction of light
          in the waveguide plane can take place [10,11].

chosen so that the Bragg condition [Eq. (5)] is satisfied at some
frequency in the acoustic band (usually the center frequency). The
condition on bandwidths [Eq. (7)] and that on matching diffraction
angles [Eq. (8)] also applies to the acoustic waveguide modulator.
The thickness dimension, b, is restrained by the waveguide properties and does not depend on the interaction length ℓ as it did in
the bulk case. The effectiveness of the interaction in the waveguide case does, however, depend on the overlap between the acoustic
field and the electromagnetic field. This is illustrated in Fig. 6.
The strain field of the SAW decays exponentially away from the surface with a decay constant proportional to the acoustic wavelength
Λ. Λ, of course, is inversely proportional to the acoustic frequency. The field of the guided electromagnetic wave depends on
the nature of the waveguide but usually can be represented as a
sinusoid with exponential tails as illustrated. The effective overlap thus will vary with acoustic frequency. For our purposes, we
will represent the effect of overlap by modifying the dimension b
with a constant $0 < \xi < 1$ such that $b_{effective} = b/\xi$. With this

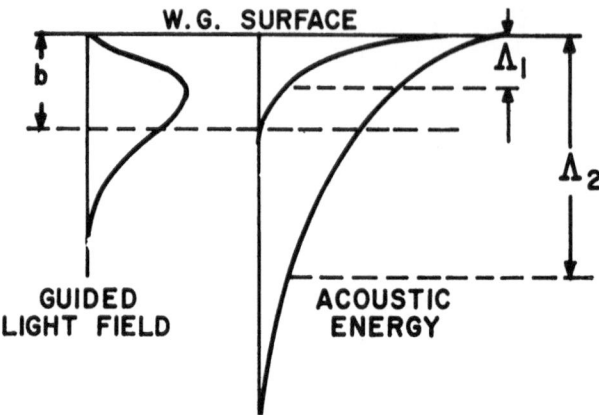

Fig. 6.   The energy distribution in the guided light wave depends
           on the waveguide film thickness and the ratio of refractive
           indices between film and substrate. The acoustic energy in
           the SAW falls off exponentially from the surface with a de-
           cay constant proportional to the acoustic wave length Λ.
           Thus, at high acoustic frequencies f_1, $\Lambda_1 = v_a/f_1$ and the
           penetration is small. At low frequencies f_2, $\Lambda_2 = v_a/f_2$
           and the acoustic energy penetrates deeply into the sample.
           The strength of the interaction depends on the overlap
           integral between the acoustic energy and the optical
           energy. This integral can be seen to vary with acoustic
           frequency.

artifice the Specific Energy is given by

$$(P/\Delta f)_2 = (45\ \lambda_o^3 f_a/n v_a^2 M_2)\,(b/\xi)\,(mW/MHz) \tag{17}$$

$$\eta = 1-\sin^2\left(\frac{1}{2}\sqrt{\frac{10^7\pi^2 n v_a^2 M_2 P_a}{2\lambda_o^2 f_a \Delta f}}\left(\frac{b}{\xi}\right)\right) \tag{18}$$

B.   Two Recent Acousto-Optic Waveguide Modulator Experiments

       Schmidt and Kaminow [10] use a waveguide made from y-cut LiNbO$_3$
diffused with 500 Å of Ti at 960°C. They estimate b = 2 μm.   At
f_o = 175 MHz, v_a = 3.3x10^5 cm/sec, and Λ = 18 μm so that the overlap factor ξ will be rather small. The configuration they use is

similar to that of Fig. 5, except only one common interdigital transducer is used. For η = 70% and Δf = 30 MHz, they measure an electrical power of P_e = 50 mW and estimate the acoustic power $P_a \leq$ 8 mW. Thus $(P_e/\Delta f)_2$ = 1.7 mW/MHz while $P_a/\Delta f$ = 0.27 mW/MHz. This is to be compared with the theoretical limit for bulk acousto-optic modulators made of LiNbO3 which is 1.3 mW/MHz. Thus, even though the overlap is relatively small, a significant improvement is realized in the waveguide modulator as compared to the bulk.

In the experiment of Nguyen and Tsai [11] the waveguide is made of out-diffused LiNbO3. b is estimated at 7 μm. With f_a = 325 MHz, Λ = 10 μm so that reasonable overlap would be expected. To obtain the large Δf of 110 MHz a series of transducers are used which act as a phased array and steer the acoustic beam to better satisfy the Bragg condition over the band. Nguyen and Tsai report P_e = 507 mW or $P_e/\Delta f$ = 1.4 mW/MHz. The acoustic power is estimated at P_a = 5.64 mW giving a specific energy of $P_a/\Delta f$ = 0.043 mW/MHz which is a very low value and is truly indicative of the advantage that can be realized in the waveguide format.

C. Electro-Optic Waveguide Modulators

We had the general expression [Eq. (14)] $(P/\Delta f)_2$ = $4 \times 10^9 (\varepsilon_0 \varepsilon \lambda_0^3 / \pi n'^6 r'^2) ab/\ell$ mW/MHz. In "bulk" modulators both a and b are limited by the diffraction expression $\sqrt{2\ell\lambda_0/n'}$. In a one-dimensional guide, however, while a will be limited by the inter-action length, b will be on the order of λ independently of ℓ. In a two-dimensional guide both a and b will be on the order of λ in-dependently of ℓ. (See Fig. 7.)

We may thus compare the drive power required for the bulk case to that required for the waveguide case using Eqs. (14) and (15).

$$\frac{(P/\Delta f)_{W.G.}}{(P/\Delta f)_{BULK}} \sim \begin{cases} \sqrt{\lambda/\ell} & \text{Planar Guide} \\ \lambda/\ell & \text{Stripe Guide} \end{cases}$$

Assuming λ = 1 μm and ℓ = 1 cm the ratio is 10^{-2} in the planar guide case and 10^{-4} in the two-dimensional guide (stripe guide) case.

Two common ways of combining waveguides with electrodes to utilize the electro-optic effect are illustrated in Figs. 7 and 9. In the semiconductor junction illustrated in Fig. 9, the electric field is developed entirely across the back-biased junction region. Since the junction can be made to be coincident with the waveguide region, the field of the guided wave and the modulating field can have good overlap as indicated in the figure.

In the second principal method, illustrated in Figs. 7, 8 and

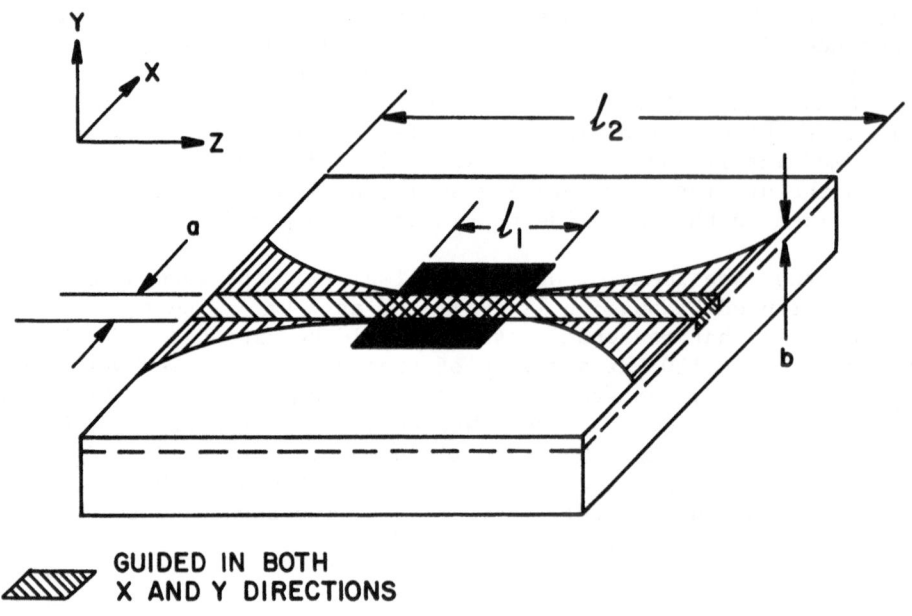

GUIDED IN BOTH
X AND Y DIRECTIONS

GUIDED ONLY IN
Y DIRECTION

Fig. 7. If light is guided in the x-z plane, then for an aperture
(a) between two electrodes the length ℓ_1 is limited by
diffraction. When the light is guided in both the x and
y direction (stripe guide), then the interaction length
is limited only by losses and the sample length ℓ_2.

10, surface electrodes are used. Here the fringing field of the
surface electrodes must overlap the optical fields. This approach
is necessary for use with most insulating crystal waveguides where
the high index film region is formed on or near the surface. In
either the junction or the surface electrode case the effect of
overlap can be accounted for by an overlap factor ξ similar to that
used in the acousto-optic waveguide case. For surface electrodes
ξ will be 0.5 or less since at least half the field will be lost in
the space above the waveguide surface. Using the overlap factor we
may write the general expression for the Specific Energy of electro-
optic waveguides

$$(P/\Delta f)_2 = (4\times10^9/\pi)(\varepsilon_0\varepsilon\lambda_0^2/n'^6r'^2)(\frac{a}{\xi_a})(\frac{b}{\xi_b})\frac{1}{\ell}$$

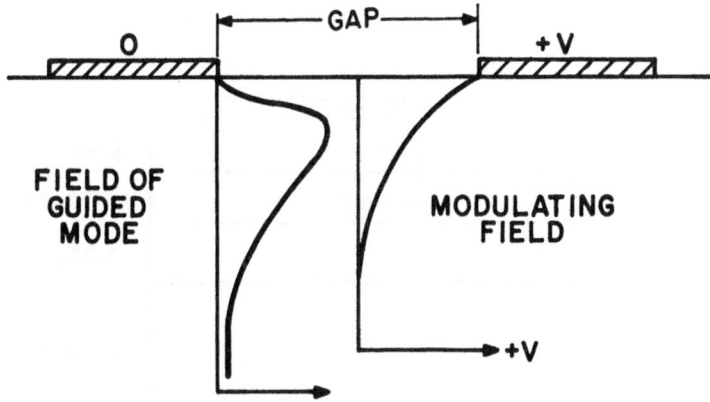

Fig. 8. The electric field due to surface electrodes, modulating
         field, decays rapidly away from the surface. The decay
         constant is a function of the gap between the electrodes.
         As in the case of the acousto-optic waveguide modulator,
         the strength of the interaction depends on the overlap
         integral between the modulating field and the field of the
         guided mode. Thus, if the gap is small compared to the
         waveguide film thickness, the overlap will be small result-
         ing in weak interaction.

The numerical factor is adjusted to give $P/\Delta f$ in mW/MHz when MKS
units are used.

                D. Experimental Electro-Optic Waveguide Modulators

        1. Planar Modulators. A good example of a semiconductor type
of waveguide modulator has been given by McKenna and Reinhart [12].
Referring to Fig. 9, light propagates in the z direction and the
junction lies in the x-z plane. If the y direction of the figure
is made parallel to the crystalline 100 direction then TE waves are
not coupled to TM waves. Thus, a simple phase modulation is obtained. If, however, y is parallel to the 110 direction as in Mc-
Kenna and Reinhart's experiment, the TE wave is coupled to the TM
wave so that a mixed TE-TM output emerges for any input. Thus a
polarization modulator results. Experimentally, $(P/\Delta f)_2$ is found
to be 0.15 mW/MHz and a good extinction ($\eta_m > 20$ dB) is obtained.
This is a planar type of modulator as there is no restraint in the
x direction.

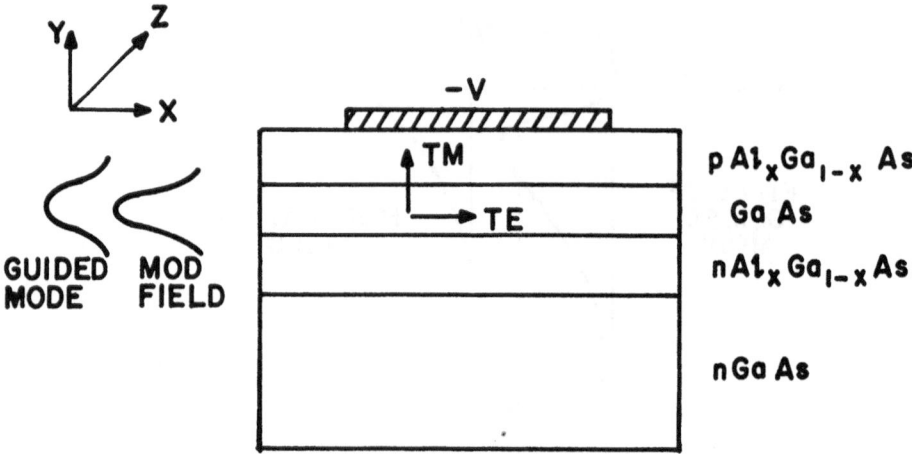

Fig. 9. Semiconducting junction modulator. When the junction is
         back-biased the entire voltage is dropped across the GaAs
         junction region. The p-Al$_x$Ga$_{1-x}$As layer and the
         n-Al$_x$Ga$_{1-x}$As layer have refractive indices lower than the
         index of the GaAs junction region. Thus, good overlap be-
         tween the field of the guided mode and the modulating
         field can be obtained.

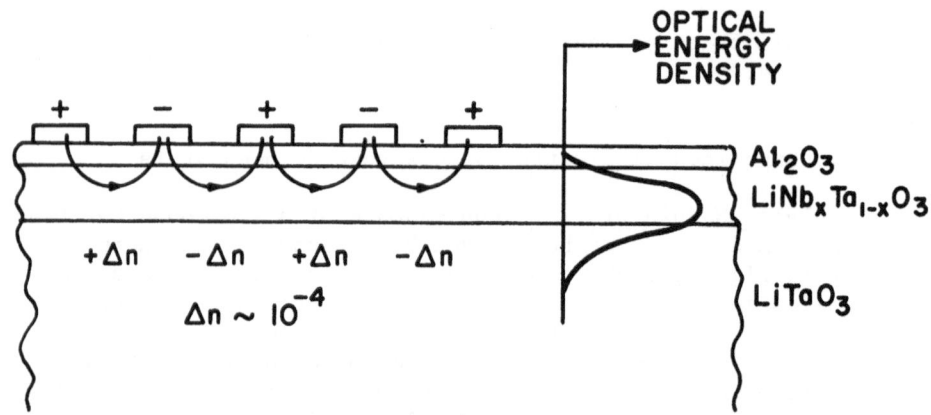

Fig. 10. Fringing fields of surface electrodes with alternating
          polarity (interdigital circuit) cause periodic changes
          in refractive index of electro-optic waveguide. Such a
          phase grating can diffract light in the waveguide plane.

We will discuss one more planar electro-optic waveguide modulator. In the work of Hammer and Phillips [13] an interdigital electrode structure is placed on the surface of a waveguide formed by diffusing Nb into $LiTaO_3$ to form the mixed crystal $LiNb_xTa_{1-x}O_3$ which has a higher index than $LiTaO_3$. As illustrated in Fig. 10, the fringing field of the interdigital structure forms a phase grating which can diffract light in the waveguide plane. The diffraction is similar to that observed in the surface acoustic wave modulators described earlier. Here, however, the spacing of the interdigital line sets the grating space which does not vary with frequency. Light enters at the Bragg angle and is diffracted in the plane of the waveguide. A number of workers have reported periodic electro-optic grating modulators [14,15,16]. The results of Hammer and Phillips [13] are shown in Fig. 11. At 6328 Å $(P/\Delta f)_2 = 1.0$ mW/MHz. For unity overlap factor ξ, the specific energy is predicted to be 0.18 mW/MHz. Thus, the effective overlap factor in this experiment is 0.18. Bulk $LiNbO_3$ modulators with appropriate safety factors require approximately 10 mW/MHz. At $\lambda_0 = 6328$ the extinction ratio is close to 20 dB at an applied voltage of 8 V. This may be compared to the hundreds of volts required by bulk modulators.

2. Examples of Stripe Guide Electro-Optic Modulators (Light Restrained in both Transverse and Lateral Directions. Perhaps the simplest example of a stripe guide modulator is the phase modulator reported by Kaminow, Stulz and Turner [17]. This is illustrated in Fig. 12. Its principle of operation is straightforward and we are content to note that the specific energy for this device is measured to be 6.8 μW/MHz.

In another experiment by Izutsu, Yamane and Sueta [18] the electrodes are shaped to form the elements of a microwave delay line. Partial synchronism between the microwave ($n_{eff} \approx 4.2$) and the light ($n_{eff} = 2.2$) is obtained. Phase modulation is obtained over a band $\Delta f = 7.5$ GHz at $\lambda_0 = 6328$ Å. The specific energy is 0.1 mW/MHz. There are no bulk modulators that have performance figures in either bandwidths or drive power that come to within an order of magnitude of the performance of this device.

3. Coupling and Interferometric Modulators. The last two groups of electro-optic waveguide modulators we will discuss have no real counterpart in conventional optics. We first consider coupling modulators as illustrated schematically in Fig. 13. Consider Fig. 13a. Two single mode strip guides run parallel to each other. The guides are placed so that their evanescent fields overlap. If the phase velocity of both guides is equal, light launched in the upper guide will transfer to the lower guide in a distance known as the critical coupling length L. If the interaction continues past a distance L the light in the lower guide will couple back into the upper guide. If the guides are arranged so that the interaction continues precisely for the distance L all the light

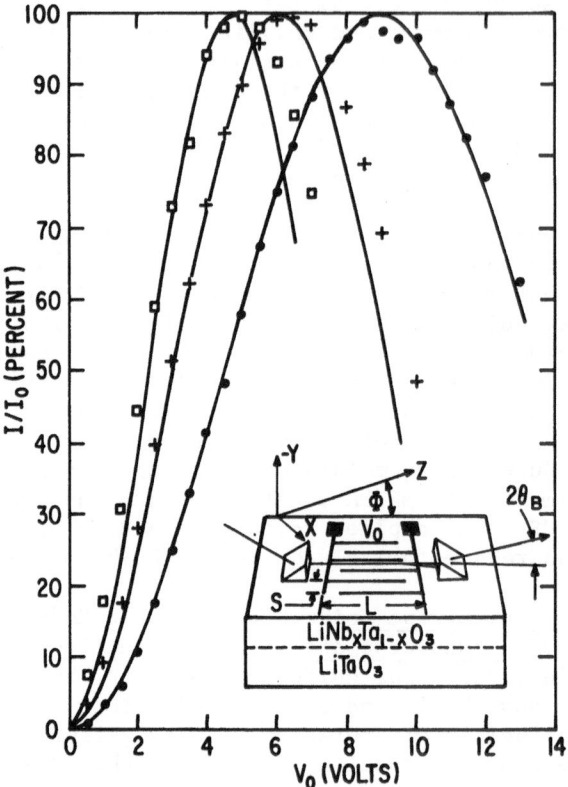

Fig. 11.   Ratio of diffracted (I) to undiffracted (I_o) light
coupled out of grating modulator using $LiNb_xTa_{1-x}O_3$ on
$LiTaO_3$ waveguides. λ_o = 4976 Å (squares), 5598 Å
(crosses), and 6328 Å (circles). Solid lines are
plots of sin^2BV normalized at I/I_o = 70%. [13]

will be transferred to and remain in the lower guide. If electrodes are arranged so that application of a voltage changes the
refractive index of one guide with respect to the second then phase
match is destroyed and light will remain in the upper guide. For
$LiNbO_3$ type guides on "C" plates the simple electrodes shown in
Fig.13a give a field component perpendicular to the plane and of
opposite sense in the two guides. For TM waves this results in a
slowing in one guide and a speeding up in the second. Hence, phase
match is destroyed. This COBRA configuration was described by
Papuchon et al. [19]. A related device constructed using semiconducting waveguides was reported by Campbell et al. [20]. Configurations using $LiNbO_3$ or $LiTaO_3$ x and y plates can also be used. These
require either a center electrode or the application of field to
only one of the two coupled guides.

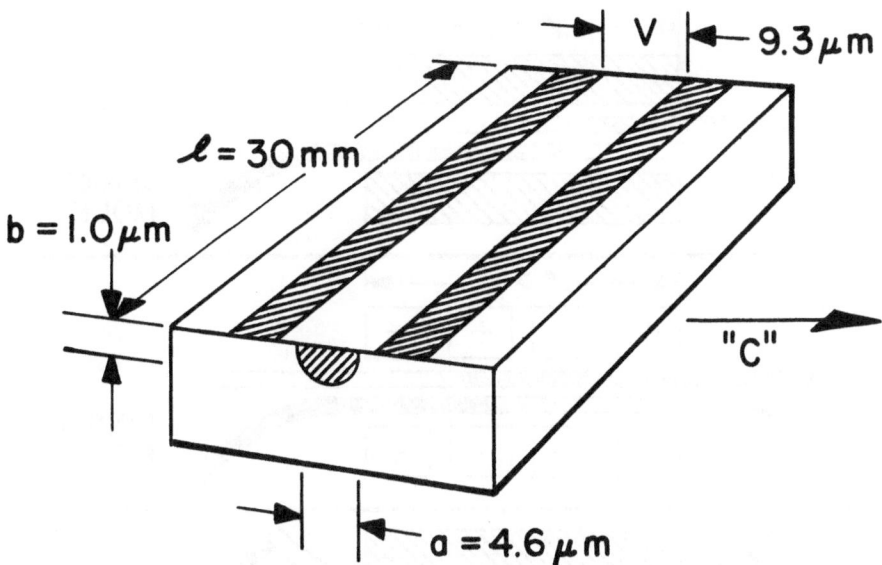

Fig. 12.   Simple phase modulator. A stripe guide of cross-section
           b = 1.0 μm, a = 4.6 μm is formed by titanium diffusion
           in LiNbO₃. Surface electrodes spaced 9.3 μm apart pro-
           vide a field which changes the effective index of the
           guide resulting in a phase shift of the guided wave which
           is proportional to the voltage. [17]

        If the actual interaction length (ℓ) departs from the critical
coupling length (L), then all of the light cannot be switched resulting in poor extinction ratios. It is difficult to control the
actual value of L. Kogelnick and Schmidt [21] have shown that
electrical control of the critical coupling can be obtained by
making L greater than a critical coupling length and using multiple
electrodes of alternating polarity. This is illustrated in Fig.
13b. They refer to this approach as the alternating Δβ method. A
detailed explanation of the operation of this device is given by
Kogelnick in another lecture of this series.

        In some very recent experiments by Cross, Schmidt and Thorn-
ton [22] a length (ℓ) of 2 cm and 12 alternating sections are used.
The drive voltage is approximately 2 V and the specific energy is
on the order of only a few microwatts/MHz.

        Finally, in the device illustrated in Fig. 13c, the two single-
mode guides are "collapsed" to form a two-mode guide in the experiment of Papuchon, Roy and Ostrowsky [23]. A more detailed view of
this experiment is given in Fig. 14a. As can be seen from the plot

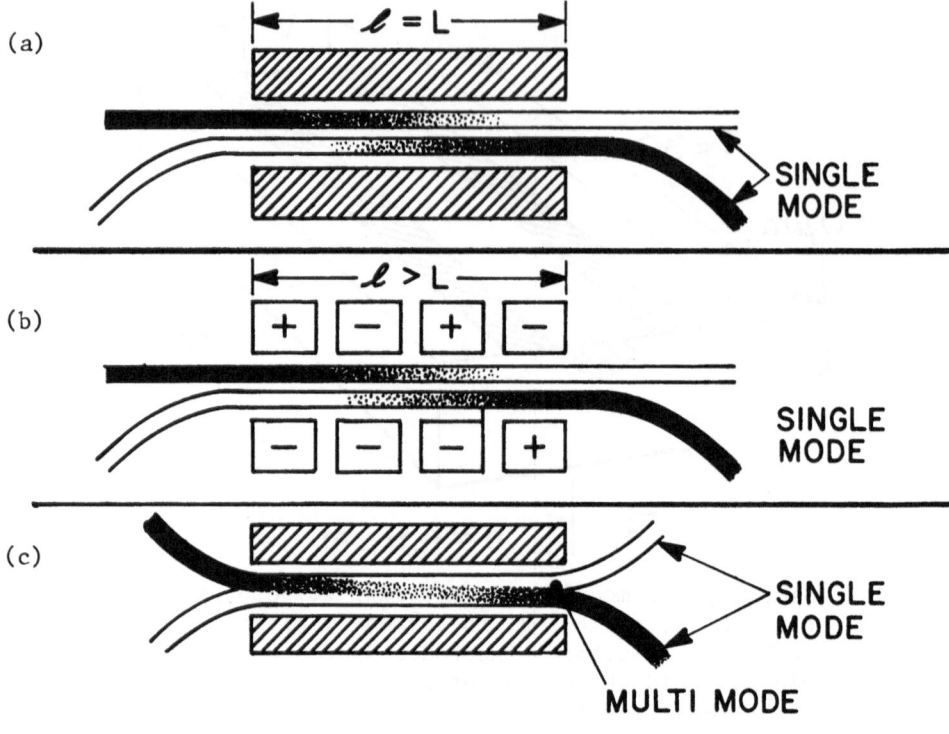

Fig. 13.   (a) Two single mode stripe guides arranged in proximity
               so that the evanescent fields overlap. If the guides
               are identical the phase velocity of light in both
               guides will be the same allowing light from the upper
               guide to couple into the lower. Application of a
               voltage changes the phase velocity of one guide with
               respect to the other decoupling the system. [19,20]

           (b) As in 13a but alternating polarity of electrodes
               makes the device less critically dependent on over-
               all interaction length. [21]

           (c) A two-mode guide is substituted for the two single-
               mode guides. The voltage here determines if an anti-
               node in the light energy arrives at the upper or
               lower branch. [23]

of light intensity vs. distance in Fig. 14b, the light energy distribution in the higher order mode progresses from having a maximum at one side of the guide to a maximum at the other returning
to its original distribution in the distance L_0. Light will be

coupled to the furcation at which the maximum occurs. Thus, if
when the branch is reached the maximum occurs at the top, light will
be coupled out of the upper branch and vice versa. In this device
application of a voltage changes the interaction length rather than
the phase match condition. Thus, the voltage changes the distance
required to go from a node at one edge to a node at the other guide
edge. This makes for a simple and non-critical·type of device but
requires relatively high drive voltage. In the 5 mm section
$\eta_m \sim 18.3$ dB is achieved at $V \sim 30$ V. For 70% extinction 15 V is
required. Using an estimated value of $C = 1.0$ pf and $(P/\Delta f)_2 =$
$\pi/2$ CV^2, $(P/\Delta f)_2 \sim 0.35$ mW/MHz. Thus, here, one trades off, as
compared to the stepped $\Delta\beta$ devices, voltage and power for simplicity.

The final type of device we will briefly discuss is the optical

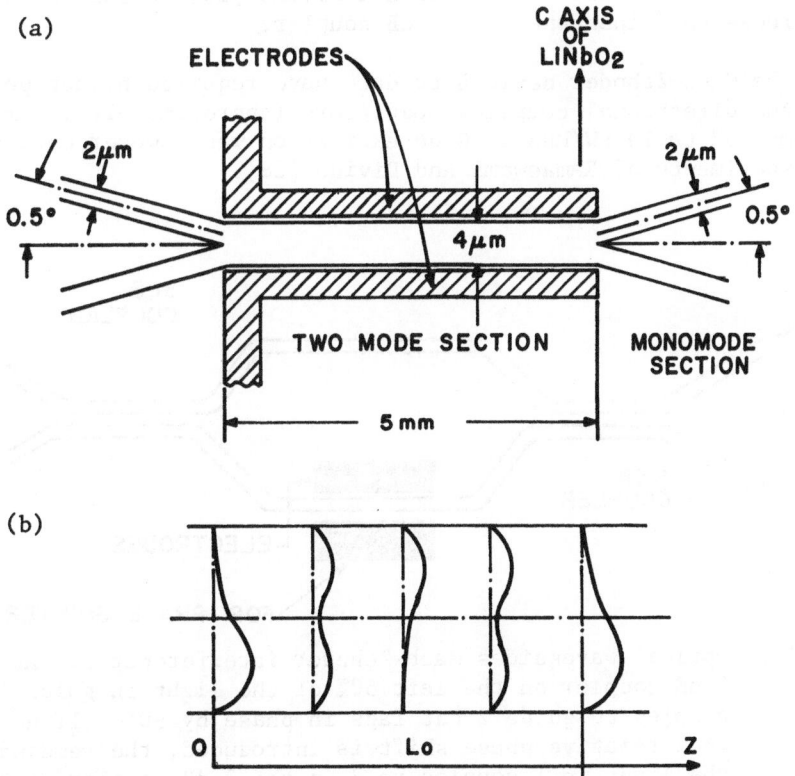

Fig. 14. Details of the two-mode coupling modulator of Papuchon,
Roy and Ostrowsky [23]. The explanation is given in the
text.

waveguide Mach–Zehnder interferometer which was first suggested by
F. Zernike, Jr. [24]. As shown in Fig. 15, this device consists of
two guides which first come together to form a 3 dB coupler, then
separate in a region where the electro–optic effect is used to give
a variable phase shift, rejoin in a 3–dB coupler configuration and
finally separate again. At the first 3 dB coupler half the light
launched in guide 1 will be transferred to guide 2. The light
coupled into guide 2 is shifted (lags) 90° out of phase with the
light in guide 1. If on arrival at the output 3 dB coupler the
light in guide 2 still lags the light in 1 by 90° the remainder of
the light in 1 will couple into guide 2. If the phase shifter is
used to shift the phase of arm 2 by an additional 90°, then all the
light will couple back to arm 1 at the output 3 dB coupler. A
number of variations of the device using furcated intersections between mono- and multi-mode guides to provide the 3 dB couplers have
been reported. See Martin [25], Burns, Lee and Milton [26], and
Ohmachi and Noda [27] who have all reported such devices.

      In order to get good extinction it is necessary to have ac-
curate 3 dB couplers. Ramaswami and Divino [28] do this by using
electrodes to "fine tune" the 3 dB coupler.

      The Mach–Zehnder devices to date have required higher power
than the directional coupling modulators (approximately 1.0 mW/MHz)
as compared to 10 µW/MHz). Good extinction was however obtained in
the experiments of Ramaswami and Divino [28].

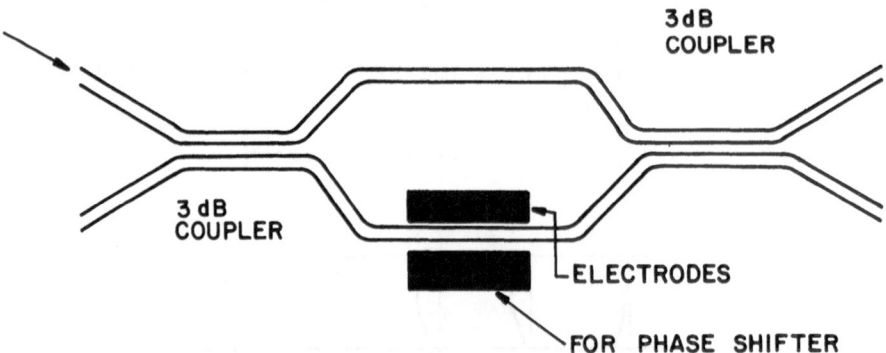

Fig. 15. Optical waveguides Mach–Zehnder interferometer. At the
          3 dB coupler on the left 50% of the light in guide 1
          couples to guide 2 but lags in phase by 90°. If no fur-
          ther relative phase shift is introduced, the remainder of
          the light in 1 couples to 2 in the 3 dB coupler at the
          right. If the electro-optic phase shifter changes the
          relative phase by 90°, all the light couples back to
          guide 1 at the right 3 dB coupler.

GENERAL REFERENCES

Integrated Optics, T. Tamir, Ed., Springer-Verlag, N.Y., 1975.

REFERENCES

1. M. Born, E. Wolf, Principles of Optics, 2nd Ed. (MacMillan Co., New York, 1964).

2. A. Yariv, IEEE J. Quant. Electronics QE-9, 919 (1973).

3. I. P. Kaminow, J. R. Caruthers, E. H. Turner, L. W. Stulz, Appl. Phys. Lett. 22, 540 (1973).

4. J. M. Hammer, Chapter IV in Integrated Optics, T. Tamir Ed., Springer-Verlag, N.Y., 1975.

5. J. Pankove, Optical Processes in Semiconductors, Dover, New York, 1975, p. 29.

6. J. F. Nye, Physical Properties of Crystals, (Oxford at the Clarenden Press, London 1957), pp. 241 et seq.

7. Ibid, pp. 244 et seq.

8. D. A. Pinow, IEEE J. Quant. Electr. QE-6, 223 (1970).

9. E. I. Gordon, Proc. IEEE 54 1391 (1966).

10. R. V. Schmidt and I. P. Kaminow, IEEE J. Quant. Electr. QE-11 57 (1975).

11. L. T. Nguyen and C. S. Tsai, Appl. Optics 16 1297 (1977).

12. J. McKenna and F. K. Reinhart, J. Appl. Phys. 47 2069 (1976).

13. J. M. Hammer and W. Phillips, Appl. Phys. Lett. 24 545 (1974).

14. J. F. St. Ledger and E. A. Ash, Electron Lett. 4 99 (1968).

15. D. P. Gia Russo and J. H. Harris, Appl. Opt. 10 2786 (1971); also J. H. Polky and J. H. Harris, Appl. Phys. Lett. 21 307 (1972)

16. J. M. Hammer, Appl. Phys. Lett. 18 147 (1971); also J. M. Hammer, D. J. Channin, M. T. Duffy and C. C. Neil, IEEE J. Quant. Electr. QE-11 138 (1975).

17. I. P. Kaminow, L. W. Stulz and E. H. Turner, Appl. Phys. Lett. 27 555 (1975).

18.   M. Izutsu, Y. Yamane and T. Sueta, IEEE J. Quant. Electr.
      QE-13 287 (1977).

19.   M. Papuchon, Y. Combemale, X. Mathieu, D. B. Ostrowsky,
      L. Reiber, B. Sejourne and M. Werner, Appl. Phys. Lett. 27 289
      (1975).

20.   J. C. Campbell, F. A. Blum, D. W. Shaw and K. L. Lawley, Appl.
      Phys. Lett. 27 203 (1975).

21.   H. Kogelnik and R. V. Schmidt, IEEE J. Quant. Electr. QE-12
      396 (1976).

22.   P. S. Cross, R. V. Schmidt and R. L. Thornton, Topical Meeting
      on Integrated and Guided Wave Optics, Salt Lake City, Utah -
      Digest, Paper Tu B1 (1978).

23.   M. Papuchon, Am. Roy and D. B. Ostrowsky, Appl. Phys. Lett.
      31 266 (1977).

24.   F. Zernike, Jr., Topical Meeting on Integrated Optics, New
      Orleans, LA, Digest, Paper WA-5 (1974).

25.   W. E. Martin, Appl. Phys. Lett. 26 562 (1975).

26.   W. K. Burns, A. B. Lee and A. F. Milton, Appl. Phys. Lett.
      29 790 (1976).

27.   Y. Ohmachi and J. Noda, Appl. Phys. Lett. 27 544 (1975).

28.   V. Ramaswami and M. D. Divino, Topical Meeting on Integrated
      and Guided Wave Optics, Salt Lake City, Utah, Technical Digest
      Paper TuA4 (1978).

29.   J. P. Kaminow in Handbook of Lasers, R.J. Pressley, Ed.,
      (Chemical Rubber Company, Cleveland, 1971).

COUPLED-WAVE DEVICES

Herwig Kogelnik

Bell Laboratories

Holmdel, New Jersey 07733

A scan of recent reviews of integrated optics[1-4] or of a
recent text book on this subject[5], makes it apparent that
coupled-wave devices play a major role in this new technology.
The reason for this is that optical effects of interest are
usually small over distances of the order of a wavelength.
Coupled-wave phenomena extend over many wavelengths and can lead
to a considerable enhancement of these effects. In integrated
optics we encounter a large variety of coupled-wave phenomena
and devices. In the following we propose to discuss a selection
of these, including directional couplers, switched directional
couplers, stepped $\Delta\beta$ couplers, bistable $\Delta\beta$ couplers, tunable
coupler filters, the tunable Harris filter, electrooptic Bragg
deflectors, acousto-optic Bragg deflectors, TE-TM mode converters,
corrugated waveguide filters, distributed Bragg reflector lasers
and distributed feedback lasers. The theoretical basis for the
treatment of coupled-wave phenomena is the coupled-wave formalism
which is summarized in References 6 and 7 and which is reviewed
in this symposium by A. Yariv. This formalism provides
expressions for the coupling constants of various devices which
we will use for our discussion.

In coupled-wave devices we deal with guiding structures which
support two waves or modes which propagate freely and uncoupled as
long as the structure is not perturbed. We follow here the
notation of Reference 7 and call these two waves R and S ("reference"
and "signal"). A perturbation of the original structure
leads to a coupling of the two waves and to an exchange of energy
between them. An important requirement for a significant interaction between the two waves is their synchronism or "phase

281

matching". In the simplest case this requires the equality

$$\beta_R = \beta_S \tag{1}$$

of the propagation constants β_S and β_R of the two waves.

We distinguish between two different types of coupled-wave interactions, codirectional interactions and contradirectional interactions, and we classify the coupled-wave devices accordingly. Codirectional interactions occur between two forward (or two backward) waves, and contradirectional interactions occur between one forward and one backward wave. Figure 1a shows a typical variation along the direction of propagation z of the wave amplitudes $R(z)$ and $S(z)$ for the codirectional or "forward scattering" case and Figure 1b indicates the typical situation for the contradirectional or "backward scattering" case.

1. CODIRECTIONAL DEVICES

The exchange of energy between two codirectional waves of complex amplitude R and S is described by coupled-wave equations of the form[6,7]

$$R' - j\delta R = -j\kappa S,$$

$$S' + j\delta S = -j\kappa R. \tag{2}$$

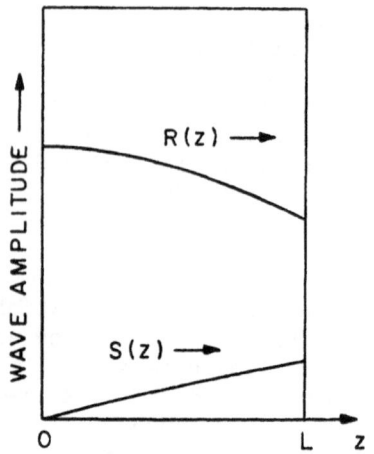

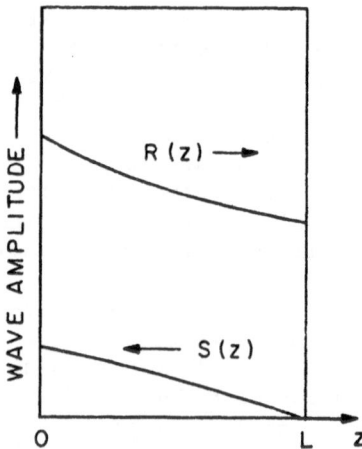

a) FORWARD SCATTERING   b) BACKWARD SCATTERING

Figure 1

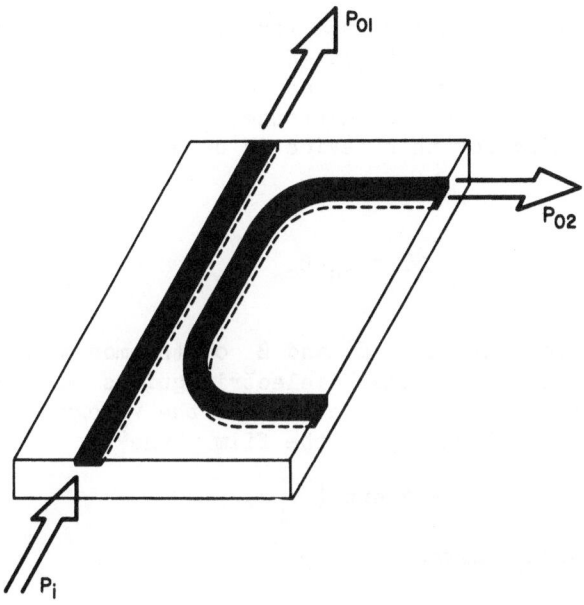

OPTICAL DIRECTIONAL COUPLER

Figure 2

where the prime indicates differentiation with respect to the propagation distance z. Here κ is the coupling constant and δ is a normalized frequency which measures the deviation from synchronism. For the boundary conditions $R(0)=1$ and $S(0)=0$ the solutions of the coupled-wave equations are

$$S(z) = -j\kappa \sin (z\sqrt{\kappa^2+\delta^2})/\sqrt{\kappa^2+\delta^2},$$

$$R(z) = \cos (z\cdot\sqrt{\kappa^2+\delta^2}) + j\delta \sin (z\sqrt{\kappa^2+\delta^2})/\sqrt{\kappa^2+\delta^2}. \quad (3)$$

For the case of synchronism ($\delta=0$) we have a sinusoidal exchange of energy between R and S,

$$S(z) = -j \sin (\kappa z), \quad R(z) = \cos (\kappa z) \cdot \quad\quad (4)$$

1.1 Directional Coupler

The directional coupler is a familiar microwave device. Figure 2 shows a typical integrated optics version. It consists of two strip guides approaching each other, running close and parallel over the interaction distance, and then separating again. Synchronism for this device is determined by the difference

$$2\delta = \beta_R - \beta_S \tag{5}$$

of the propagation constants β_R and β_S of the modes in the two strip guides. For rectangular dielectric guides the coupling constant κ is an exponential function of the waveguide separation c and the decay constant γ_y in the film plane[8]

$$\kappa \, \alpha \, \exp(-\gamma_y c). \tag{6}$$

Typical conversion lengths

$$\ell = \pi/2\kappa \tag{7}$$

range from fractions of 1 mm to several millimeters. To achieve full transfer of energy from one guide to the other one has to fabricate an interaction length equal to ℓ or an odd multiple thereof (see equation (4)). Experimental directional couplers have been prepared in a variety of structures and materials systems[1,5], including ferroelectric crystals and semiconductors.

1.2 Switched Directional Coupler

Several proposals and attempts have been made to control a directional coupler electrically in order to switch the guided light from one waveguide to the other (for a history see References 9 and 1). Figure 3 shows a sketch of a switched

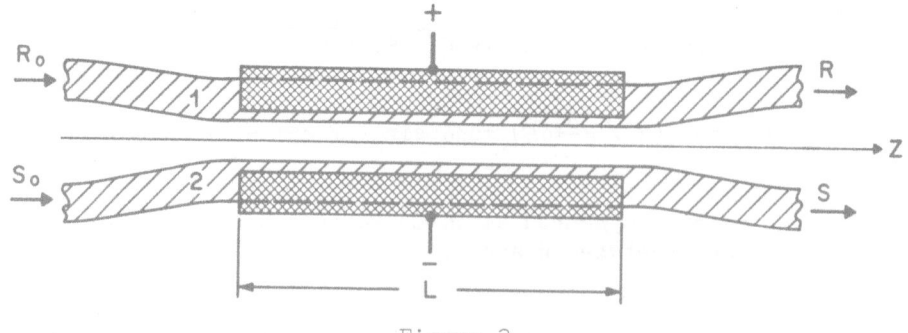

Figure 3

directional coupler which is typically fabricated in an electro-
optic material such as LiNbO$_3$. The electrodes, shown in the
COBRA configuration[10], allow control of the coupler via an
applied voltage. In an electrooptic material an applied electric
field E induces a refractive index change $\Delta n'$ of magnitude[9]

$$\Delta n' = - n'^3 r' E/2, \tag{8}$$

where n' is the effective refractive index of the material and
r' the effective electrooptic coefficient. The field is applied
with opposite polarity to the two guides and induces a phase-
mismatch

$$2\delta = \Delta\beta \approx \frac{2\pi}{\lambda} \Delta n = \frac{2\pi}{\lambda} n'^3 r' E \quad , \tag{9}$$

where λ is the free-space wavelength of the light. As the
induced index changes are small, there is, to first order, no
effect of the applied field on the coupling coefficient κ.

The device principle is as follows: One aims to fabricate
a coupler with an interaction length L which is equal to an odd
multiple of the conversion length ℓ. This leads to complete
transfer of the light from one guide to the other when the guides
are phasematched (see equation (4)). By application of a voltage
one can induce a phase mismatch $2\delta = \Delta\beta$ which spoils the interaction and leads to zero transfer of light (see equation (4)).

Figure 4 provides a graphical illustration of the device
operation. Here the mismatch-length product δL (or the voltage)
are plotted on the horizontal axis and the interaction length to
conversion length ratio L/ℓ is plotted on the vertical axis.
Based on equation (4) one can now mark the conditions for
complete transfer (Ⓧ) and zero transfer (⊜), which leads to
the isolated points and circles shown. For any given device an
applied voltage will move the state of the switch on this diagram
along a horizontal line.

1.3 Stepped -$\Delta\beta$ Coupler

In practice it turns out to be difficult to fabricate the
interaction length L and the conversion length ℓ of switched
directional couplers with sufficient accuracy to insure
essentially complete transfer of light ($\geq$ 20 db). No electrical
control can be provided to compensate for fabrication errors and
to adjust for complete crossover. The stepped -$\Delta\beta$
configuration[11,12] makes this electrical adjustment of the
crossover state possible. Figure 5 shows a sketch of the stepped
-$\Delta\beta$ coupler structure. Note that the electrodes are split into
two sections and that voltages of opposite polarity are applied
to the two sections inducing mismatch of opposite sign. The

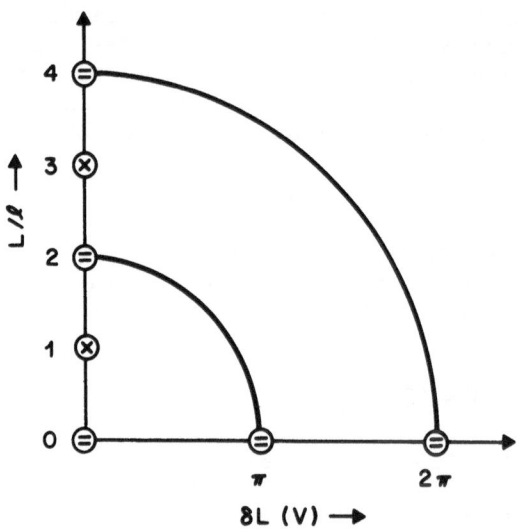

Figure 4

stepped $-\Delta\beta$ coupler is easily analyzed by casting the coupled-
wave solutions (4) into matrix form, and multiplying the
appropriate matrices of the individual sections[11]. One finds
that the condition for the crossover state of this coupler is

$$\frac{\kappa^2}{\kappa^2+\delta^2} \; \sin^2 \frac{L}{2} \sqrt{\kappa^2+\delta^2} = \frac{1}{2} \; , \qquad (10)$$

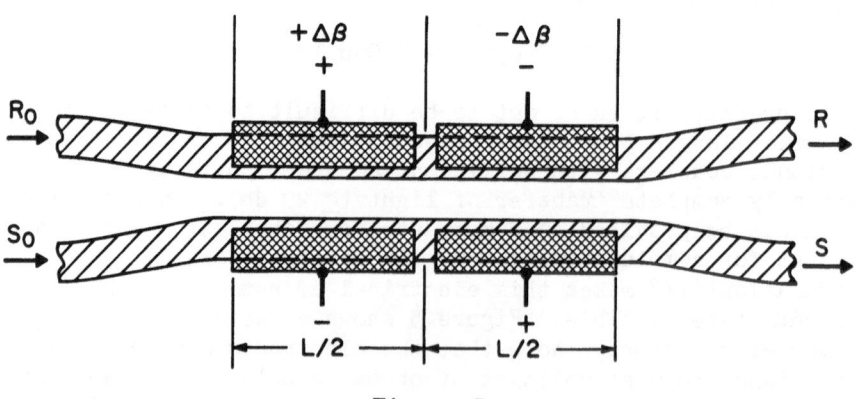

Figure 5

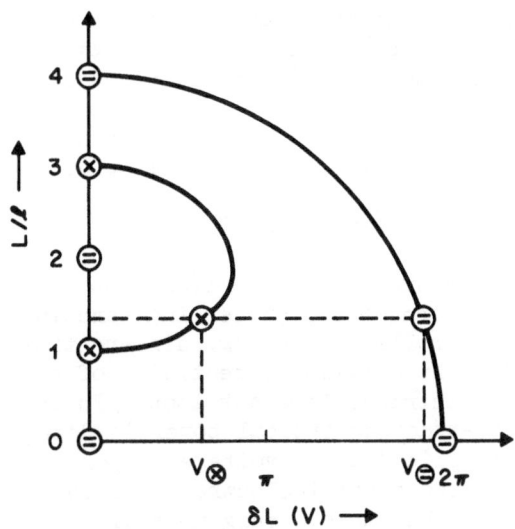

Figure 6

and that the zero-transfer state is obtained if

$$L/\ell = 2(2\gamma+1),\qquad\qquad\qquad(11)$$

or

$$(L/\ell)^2 + (2\delta L/\pi)^2 = (4\gamma)^2,$$

where γ is an integer. The switching diagram of Figure 6 gives a graphical representation of these switching conditions. Note that there is a curve representing crossover states for couplers longer than one conversion length ℓ but shorter than 3 ℓ. For any coupler in that range we have the capability of an electrical adjustment of the crossover state as well as the zero-transfer state.

The stepped $\Delta\beta$ configuration has allowed the construction of optical switches[12] with a crossover ratio of 400:1 (26 db) using Ti diffused strip guides in LiNbO$_3$. The width and spacings of the guides were about 3 μm and the interaction length about 3 mm. Stepped $\Delta\beta$ couplers have also been used in the demonstration of an experimental 4x4 switching network in which 5 such switches were integrated on a single LiNbO$_3$ chip.[13]

Experimental stepped − $\Delta\beta$ couplers have also been made in GaAs based semiconductor compounds.[14,15] Couplers with more

than two sections of alternating Δβ also allow electrical
adjustability and are useful when the interaction length is
several coupling lengths long.[11] Six sections of alternating
Δβ have recently been used in the construction of an efficient
switch and amplitude modulator[16] which operates at data rates
in excess of 100 Mbit/sec with drive voltages as low as 3V.

1.4   Bistable Δβ-Coupler

The bistable Δβ-coupler is an optical 4-port bistable
switch which can perform signal processing functions such as
remote optically controlled switching, differential gain and
optical limiting.[17] It consists essentially of an alternating
Δβ coupler and an electrical feedback loop. In this loop a
portion of the zero-transfer optical power is detected (e.g. by an
avalanche photomultiplier) and a voltage proportional to this
power is applied to the control electrodes of the Δβ coupler.
One of the functions that this arrangement allows is the optically
controlled switching between the crossover and the zero-transfer
state of the coupler. An experimental device of this type
realized[17] in Ti:LiNbO$_3$ had a switching time of about 300 μsec
and an optical switching energy of 3 pJ.

1.5   Tunable Coupler Filter

The tunable coupler filter[18] is essentially a directional
coupler with wavelength-dependent transfer characteristics.
Electrical control of both the filter center wavelength and the
crossover efficiency is provided. This new structure[18]
combines the Δβ-coupler idea with a proposal[19] for a directional
coupler consisting of two nonidentical waveguides which have
intersecting dispersion characteristics as sketched in Figure 7.
The two waveguides are fabricated to different widths (1.5 and
3 μm, with a 3.5 μm spacing) with different effective guide
index $N_1(\lambda)$ and $N_2(\lambda)$. This leads to a wavelength dependent
mismatch

$$\delta(\lambda) = \frac{\pi}{\lambda} (N_2 - N_1).$$ (12)

The mismatch δ can also be controlled electrically via the
electrooptic effect as the device is made in Ti: LiNbO$_3$. The
device fabricated[18] with an interaction length of 1.5 cm, had a
measured filter bandwidth of 200 Angstroms and was tunable at the
rate of 110 Angstroms/Volt.

1.6   The Electrooptic Bragg Deflector

The electrooptic Bragg deflector is a thin-film amplitude
modulator (for a review see Reference 9). As opposed to the

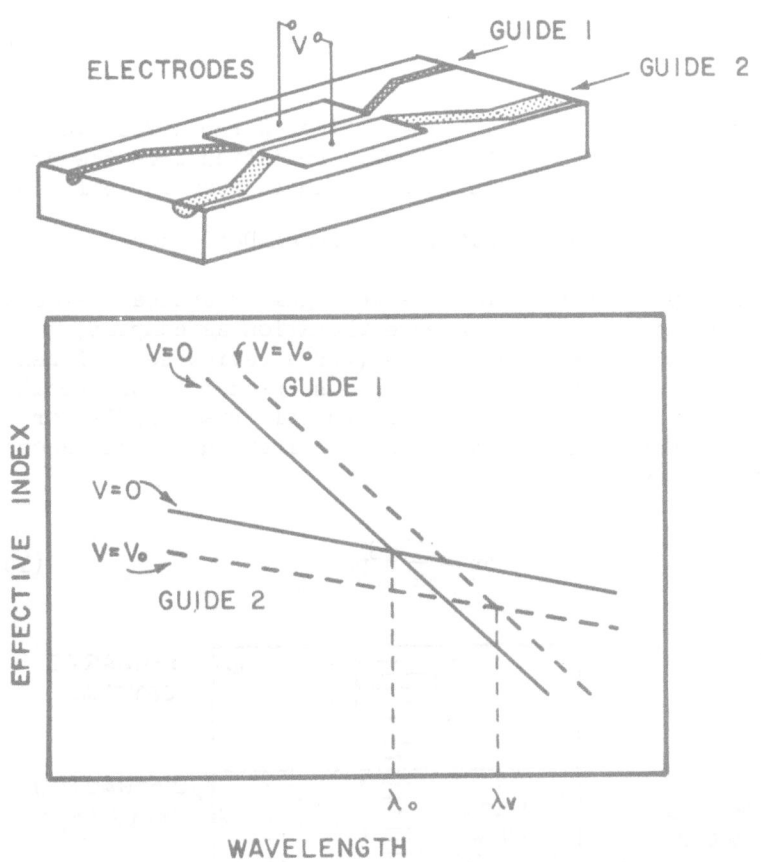

Figure 7

strip guides used in the devices discussed above, this
modulator uses a planar film guide in an electrooptic material.
Spatially periodic refractive index variations of amplitude Δn
are induced electrooptically via a periodic set of electrodes.
The device principle is very similar to that of the acoustooptic
Bragg deflector discussed in the next section and sketched in
Figure 8. When light is incident near the Bragg angle on the
induced grating a portion of it is scattered and a coupled-
wave interaction between the scattered and incident light takes
place following Equations (3-4   ). The coupling coefficient
for this interaction is[20]

$$\kappa \approx \frac{\pi}{\lambda} \Delta n. \qquad\qquad (13)$$

We have synchronism when the Bragg condition

$$\sin \Theta_B = \lambda/2N\Lambda \qquad (14)$$

is obeyed, where Λ is the grating period, N the effective film index and Θ_B the angle of incidence. Deviation from the Bragg condition leads to asynchronism and reduced transfer efficiency.

1.7 The Acoustooptic Bragg Deflector

Acoustooptic interactions between guided optical waves and acoustic surface waves are under exploration as compact, miniature light deflection devices (for a review see Reference 21). The device is sketched in Figure 8. The principle of operation is very similar to that of the electrooptic Bragg deflector except that, here, the index grating is induced via the acousto-optic effect. We have

$$\Delta n = \frac{1}{2} n^3 pS, \qquad (15)$$

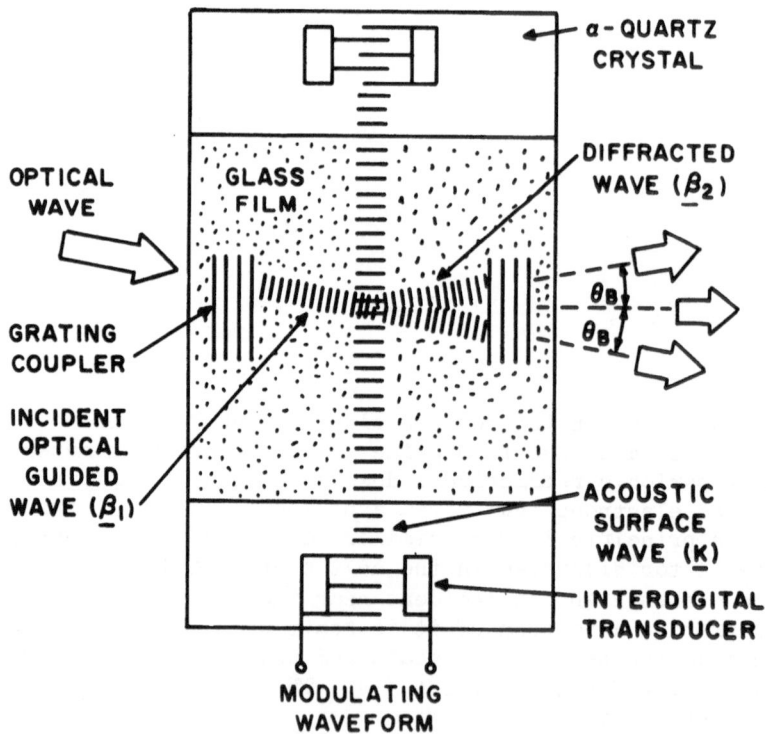

Figure 8

where p is the photoelastic constant and S the acoustic strain
amplitude. Typical operating characteristics[21] for a deflector
made in Ti: LiNbO$_3$ are an acoustic frequency of 175 MHz, a
bandwidth of 35 MHz and 50 mWatt drive power for a 70% deflection
efficiency.

1.8 TE-TM Mode Converter

TE to TM mode converters, in combination with polarizer
elements, can serve as optical amplitude modulators (for a review
see References 1 and 9). The TE-TM mode conversion process can
be regarded as another case of a codirectional coupled-wave
interaction.[6,7] We are dealing here with a single waveguide in
which a TE and a TM mode propagate uncoupled in the absence of a
perturbation. Even scalar perturbations of the refractive index
leave the two modes uncoupled. However, a tensor perturbation
$\Delta\varepsilon_{ij}$ of the dielectric constant

$$\Delta\varepsilon_{ij} = \begin{pmatrix} \Delta\varepsilon_1 & \eta & 0 \\ \eta & \Delta\varepsilon_2 & 0 \\ 0 & 0 & \Delta\varepsilon_3 \end{pmatrix} \tag{16}$$

leads to a coupling of the modes via the off-diagonal elements η.
This perturbation can be caused by electrooptic or acoustooptic
effects. The coupling constant for this process is of the form[6,7]

$$\kappa = \omega \int_{-\infty}^{+\infty} dx\ \eta\ E_{TM} E_{TE}^*, \tag{17}$$

where E_{TM} and E_{TE} are the field distributions in the guide of the
transverse electric field components of the TM and TE modes. The
phase mismatch δ is determined by the corresponding propagation
constants β_{TE} and β_{TM},

$$2\delta \approx \beta_{TM} - \beta_{TE}. \tag{18}$$

Reference 36 reports recent experimental results on amplitude
modulation at 1.06 μm by TE-TM mode conversion in double-hetero-
structure AlGaAs [110] p-n junctions with half-wave voltages of
12V and extinction ratios better than 20 db.

1.9 Guided-Wave Harris Filter

The socalled Harris filter[22] is based on a colinear
acoustooptically induced TE-TM mode conversion. The propagation
constants β_{TE} and β_{TM} are chosen different and dispersive.

Synchronism is established by chosing the acoustic wavelength Λ such that

$$\beta_{TE}-\beta_{TM} = 2\pi/\Lambda. \tag{19}$$

The filter characteristics of this device result from the dispersion of $\beta_{TE}(\lambda)-\beta_{TM}(\lambda)$ which causes increasing asynchronism away from the center wavelength of the filter given by Equation (19). Changing the acoustic wavelength allows tuning of the center frequency of the filter. Integrated-optics versions of this device using surface-acoustic waves are under study.[23-25]

2. CONTRADIRECTIONAL DEVICES

The energy interchange between two contradirectional waves of complex amplitude R and S is described by coupled-wave equations of the form[6,7]

$$R'+j\delta S = -j\kappa S$$

$$S'-j\delta S = j\kappa R, \tag{20}$$

where κ is the coupling coefficient and δ a measure for the deviation from synchronism, as in the codirectional case. For a contradirectional interaction we have the boundary conditions R(0)=1 and S(L)=0, which leads to the solutions

$$S(0) = -j\kappa/\left[\sqrt{\kappa^2-\delta^2}\coth(L\cdot\sqrt{\kappa^2-\delta^2})+j\delta\right]$$

$$R(L) = \sqrt{\kappa^2-\delta^2}/\left[\sqrt{\kappa^2-\delta^2}\cosh(L\cdot\sqrt{\kappa^2-\delta^2})+j\delta\sinh\right.$$

$$\left.(L\cdot\sqrt{\kappa^2-\delta^2})\right]. \tag{21}$$

For the case of synchronism we have $\delta=0$ and a simplification to

$$S(0) = -j\tanh(\kappa L), \quad R(L) = 1/\cosh(\kappa L). \tag{22}$$

2.1 Corrugated Waveguide Filters

One way to make a guided-wave filter is to machine a corrugation of very short period Λ into the surface of a planar film guide, as shown in Figure 9. Such a corrugated waveguide of length L provides a band-rejection filter with a fractional bandwidth of about

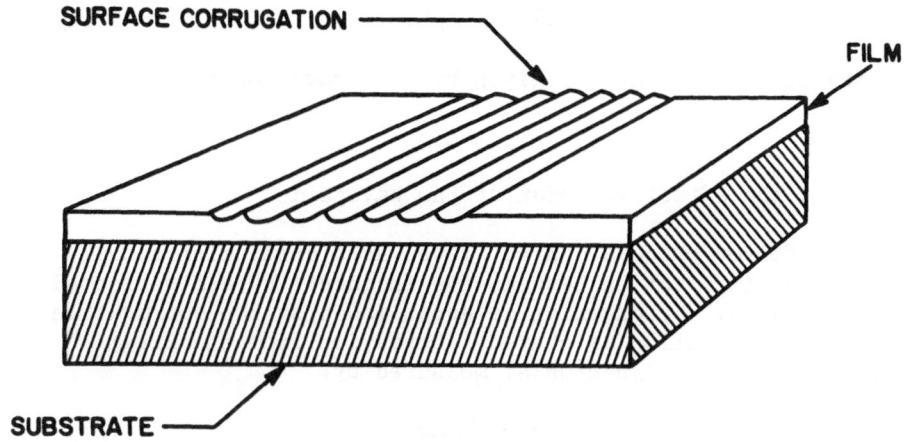

SURFACE CORRUGATION

FILM

SUBSTRATE

Figure 9

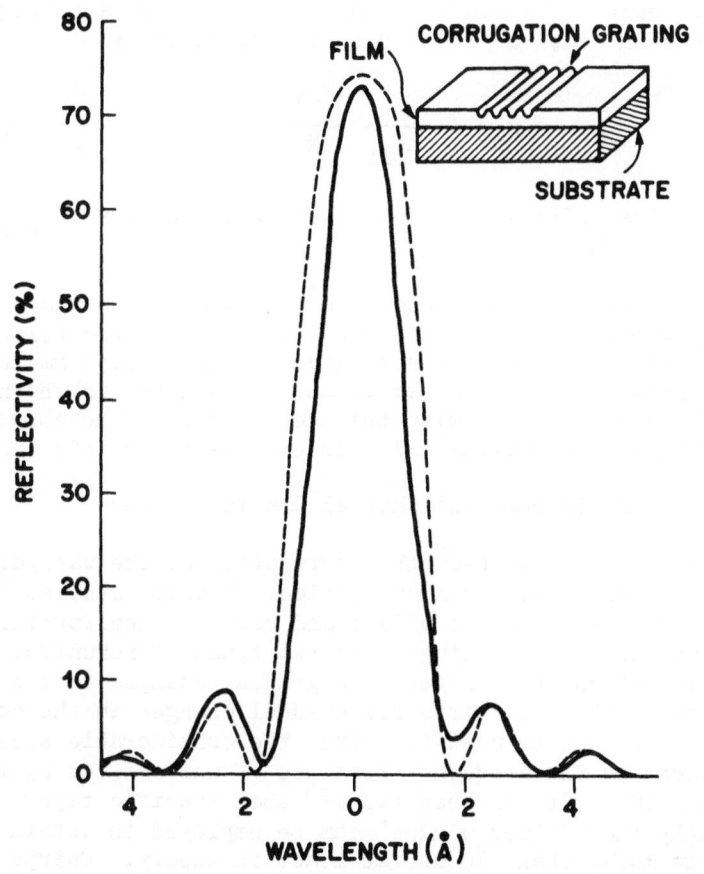

Figure 10

$$\Delta\lambda/\lambda \approx \Lambda/L, \tag{23}$$

centered at a wavelength λ_o given by the Bragg condition

$$\lambda_o = 2N\Lambda, \tag{24}$$

where N is the effective index of the waveguide.

This filter process can be viewed as a contradirectional coupled-wave interaction between the incident (R) and reflected (S) guided wave. Synchronism is established by the corrugation if the Bragg condition is obeyed. Deviation from the Bragg condition leads to asynchronism measured by[7]

$$\delta = \beta - \beta_o \approx \Delta\omega/v_g, \tag{25}$$

where $\beta_o = 2\pi N/\lambda_o$, $\beta = 2\pi N/\lambda$, v_g is the group velocity of the guided wave, and $\Delta\omega$ is the radian frequency deviation from the center frequency. The coupling coefficient κ for a corrugation of amplitude Δh is, to first order in Δh, given by[7]

$$\kappa = \frac{\pi}{\lambda} \frac{\Delta h}{h_{eff}} \frac{n_f^2 - N^2}{N}, \tag{26}$$

where n_f is the refractive index of the film guide and h_{eff} the effective film thickness.

In practice the Bragg condition Equation (24) demands ultrashort periods Λ of 1000 to 3000 Å. The measured filter response[26] of a corrugated glass guide filter of 0.57 mm length with corrugations of 2000 Å period and 460 Å depth is shown in Figure 10. The agreement with the predictions of the coupled wave theory, i.e. Equations (21), is excellent for this case.

2.2 Tapered and Chirped Grating Filters

In the previous section the corrugation of the waveguide was assumed uniform in depth and period. Several studies[26,30] have already considered the effect and use of nonuniformities in the corrugation. One distinguishes two types of nonuniformities, i.e. tapers and chirps. Tapers are gradual changes in the corrugation depth, and chirps are gradual changes in the corrugation period. Tapers can be used to reduce the considerable side-lobe level present in the frequency response of uniform corrugations (see Figure 10). It has been found[29] that specific taper functions, particularly the "Kaiser window" can be employed to obtain side-lobe levels lower than -70 db, at least in theory. Chirps can be used to broaden the filter response, which, in uniform corrugations is limited by the coupling coefficients obtainable in practice.

To allow for corrugation nonuniformities, the coupled wave equations have to be modified to the form

$$R' + j\delta R = -j\kappa S \exp(-j\phi)$$

$$S' - j\delta S = j\kappa R \exp(j\phi), \tag{27}$$

where $\kappa(z)$ describes the taper and $\phi(z)$ describes the chirp. To perform numerical evaluations it is convenient to transform these expressions via the definition of a local reflection coefficient $\rho(z)$,

$$\rho = \frac{S}{R} \exp(-j\phi). \tag{28}$$

The result[27] is a Riccati differential equation of the form

$$\rho' = j(2\delta - \phi')\rho + j\kappa(1 + \rho^2). \tag{29}$$

This can be solved for the filter response by techniques such as the Runge-Kutta method.

2.3 Distributed Bragg Reflector Lasers

Corrugated waveguides as described in section 2.1 can be regarded as wavelength dependent reflectors. Two such elements can be used to form a thin-film resonator for a guided-wave laser such as double heterostructure semiconductor laser.[31,32] This is called the "Distributed Bragg Reflector" laser. If operated at the center wavelength the threshold condition for this laser is the equality of the reflectivity of the corrugation R and the excess laser gain described by the gain constant α

$$R = \tanh^2 \kappa L_R = \exp(2\alpha L_L), \tag{30}$$

where L_R is the reflector length and L_L the length of the laser medium.

2.4 Distributed Feedback Lasers

Distributed feedback laser structures[33] promise to provide compact and low-loss optical cavities for semiconductor lasers which allow longitudinal mode control, frequency selection and reduced temperature sensitivity of the output wavelength. These structures are essentially an integration of a corrugated wave-

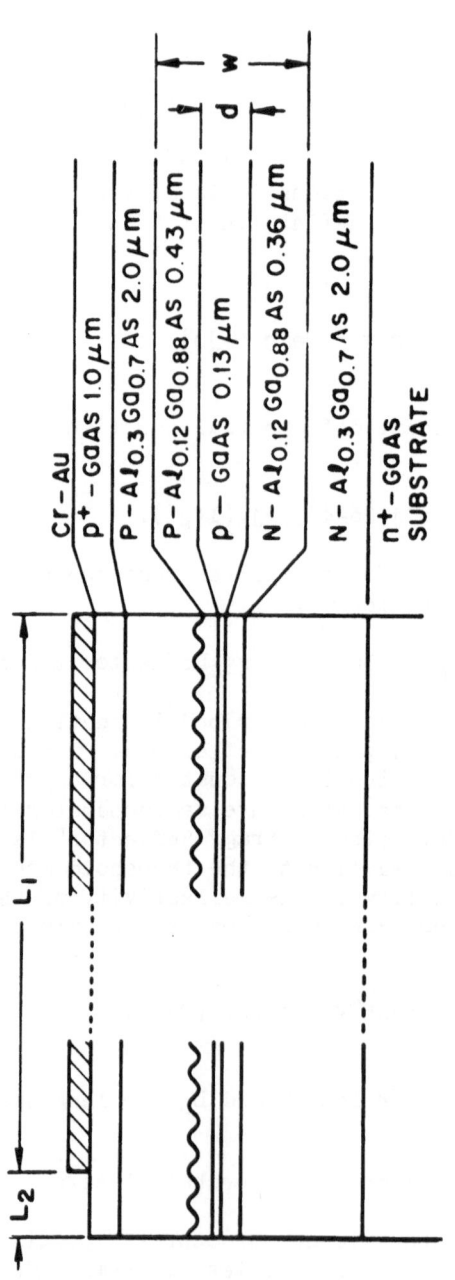

Figure 11

guide filter with a waveguide medium possessing laser gain. cw
operation at room temperature has been achieved in separate
confinement semiconductor DFB lasers[34] (for a review of recent
work see Reference 2). Figure 11, taken from Reference 35 shows
a sketch of such a DFB structure.

The DFB process is a contradirectional coupled-wave interaction in the presence of laser gain and with a new kind of
boundary condition. To allow for gain, the coupled-wave equations
are written in the form[33]

$$-R' + (\alpha - j\delta)R = j\kappa S \qquad\qquad (31)$$

$$S' + (\alpha - j\delta)S = j\kappa R,$$

where α is the gain constant. The new boundary conditions occur
because we are now dealing with an oscillator on which no light
is incident at the device terminals at $z = -L/2$ and $z = L/2$.
Correspondingly the boundary conditions are

$$R(-L/2) = S(L/2) = 0. \qquad\qquad (32)$$

The solution of Equations (31) and (32) predicts the laser
threshold and the wavelength selectivity of DFB lasers.[33]

REFERENCES

1. P. K. Tien, Rev. Mod. Phys, 49, 361, April, 1977.

2. A. Yariv and M. Nakamura, IEEE J. Quantum Electronics, QE-13, April, 1977.

3. H. Kogelnik, Fibers and Integrated Optics, 1, 227, 1978.

4. H. Kogelnik, IEEE Transactions MTT-23, 2, January, 1975.

5. Integrated Optics, T. Tamir Editor, Springer, Berlin, 1975.

6. A. Yariv, IEEE J. Quantum Electronics, QE-9, 919, September, 1973.

7. H. Kogelnik, in Integrated Optics, T. Tamir ed., Springer, Berlin, 1975.

8. E. A. J. Marcatili, Bell Syst. Techn. J. 48, 2071, September, 1969.

9. J. M. Hammer, in Integrated Optics, T. Tamir ed., Springer, Berlin, 1975.

10. M. Papuchon et al, Appl. Phys. Lett. 27, 289, September, 1975.

11. H. Kogelnik and R. V. Schmidt, IEEE J. Quantum Electronics, QE-12, July, 1976.

12. R. V. Schmidt and H. Kogelnik, Appl. Phys. Lett. 28, 503, May, 1976.

13. R. V. Schmidt and L. L. Buhl, Electronics Lett. 12, 575 October, 1976.

14. F. J. Leonberger and C. O. Bozler, Appl. Phys. Lett. 31, 223, 1977.

15. J. C. Shelton, F. K. Reinhart and R. A. Logan, IEEE Trans. ED-24, 1198, 1977.

16. R. V. Schmidt and P. S. Cross, Optics Letters 2, 45, February, 1978.

17. P. S. Cross, R. V. Schmidt, R. L. Thornton and P. W. Smith, IEEE J. Quantum Electronics, August, 1978.

18. R. C. Alferness and R. V. Schmidt, Appl. Phys. Lett., July, 1978.

19. H. F. Taylor, Optics Communications $\underline{8}$, 421, August, 1973.

20. H. Kogelnik, Bell Syst. Tech. J. $\underline{48}$, 2909, November, 1969.

21. R. V. Schmidt, IEEE Trans. Sonics and Ultrasonics $\underline{SU-23}$, 22, January, 1976.

22. S. E. Harris and R. W. Wallace, J. Opt. Soc. Am. $\underline{59}$, 744, 1969.

23. Y. Ohmachi, Electronics Lett. $\underline{9}$, 541, November, 1973.

24. H. Hayashi and Y. Fujii, IOOC '77, Tokyo, Japan, 1977.

25. C. S. Tsai et al., IOOC '77, Tokyo, Japan, 1977.

26. D. C. Flanders et al, Appl. Phys. Lett. $\underline{24}$, 194, 1974.

27. H. Kogelnik, Bell Syst. Techn. J. $\underline{55}$, 109, January, 1976.

28. M. Matsuhara, K. O. Hill and A. Watanabe, J. Opt. Soc. Am. $\underline{65}$, 804, July, 1975.

29. P. Cross and H. Kogelnik, Optics Lett. $\underline{1}$, 43, July, 1977.

30. C. S. Hong et al, Appl. Phys. Lett. $\underline{31}$, 276, August, 1977.

31. S. Wang, J. Quantum Electronics $\underline{QE-10}$, 413, 1974.

32. F. K. Reinhart, R. A. Logan and C. V. Shank, Appl. Phys. Lett. $\underline{27}$, 45, 1975.

33. H. Kogelnik and C. V. Shank, J. Appl. Phys. $\underline{43}$, 2327, May, 1972.

34. M. Nakamura et al, Appl. Phys. Lett. $\underline{27}$, 403, 1975.

35. H. C. Casey, S. Somekh and M. Illgems, Appl. Phys. Lett. $\underline{27}$, 143, 1976.

36. J. McKenna and F. K. Reinhart, J. Appl. Phys. $\underline{47}$, 2069, May, 1976.

EXOTIC USES OF GUIDED WAVE OPTICS

D. B. Ostrowsky

Laboratoire Electrooptique
Université de Nice - Parc Valrose
06034 Nice Cédex - France

1 - INTRODUCTION

Practically all of the papers presented in this book deal with those aspects of guided wave optics that are related to optical fiber communication systems, real or envisaged. While such systems have been the inspiration for nearly all the work on guided wave optics, the rapidly developing technology can, and I believe will, have an important impact in many other fields and in ways that the people who developed the technology never imagined. In this paper I shall briefly outline what I consider to be some interesting examples of such technological "fallout". I shall begin by discussing fiber devices, go on to mention some integrated optical devices and conclude by outlining two examples of the use of guided wave optics as an experimental tool.

2 - FIBER DEVICES

The possibility of guiding light several kilometers without catastrophic loss creates many possibilities besides those offered by communication systems.

Essentially, many weak effects can now be integrated to give appreciable totals over the enormous interaction length involved. In this section I shall outline the way this property can be used in two types of fiber sensors and mention, for completeness, the way fibers can be used to realize lasers.

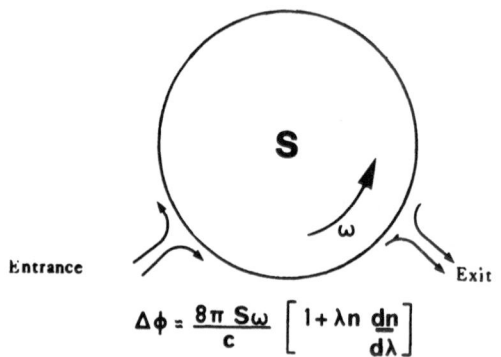

$$\Delta\phi = \frac{8\pi\,S\omega}{c}\left[1 + \lambda n\,\frac{dn}{d\lambda}\right]$$

Figure 1 - Fiber gyroscope schematic

2.1 - Movement sensors : Fiber gyroscopes

A laser gyroscope consists, essentially, of a dispositive in which an entering beam is divided into two beams, one running clockwise and one running counterclockwise, over the same path (Fig. 1). The two beams are then recombined and their relative phase change is measured (1,2). The relative phase change is approximately given by :

$$\Delta\phi = \frac{8\pi S\omega}{c}\left[1 + \lambda n\frac{dn}{d\lambda}\right]$$

where c is the speed of light, λ the vacuum wavelength, n the index of refraction and S the enclosed surface. The essential point to note is that the phase change is directly proportional to the surface, S, enclosed by the beam path. One can immediately see the advantage provided by a single mode fiber optic version of such a device. Without increasing the size of the device one can add a large number, N, of spirals and the total enclosed surface will then be given by NS where S is the elementary surface enclosed.

The number of spirals one can add will eventually be limited by the fiber loss which will reduce the fringe displacement signal below the level accepted by the detector system. With fibers in the 2-5 db/km range and 10 mwatts of injected power one can hope to detect angular rotations of 10^{-9} rad/sec which is an improvement of several orders of magnitude over existing devices.

2.2 - Fiber Parameter Sensors

The essential configuration of these devices is shown in

Figure 2. One constructs a fiber interferometer in which one arm is placed in the environment to be measured, and the other in a controlled environment. The exit beams are recombined and the fringe displacement detected. Since the fiber index is a function of such parameters as temperature and pressure the variation in these parameters, integrated over the fiber length, will be measured.

As a rough example consider such an interferometer used as a temperature sensor. In the perturbed fiber the phase change of the emerging light will be mainly due to the change in the index of refraction (since the length change of the fiber will give a contribution an order of magnitude smaller). This change will be given by

where λ_0 is the vacuum wavelength of the light, Δn the index change, and L the perturbed length.

If we call the minimum detectable phase change $\Delta\phi_m$ we need a minimum Δn_m given by

$$\Delta n_m \geq \frac{\lambda}{L} \frac{\Delta\phi_m}{2\pi}$$

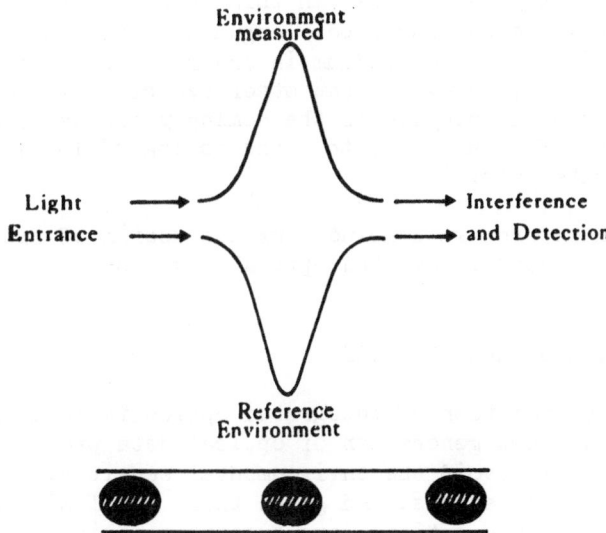

Figure 2 - a) Fiber parameter sensor
            b) Side observation of polarized light in a
               birefringent fiber

If we take $\Delta\phi_m = \frac{\pi}{4}$ and $\lambda = 1$ micron we have

$$\Delta n_m \gtrsim \frac{10^{-7} \, (\mu)}{L \, (\mu)}$$

Since

$$\Delta n_m \approx \Delta T \times 10^{-5} \, (^{\circ}C)^{-1}$$

We obtain

$$\Delta T_m \approx \frac{2 \times 10^{-2} \, (^{\circ}C \cdot \mu)}{L \, (\mu)}$$

or for a measurement length of 10 meters, 2 millidegrees C which is a rather good sensitivity.

We should also keep in mind that Dr. Stolen has shown how to simplify such devices even further. By using a birefringent fiber excited with a mixture of the two mode polarizations one can use just one fiber as the sensor and reference by counting the beat fringe displacements, as shown on the bottom of figure 2, since the birefringence has a similar sensitivity to the various parameters to be measured.

2.3 - Fiber lasers

Fiber lasers are mentioned here only for the sake of completeness since Dr. Stolens lectures in this volume give very detailed examples of how one can use fiber length to compensate for a weak optical gain in an excited medium in order to obtain laser action. We note in passing, however, that other examples of such lasers will certainly be developped in the coming years based on other nonlinear effects (Kerr effect, etc.) and doping of low loss fibers with active materials.

Here, therefore, we conclude our discussion of fiber devices and go on to discuss integrated optical devices.

3 - INTEGRATED OPTICAL DEVICES

The rapid evolution of integrated optics is leading to the development of a new generation of optical data processing devices. These devices will avoid the environmental sensitivity which have plagued their predecessors, and given their small size and projected low cost, should make optical data treatment a viable alternative in a variety of situations. In this section we shall outline the principles of several such devices.

3.1 - Integrated optical spectral analyzer

Figure 3 presents a schematic of this device. Light coupled into a lithium tantalate or niobate guide from an injection laser is collimated by a series of geodesic lenses and diffracted by an acoustic surface wave modulated by a microwave signal which is to be processed. Since the deflection angle depends on the frequency this permits an analysis of the microwave signal by a detector array. The device is therefore a real time signal analyzer capable of handling complex signals of the type encountered in various electronic counter measure schemes. It should provide a

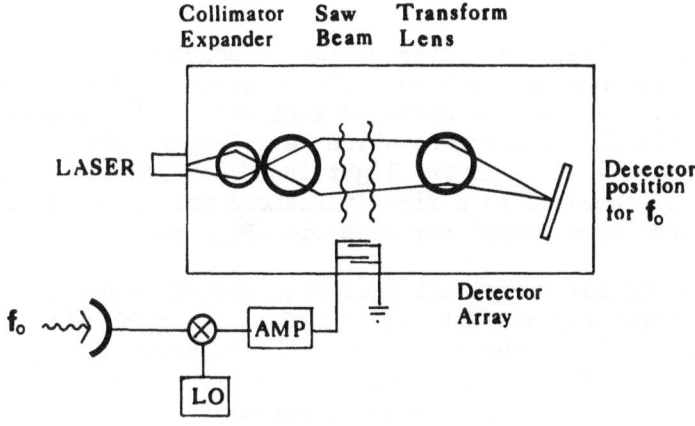

Figure 3 - Integrated optical spectral analyzer

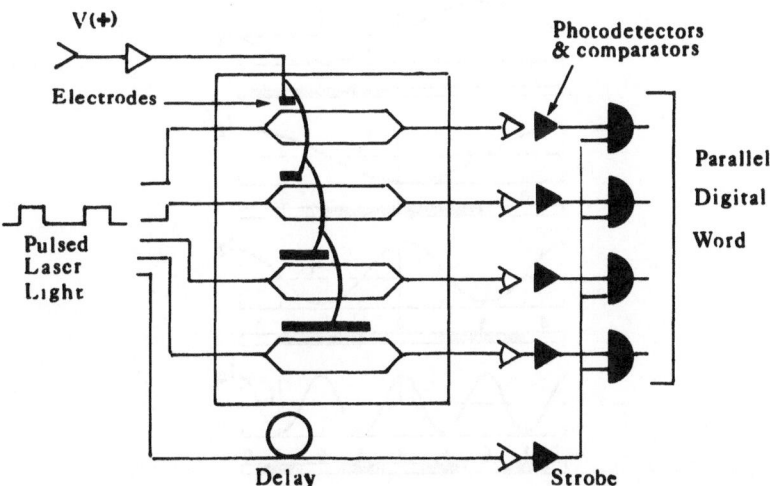

Figure 4 - Integrated optical A/D converter

very compact means of treating Gigahertz bandwidth signals at a
cost that is projected to be an order of magnitude lower than
that of competitive techniques (3).

3.2 – Integrated optical A/D converters

A/D converters are extremely important interfaces between the
analog signals generated by various sensors and digital data handling devices. Integrated optics offers a means of realizing devices
that should operate considerably faster than their existing, purely
electronic counterparts (4). A schematic of such a device is shown
in Figure 4.

An array of identical interferometric modulators is realized on
an electrooptic crystal such as lithium niobate. Each waveguide is
excited by a laser and the analog signal voltage is simultaneously
applied to all the electrodes, which have lengths given by
$L_n = 2^{n-1} L_1$, n = 1, 2, 3. The light from each modulator is detected,
amplified and compared to a fixed threshold intensity, thereby generating a "one" or a "zero" for each bit (Fig. 5).

An experimental verification of this principle has been given
using 1-3V input signals for bandwidths of up to 200 Mhz. By using
100ps laser pulses at the input for both the treatment and pulse
strobe signal such a device should be capable of treating 1 Gword/s
with 6 bit precision, a considerable speed advantage over existing
devices.

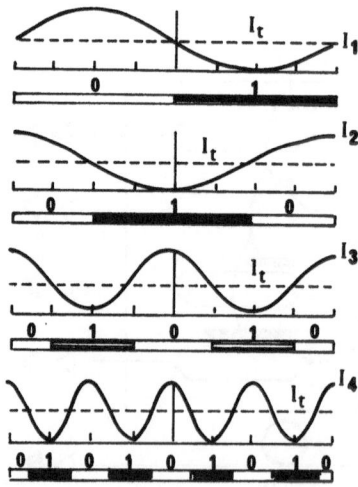

Figure 5 – Comparator signal with the integrated optical A/D
           converter

3.3 - Integrated optical logic devices

Integrated optical logic devices offer the possibility of gate delay times that are about an order of magnitude lower (20-40 psec) then the fastest electronic circuitry available (5) with equivalent power dissipation levels (~ 50 pj).

One possible configuration for a simple device is shown in Fig. 6. The device is an interferometric modulator fabricated in an electrooptic material such as lithium niobate. Light coupled into the straight section is equally divided into the two arms. The phase shift of the light in each of the two symmetric arms is controlled by the applied voltages. If the relative phase shift is 2πN radians, where N = 0, $\pm$ 1, $\pm$ 2..., the light beams arriving at the output will interfere constructively and excite the output guide. If the relative phase shift is an odd number of π radians however, the light will interfere destructively at the output and no light will enter the output guide. Therefore, if one designs the device so that a particular voltage, Vo, produces a π phase shift, light will be transmitted only if the number of inputs at Vo (the binary "one") is even. With a CW optical input provided it acts, therefore, as an even parity generator. By applying a DC bias of Vo to an extra electrode the device becomes an odd parity generator. Interconnections of such devices can provide practically any desired logic function.

When using such devices the circuit speed will be determined by the electronics at the detection end where the optical signal is reconverted to an electrical signal. Whenever it is feasible, therefore, to perform a series of operations with the data maintained in the optical form before reconversion, the optical circuit can provide a considerable speed advantage.

3.4 - Integrated Optical Bistable Devices

Optical bistable devices have interested research workers for a number of years (6,7). The early proposals for such devices, however, were based on the use of purely optically-induced non-linearities which require extremely high optical power densities. More recently Smith (8) suggested the use of an electrooptically

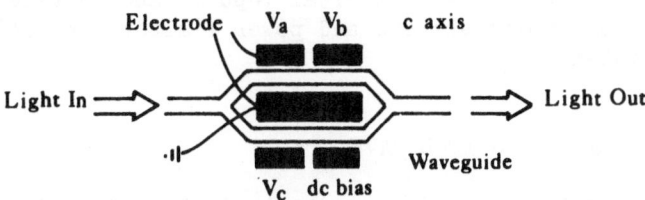

Figure 6 - Integrated optical interferometric optical logic device

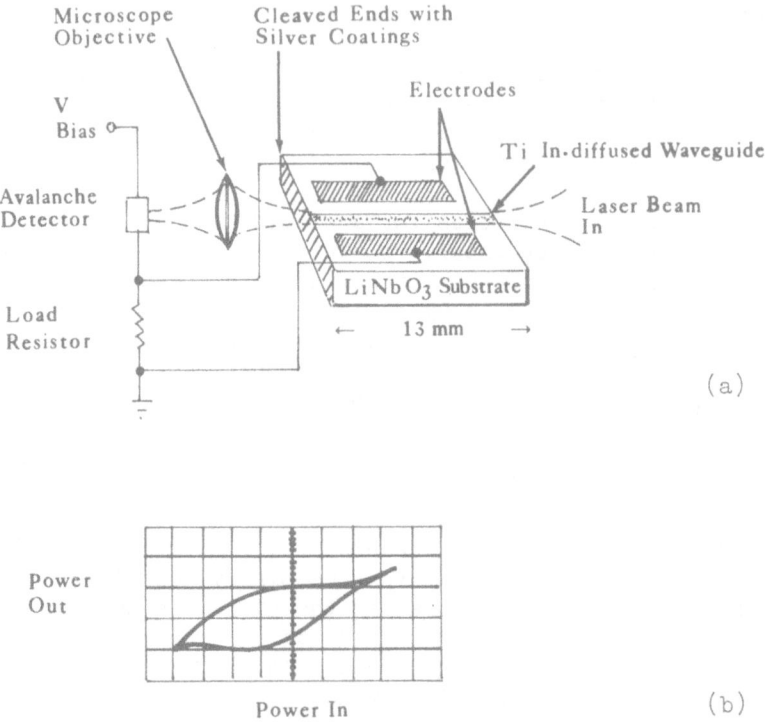

Figure 7 - a) Integrated optical bistable device
            b) Observed hysteresis

induced non-linearity to overcome this problem and has now demonstrated an integrated optical version of such a device, which is
shown in Fig. (7a). The light at the output of the cavity formed by
the silvered cleaved crystal faces is detected and amplified by the
avalanche detector. The electrical signal is then fed back to the
electrodes which control the optical length of the cavity via the
electrooptic effect. The hysteresis, and hence bistability, observed
with this configuration is shown in the lower part of Fig. (7b).
With this integrated optical device, nanosecond switching times with
picojoule energy dissipation become possible with switching controllable by either optical or electrical inputs. Such devices could be
used as efficient pulse shapers and power limiters in optical communication systems.

4 - SCIENTIFIC USE OF GUIDED WAVE OPTICS

     In addition to the various technological utilizations of guided
optical wave devices we shall, hopefully, see an evergrowing use of

the associated techniques by scientists in various fields. In the
following sections we shall describe two such examples of technological "spinoff."

4.1 – Absorption Spectroscopy Using Integrated Optics

Absorption spectroscopy is a technique commonly used to measure
the concentration of various chemicals in nonabsorbing solvents.

This technique, however, does not work for solutions which
scatter light since it is practically impossible to separate attenuation due to scattering from that due to absorption.

Integrated optics, however, provides a means of measuring absorption in certain cases with an immunity to scattering (9). It has
been shown, for example, that it is possible to measure the absorption due to small bilirubin molecules in whole blood while avoiding
scattering by red blood cells. The device used to do this is shown
in Figure 8. The guided light in this configuration does not "see"
the red blood cells since they are about two orders of magnitude
larger in diameter than the penetration distance of the evanescent
wave into the sample. Experiments performed with this device have
shown that scattering due to red blood cells was less than 0.01db/cm.
Clearly this technique will be valuable for any samples with relatively high optical density and scatterers which are a wavelength
or larger in diameter.

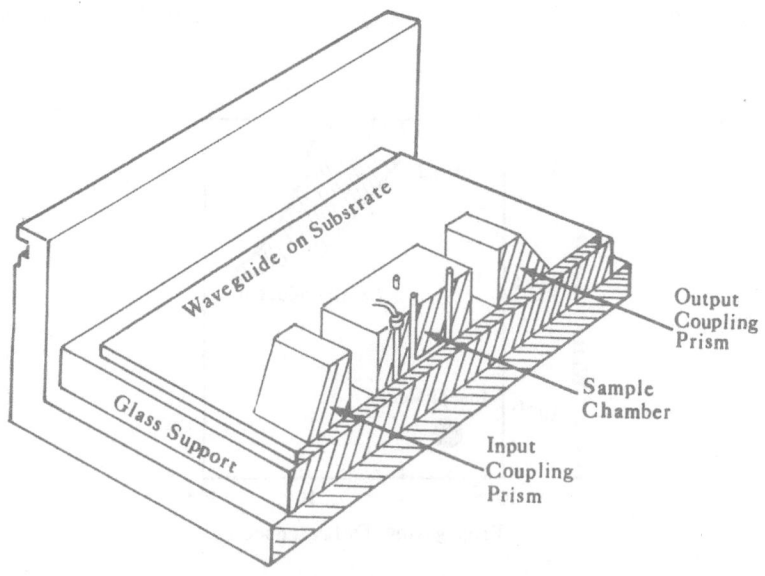

Figure 8 – Waveguide spectrometer

4.2 - Superconducting Weak Links Induced by Optical Guided Waves

Superconducting weak links are a type of Josephson junction. These junctions consist of two superconducting regions separated by a narrow region exhibiting no, or weak, superconductivity. Such junctions exhibit macroscopic quantum interference effects which give rise to such phenomena as rapid current oscillations ($\nu = 484$Ghz/mv) when a DC voltage is maintained across them. They are used in magnetometers and show promise for use as extremely rapid computer switching elements. If one considers a plot of power dissipation versus gate delay time, as shown in Figure 9, we see that such superconducting devices offer a considerable power dissipation advantag over electrical, or projected integrated optical devices such as those discussed in this article.

Several authors have discussed the possibility of using electrons (10, 11), microwaves, or light to locally weaken the superconductivity of a superconducting film and thereby realize dynamically controllable weak links. The experiment I shall now describe demonstrates that optical guided waves provide an excellent means of accomplishing this(12)

The essential problem is to create a very narrow region of weak superconductivity to allow tunneling of the superconducting wave functions through this region. To do so we have used the configuration shown in Figure 10. A narrow lead (10μ) strip is formed upon a two micron wide optical waveguide realized by ion exchange in a glass substrate. When the guide is excited with the sample held at cryogenic temperatures (below $\sim$ 7°K) as shown in Fig. 11, the evanescent portion of the guided wave locally weakens the superconduc-

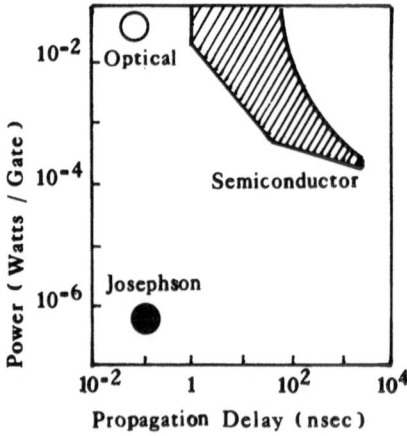

Figure 9 - Power dissipation versus gate propagation delay for various classes of devices

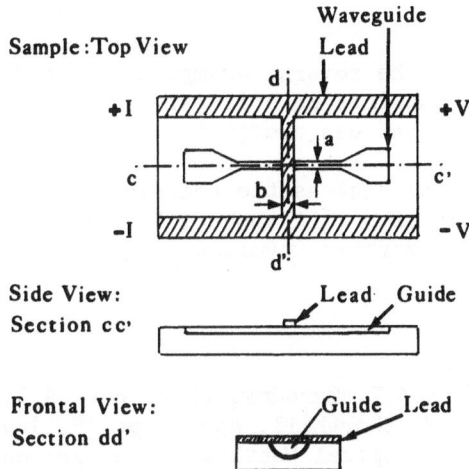

Figure 10 - Sample configuration of an optically induced
superconducting weak link (Josephson junction)

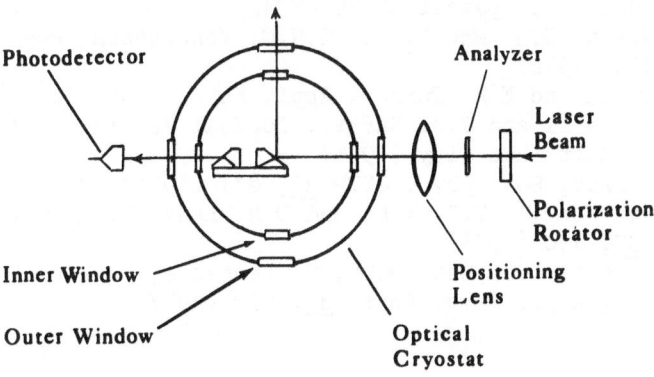

Figure 11 - Experimental arrangement for the optically induced
weak link

tivity in the contacted region and, for a certain range of optical
power levels (corresponding to several μwatts of absorbed power)
Josephson junction behavior is observed. At lower optical power
levels the entire lead strip is superconducting and at higher levels
the strip becomes normally resistive. This demonstrates, therefore,
that it is possible to realize dynamically controlled junctions by
this means.

5 – CONCLUSION

It is hoped that the several examples of "exotic" uses of guided wave optics outlined in this paper give some idea of the scope of the fields that can profit from these techniques.

I hope it is clear that as the techniques become easier to manipulate many more uses will appear, limited only by the originality of the research workers involved.

REFERENCES

1) – H.R. Bilger and A.T. Zarodny, Phys. Rev. A 5, 591, (1972)
2) – V. Vali and R.W. Shorthill, Appl. Opt 15, 1099 (1976)
3) – T. Galliorenzi, Topical Meeting on Integrated and Guided Wave Optics, Salt Lake City (1978)
4) – H.F. Taylor, Topical Meeting on Integrated and Guided Wave Optics, Salt Lake City (1978)
5) – H.F. Taylor, Appl. Opt 17, 1493 (1978)
6) – H. Seidel, U.S. patent n° 3610731
7) – H.M. Gibbs, S.L. McCall and T.N.C. Venkatesan, Phys. Rev.Lett. 36, 1135 (1976)
8) – P.W. Smith and E.H. Turner, Appl. Phys. Lett. 30, 280 (1977)
9) – G.L. Mitchell and J.H. Harris, Topical Meeting on Integrated Optics, Salt Lake City (1976)
10) – A.F. Volkov, Sov. Phys. JETP 33, 811 (1971)
11) – Ting-Wah Wong, J.T.C. Yeh, and D.N. Langenberg, Phys. Rev. Lett. 37, 150 (1976)
12) – A. Gilabert, D.B. Ostrowsky, C. Vanneste, M. Papuchon and B. Puech, Appl. Phys. Lett. 31, 590 (1977)

NONLINEAR EXPERIMENTS WITH Ti DIFFUSED LiNbO$_3$ OPTICAL WAVEGUIDES

A. Neyer, W. Sohler and H. Suche

Institut für Physik, Universität Dortmund

D-4600 Dortmund, Germany

Linear optics is based upon the fundamental relation between the polarization $\vec{P}$ of a crystal and an applied electric field $\vec{E}$:

$$\vec{P}(t) = \chi\, \vec{E}(t) \qquad\qquad \chi : \text{susceptibility}$$

However, if the amplitude of this field grows, higher order terms have to be taken into account, which are responsible for nonlinear effects [1]:

$$P_i(t) = \chi_{ij}\, E_j(t) + \chi_{ijk}^{(2)}\, E_j(t)\, E_k(t) + \ldots$$

The nonlinear susceptibility tensors have generally very small components, so that a restriction to the second order term (a third rank tensor) is in most cases sufficient. By symmetry considerations one can show that $\chi_{ijk}^{(2)}$ is nonvanishing only in crystals which lack an inversion center. Thus only such crystals have an electrooptic (Pockels-) effect or can be used for second order parametric interaction of optical waves. In the first case a modulating RF-field with frequency Ω and an optical field with frequency ω interact e.g. in a phase modulator and give rise to nonlinear polarization terms of frequency $\omega\pm\Omega$. These are the source terms for the modulation sidebands. In the case of parametric coupling high optical fields interact. The simplest process is the interaction of two fields of the same time dependence ω. Again a nonlinear polarization occurs which is now the source term of an optical field of doubled frequency. This is called second harmonic generation (SHG) which we have studied in detail in optical waveguides. Mathematically, the nonlinear interactions are treated by coupled mode theory [1,2].

Integrated optics is very attractive to study nonlinear proces-

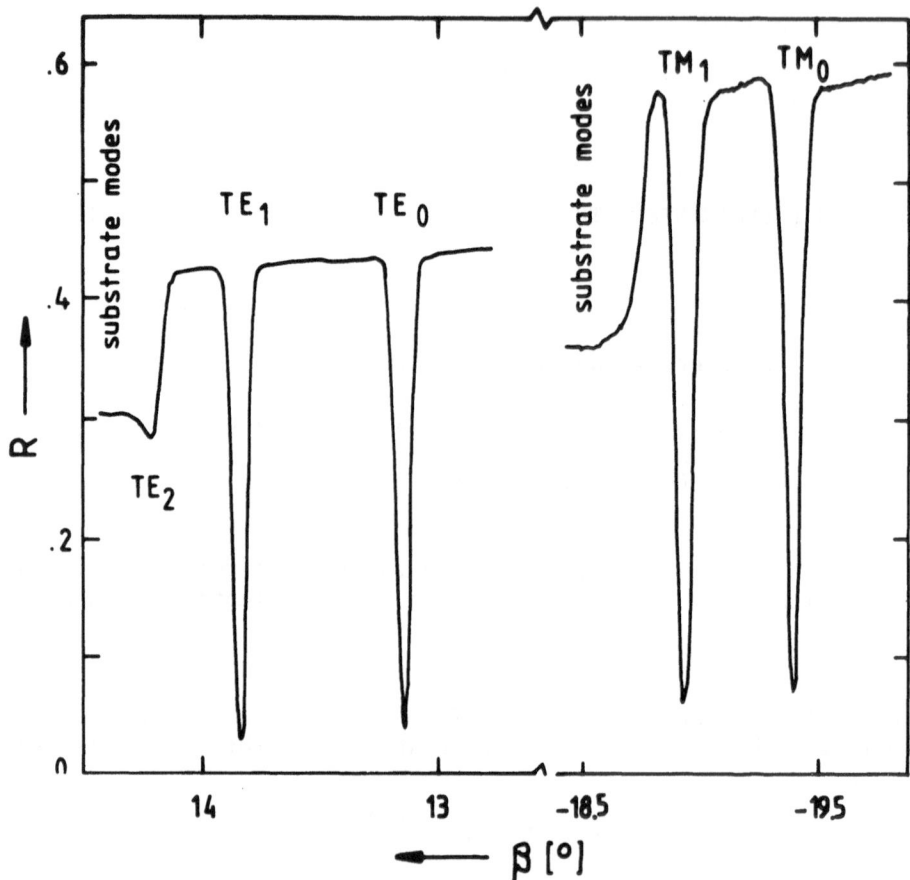

Fig.1: Prism coupling into a Ti:LiNbO$_3$ optical waveguide: internal reflectivity R of the input prism plotted versus the angle of incidence β at the outer prism face (see also the schematical drawing in Fig.3). From the resonances the coupling efficiencies in the various modes can be derived (e.g. 91% in the TE$_1$-mode [3]). λ = 0.63μ.

ses. Especially, the necessary high optical power density is easily achieved in waveguides e.g. by prism coupling (see Fig.1). It can be maintained and guided over large interaction lengths without limitations by diffraction and beam walk off effect as in the bulk. The necessary high RF-fields require only small voltages and power due to the small interaction volume of integrated devices.

    As an example we present a cutoff modulator in a Ti diffused LiNbO$_3$ optical strip guide [4]. In the upper part of Fig.2 a schematical drawing is shown. The actual modulator is the 1mm long, 3μ wide strip guide. Its index of refraction can be lowered by the electro-

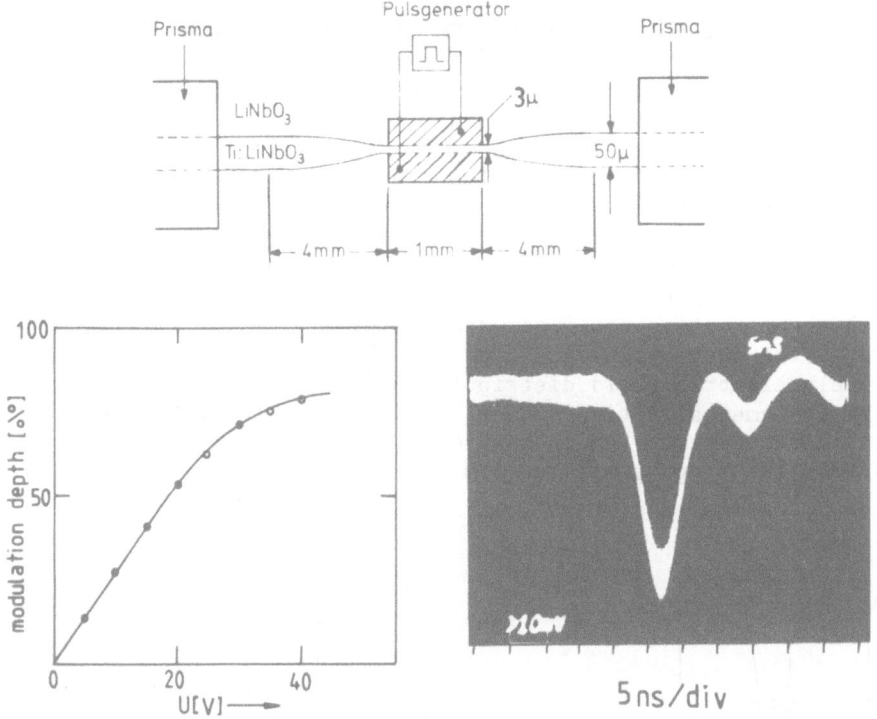

Fig.2: Cutoff modulator with a Ti:LiNbO$_3$ strip guide. Upper part: schematical drawing of the modulator. Lower part: modulation depth versus switching voltage (left); optical response to high speed modulation with 5V pulses (right).

optic effect in the region between the electrodes. If the dimensions and index of refraction of the guide are properly chosen, its cutoff frequency can be electrooptically increased so much that the propagation of an optical mode of given frequency is prevented. The mode couples out into the substrate. In this simple way an intensity modulation is obtained. The cutoff modulator is a typical device of integrated optics which has no counterpart in bulky, conventional optics. Its specific advantage is the small electrode structure and therefore small capacity resulting in a high speed (several GHz bandwidth) modulation capability. First experimental results are shown in the lower part of Fig.2. On the left the modulation depth is plotted versus the switching voltage. On the right an example of high speed operation with only 5V pulses is shown. (The effective switching voltage was 10V, as the modulator as pure capacity reflects nearly perfectly the incoming electrical pulse). The observed risetime of about 3nsec was determined by the restricted bandwidth of preamplifier and oscilloscope and not by the modulator.

    As a nonlinear experiment at optical frequencies we present SHG
in Ti diffused LiNbO$_3$ waveguides. The fundamental beam of frequency
ω (near infrared) and TM polarization is coupled into the waveguide
where it excites a harmonic wave (see the schematical drawing of
Fig.3). In the case of phasematching the harmonic wave adds up with
the correct phase in every place of the interaction region and can
thus grow to a large intensity. The conversion efficiency η of SHG
in a waveguide is given by [5]:

$$\eta \sim \chi^2_{eff} \ L^2 \ S \ P/A \quad \text{with} \quad S^{(m,n)} = \int_{-\infty}^{+\infty} H_y^{(m,\omega)}(x) \ H_y^{(m,\omega)}(x) \ E_y^{(n,2\omega)}(x) \ dx$$

χ_{eff} is the effective nonlinear susceptibility, L the interaction
length, S the overlap integral, H_y and E_y the transverse magnetic respectively electric field distribution of the interacting modes and
P/A the fundamental optical power density. Remarkable is the true
quadratic dependence of η on L which allows to take advantage of large interaction lengths without limitations by diffraction or beam

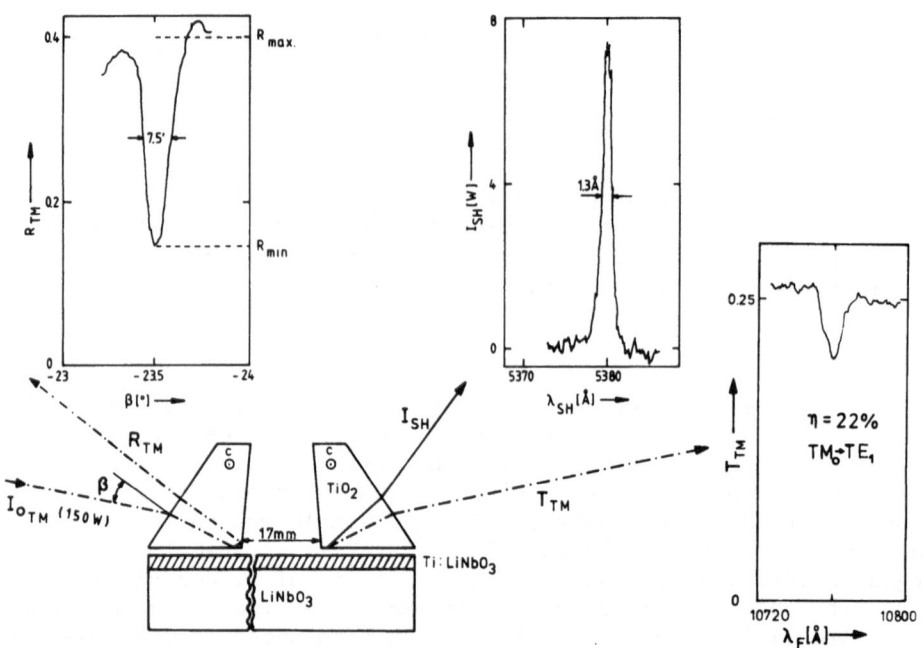

Fig.3: Schematical experimental arrangement for second harmonic generation in a Ti diffused LiNbO$_3$ optical waveguide together with results of measurements. Left inset: internal reflectivity of the
coupling prism versus the angle of incidence. Lower right inset:
transmission of the fundamental beam near the phasematch wavelength.
From the resonance a conversion efficiency of 22% is derived. Upper
right inset: Harmonic power versus wavelength.

walk off. This attractive feature of guided wave optics has been fully exploited for nonlinear interactions in optical fibers |6|. Our results of SHG in a 17mm long $Ti:LiNbO_3$ planar waveguide, produced by indiffusion of 380 Å Ti at $1040^{\circ}C$ for 5h, are shown in Fig. 3. In the lower right inset the transmission T of the fundamental beam is plotted versus the wavelength. In the case of phasematching the transmission is lowered due to coupling of energy to the corresponding harmonic mode. From the resonance a conversion efficiency of 22% is derived [7]. The power of the second harmonic output beam is plotted versus its wavelength in the upper right inset (λ_F and ß const.). The high conversion efficiency was obtained due to the advantages of integrated optics. With conventional optics at most 6% conversion efficiency would be possible at the same input power level. It was 45W, derived from the reflectivity curve in the left inset. To avoid optical damage in the waveguide, the experiment was performed in a pulse mode operation (5pps) with a tunable optical parametric oscillator as the fundamental light source.

Phasematching in optical waveguides can be obtained as in the bulk by adjusting the propagation direction or the wavelength of the pump beam. However, there are further possibilities of phasematching in waveguides which do not exist in the bulk. Taking advantage of mode dispersions, phasematching becomes possible for various modes at discrete wavelengths at practically the same crystal orientation. This was demonstrated with a deeper waveguide supporting 5 TE modes in the visible and 2 TM modes in the near infrared. Therefore 10 combinations of phasematched mode coupling should be possible; 9 of them were observed. The results are given in Table 1. It is remarkable that the fundamental wavelength may be smaller than in the bulk case as a consequence of the mode structure in guided wave optics. The SH-intensity is mainly determined by the overlap integral S.

Table 1: Phasematch wavelengths (harmonic wave) of a SHG-experiment in a $Ti:LiNbO_3$ optical waveguide made by indiffusion of 1180 Å Ti at $1040^{\circ}C$ for 5h. Phasematch wavelength for the bulk case: 5365 Å. (v=very, h=high, m=medium, l=low). (I=intensity).

	λ_{SH}(Å)	I			λ_{SH}(Å)	I
$TM_0 \rightarrow TE_0$	5760	v h		$TM_1 \rightarrow TE_0$	5495	v l
$\rightarrow TE_1$	5527	h		$\rightarrow TE_1$	5428	m
$\rightarrow TE_2$	5428	l		$\rightarrow TE_2$	5377	m
$\rightarrow TE_3$	5360	m		$\rightarrow TE_3$	5314	v l
$\rightarrow TE_4$	5303	v l		$\rightarrow TE_4$	not observed	

Acknowledgement

We thank the Deutsche Forschungsgemeinschaft for financial support and the Nato for two scholarships (A.N. and H.S.).

References

[1]: A. Yariv, Quantum Electronics (John Wiley & Sons, Inc., New York, 1975)

[2]: e.g.: A. Yariv, "Coupled Modes", this summer school.

[3]: W. Sohler and H. Suche, to be published in Wave Electronics.

[4]: A. Neyer and W. Sohler, to be published.
     The electrooptic control of waveguide cutoff has been demonstrated for the first time in GaAs by D. Hall, A. Yariv and E. Garmire, Appl. Phys. Lett. 17, 127 (1970).

[5]: A. Yariv, IEEE J. Quantum Electron. QE-9, 919 (1973).

[6]: R.H. Stolen, "Fiber Raman Lasers", this summer school.

[7]: W. Sohler and H. Suche, Appl. Phys. Lett. in press (Sept. 78).

COHERENCE EFFECTS OF THE ELECTROMAGNETIC FIELD PROPAGATING IN A MULTIMODE OPTICAL FIBER

Bruno Crosignani and Benedetto Daino

Fondazione Ugo Bordoni
Istituto Superiore Poste e Telecommunicazioni
Viale Europa, Roma, Italy

1. INTRODUCTION

The behavior of the envelope of a pulse injected into a fiber is, in general, sufficient to characterize the electromagnetic field from the point of view of telecommunications. However, it is evident that the analysis of the field itself furnishes a much more detailed description and that its study gives more information on the fiber structure. Of course, the electromagnetic field is not practically observable at optical frequencies, its statistical behavior being described by a hierarchy of correlation functions. At the lowest significant order of this hierarchy one has the mutual coherence function (1)

$$\Gamma(\underline{r},\underline{r}',z,t,t') = \; < \underline{E}(\underline{r},z,t)\cdot\underline{E}^*(\underline{r}',z,t') > \; , \tag{1}$$

the symbol <.....> indicating ensemble average. We refer to a cylindrical fiber whose axis coincides with the z axis (see Fig. 1). The measurement of this quantity will be shown to furnish a method for determining the index profile in a fiber.

Another relevant quantity is furnished by the average, performed on a time interval to be chosen according to the effects one wishes to reveal,

$$\tilde{\Gamma}(\underline{r},\underline{r}',z,t,t') = \overline{\underline{E}(\underline{r},z,t)\cdot\underline{E}^*(\underline{r}',z,t')} \qquad . \tag{2}$$

In particular, $\tilde{\Gamma}$ reduces to Γ in a stationary case if the time interval is larger than all the typical fluctuation times of the field.

319

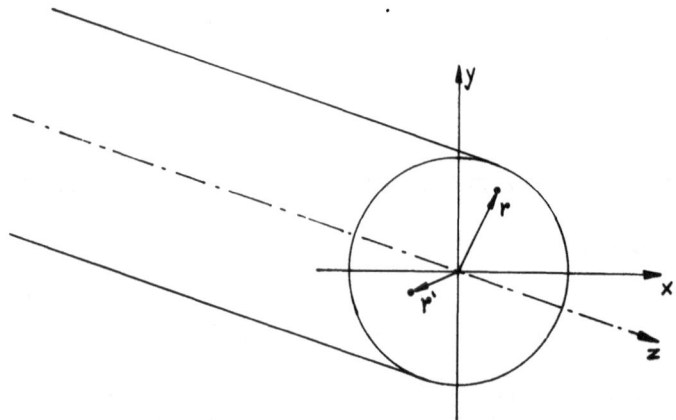

Fig. 1. Geometry of the Cylindrical Fiber.

2. MEASUREMENT OF THE NEAR-FIELD INTENSITY

For $\underline{r}=\underline{r}'$ and $t=t'$, $\tilde{\Gamma}$ is proportional to the average of the near-field intensity, that is

$$\tilde{\Gamma}(\underline{r},\underline{r},z,t,t) \propto \overline{I}(\underline{r},z,t) ,\qquad(3)$$

where $I(\underline{r},z,t)$ is the instantaneous intensity, that is the average of $\underline{E}(\underline{r},z,t)\cdot\underline{E}*(\underline{r},z,t)$ over a few optical periods (1).

When the fiber is considered from the point of view of tele-communications, the relevant quantity is the power from the whole fiber final section, which is an integral of the near-field inten-sity.

$$P(z,t) = \int \overline{I}(\underline{r},z,t)d\underline{r} .\qquad(4)$$

Let us look for the kind of information one can obtain from a detailed analysis of $\overline{I}(\underline{r},z,t)$ which is not contained in $P(z,t)$. To this end, let us write the electromagnetic field $\underline{E}(\underline{r},z,t)$ inside the fiber as the superposition of a certain number N of guided modes $\underline{E}_n(\underline{r})$, whose propagation constants are indicated by $\beta_n(\omega)$, with suitable expansion coefficients $c_n(\omega,z)$

$$\underline{E}(\underline{r},z,t) = \sum_{n=1}^{N} \underline{E}_n(\underline{r}) \int c_n(\omega,z)e^{-i\beta_n(\omega)z+i\omega t} d\omega .\qquad(5)$$

This can be rewritten identically in terms of the mode-field ampli-tudes $a_n(z,t)$ as

$$\underline{E}(\underline{r},z,t) = \sum_n \underline{E_n}(\underline{r})a_n(z,t) \quad , \tag{6}$$

where the z-dependence of the c_n's can take into account the possibility of mode coupling and of differential attenuation (2).

For evaluating the near-field intensity in a rigorous way, one has to introduce the flux of the complex Poynting vector $\underline{S}$,

$$\underline{S} = (1/2) \overline{\underline{E} \times \underline{H}^*} \quad , \tag{7}$$

through a small area σ of the final fiber section, which furnishes the power carried by the field through this area,

$$P^\sigma(z,t) = \text{Re}\int_\sigma \underline{e_z} \cdot \underline{S} \, d\underline{r} \quad , \tag{8}$$

where $\underline{e_z}$ is a unit vector in the direction of the z axis and Re means "real part of". Of course, the power carried through the whole fiber section is obtained by letting σ go to infinity. In this case, since the guided modes are mutually orthogonal, no correlation term is present in $P(z,t)$, which turns out to be the sum of the powers carried by each mode through the whole fiber section,

$$P(z,t) = \sum_n P_n(z,t) \quad , \tag{9}$$

where

$$P_n(z,t) \propto \overline{|a_n(z,t)|^2} \quad . \tag{10}$$

If one conversely considers $P^\sigma(z,t)$, its expression contains, besides the P_n's, cross-correlation terms between the various modes of the kind

$$\overline{a_n(z,t)a_m^*(z,t)} \quad . \tag{11}$$

These terms are important because they depend on the phases of the various complex mode-field amplitudes, which are clearly absent from the P_n's. The analysis of the cross-correlation terms furnishes a simple method for measuring modal dispersion.

The a_n's can be evaluated easily if one neglects material dispersion, that is if one approximates the propagation constant $\beta_n(\omega)$ as

$$\beta_n(\omega) \simeq \beta_n(\omega_o) + (\omega-\omega_o)/V_n \quad , \tag{12}$$

where

$$V_n = (d\beta_n/d\omega)_{\omega_o}^{-1} \tag{13}$$

is the group velocity of the n-th mode, and assumes for sake of simplicity, coupling and attenuation to be absent. Under the previous assumptions, one obtains

$$\underline{E}(\underline{r},z,t) = \sum_n \underline{E}_n(\underline{r})e^{(-i\beta_n(\omega_0)z+i\omega_0 z/V_n)} a_n(0,t-z/V_n) \ , \tag{14}$$

with

$$a_n(0,t-z/V_n) = (1/2p) \int \underline{e}_z \cdot \underline{E}(\underline{r},z=0,t-z/V_n) \times \underline{H}_{-n}^*(\underline{r})d\underline{r} \ , \tag{15}$$

having taken advantage of the orthogonality relation (2)

$$\int \underline{e}_z \cdot \underline{E}_n(\underline{r}) \times \underline{H}_m^*(\underline{r})d\underline{r} = 2p\delta_{mn} \ . \tag{16}$$

Assuming the boundary field at z=0 to be of the form

$$\underline{E}(\underline{r},z=0,t) = \underline{A}(\underline{r})e^{i\omega_0 t} f(t) \ , \tag{17}$$

the a_n's are proportional to f, that is

$$a_n(z,t) \propto f(t-z/V_n) \ . \tag{18}$$

Accordingly, the non-diagonal terms, which can be put into evidence by a near-field intensity measurement, are of the form

$$\overline{a_n(z,t)a_m^*(z,t)} \propto \overline{f(t-z/V_n)f^*(t-z/V_m)} \ . \tag{19}$$

Let us first consider the case of a stationary source possessing a coherence time

$$t_c = 2\pi/\delta\omega \ , \tag{20}$$

$\delta\omega$ being the bandwidth of the carrier. One then has

$$\overline{a_n(z,t)a_m^*(z,t)} \propto \overline{f(0)f^*(|z/V_n - z/V_m|)} \ , \tag{21}$$

which vanishes whenever the difference between the group delays $\tau_n=z/V_n$ and $\tau_m=z/V_m$ exceeds the coherence time, that is

$$|z/V_n - z/V_m| = |\tau_n - \tau_m| > t_c \ . \tag{22}$$

The progressive disappearance of these terms can be put into evidence (obtaining in this way an estimate of the order of magnitude of the difference between the nearest group delays) by detecting $P^\sigma(z,t)$ and noting the progressive vanishing of local rapid variations in the intensity (speckle-pattern) for decreasing values of t_c. As a matter of fact, one has

$$P^{\sigma}(z,t) = \sum F_{nn} \overline{|a_n(z,t)|^2} +$$

$$\sum_{n\neq m} \sum F_{nm} \overline{a_n(0)a_m^*(|z/V_n - z/V_m|)} \quad , \tag{23}$$

where the F_{nm} are suitable coefficients (3) such that the non-diagonal terms give rise to contributions which add or subtract locally on the fiber section to the background constituted by the diagonal terms (which are z-independent), thus giving rise to a speckle-pattern.

A source with a coherence time, that is a bandwidth, adjustable in a large range has been obtained in our case by employing a tunable dye-laser, which is able to span wavelength intervals as large as 200 Å, and taking an average over a time interval larger than the sweeping time of the laser. In fact, this is equivalent, for our practical purposes, to having a source with a bandwidth equal to the one spanned by the laser. The near-field intensity pattern, obtained with the experimental arrangement of Fig. 2 is shown in Fig. 3 for various values of the source bandwidth δf.

From the analysis of these figures one can infer a minimum group-delay difference of the order of 1 nsec.

Let us now consider the general case in which the source is not necessarily stationary. If, for example, the source exhibits slow phase variations (compared with the averaging time used in the experiment), that is

$$f(t) = e^{i\phi(t)}F(t) \quad , \tag{24}$$

F(t) representing the fast fluctuations of the source, one has

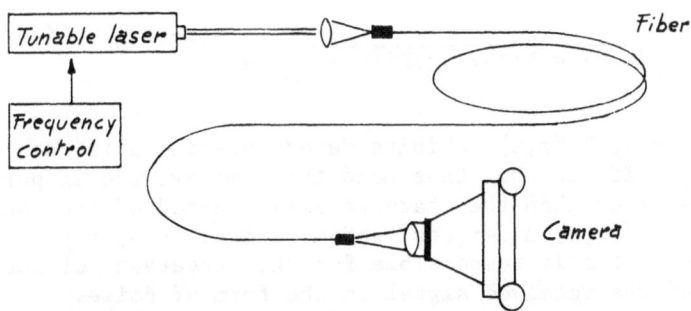

Fig. 2. Experimental Arrangement for Detecting Speckle-Patterns.

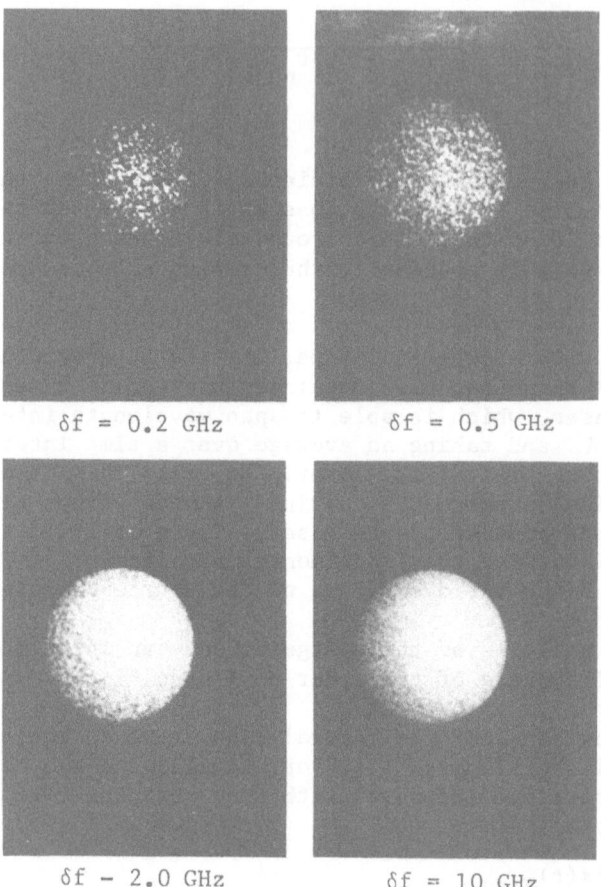

$\delta f = 0.2$ GHz $\delta f = 0.5$ GHz

$\delta f - 2.0$ GHz $\delta f = 10$ GHz

Fig. 3. Speckle-Patterns for Various Values of $\delta f = \delta\omega/2\pi$.

$$\overline{a_n(z,t)a_m(z,t)} \propto F(t-z/V_n)F^*(t-z/V_m) \, e^{i\phi(t-z/V_n)-i\phi(t-z/V_m)}$$

(25)

Accordingly, $P^\sigma(z,t)$ exhibits deterministic variations depending on $\phi(t)$. If, on the other hand the time average is performed on a scale shorter than the characteristic period of the source fluctuations, $P^\sigma(z,t)$ undergoes random fluctuations, a process which can be partially responsible for the appearance of amplitude modulation of the received signal in the form of noise.

Referring to the case described by Eq.(25), the tunable dye-laser already described in connection with the speckle-pattern experiment is able to furnish an optical carrier linearly modulated in frequency, so that

$$\phi(t) = \alpha t^2 \tag{26}$$

and

$$a_n(z,t)a_m^*(z,t) \propto e^{i\Omega_{nm}t} \tag{27}$$

with

$$\Omega_{nm} = 2\alpha \left| z/V_n - z/V_m \right| \quad . \tag{28}$$

Equation (27) implies that the time-dependent part of $P^\sigma(z,t)$ is the sum of various sinusoidal terms. The experimental set-up for observing this behavior is shown in Fig. 4 and the corresponding output, taken at various radial positions on the fiber exit-face, in Fig. 5, for various values of the modulation rate. The analysis of the variations of $P^\sigma(z,t)$ furnishes a powerful method for measuring group-delay differences as small as a few picoseconds (4).

3. MEASUREMENT OF THE COMPLEX DEGREE OF COHERENCE

Let us now consider the measurement of spatial coherence, that is the measurement of the complex degree of coherence given, in a stationary case, by

$$\gamma(\underline{r},\underline{r};z) = \tilde{\gamma}(\underline{r},\underline{r};z) = \overline{\underline{E}(\underline{r},z,t)\cdot\underline{E}^*(\underline{r},z,t)}/I(\underline{r})^{1/2}I(\underline{r}')^{1/2}, \tag{29}$$

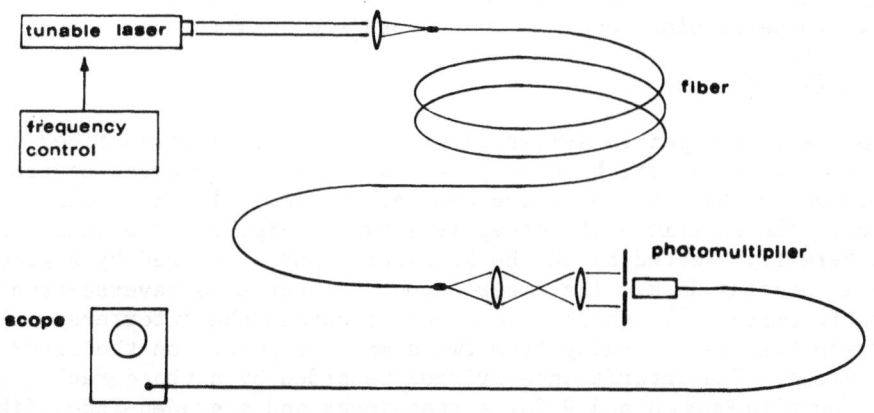

Fig. 4. Experimental Set-Up for Measuring $P^\sigma(z,t)$.

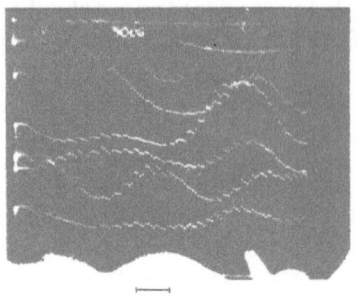

1.5 GHz

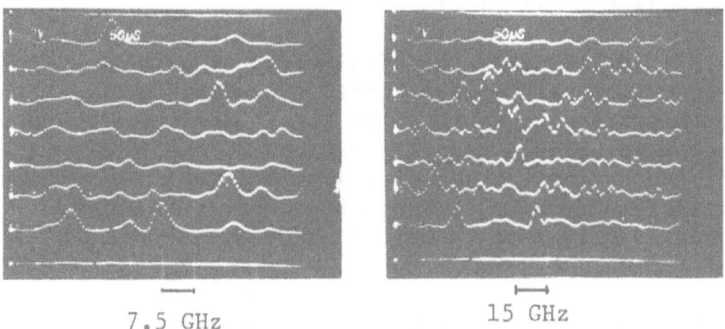

7.5 GHz   15 GHz

Fig. 5. Scope Displays of $P^\sigma(z,t)$ at Various Radial Positions.

with

$$I(\underline{r}) = \overline{\left| E(\underline{r},z,t) \right|^2} \quad . \tag{30}$$

Under certain conditions the investigation of γ leads to a simple method for observing the index profile of an optical fiber (5). In particular, one has to measure the quantity

$$\gamma(x,y,-x,y,z) \tag{31}$$

relative to two points symmetric with respect to a given fiber diameter (see Fig. 6), by employing a spatially incoherent source, which means that the modes are completely uncorrelated at the fiber input. The experimental set-up is shown in Fig. 7. The fiber is incoherently excited by the He-Ne laser light scattered by a ground-glass rotating disk. The fiber output is sent to a reverse-front interferometer (6), which allows one to obtain the interference between the fields coming from two symmetric points on the fiber exit-face. The interference fringes recorded by a photographic film are shown in Figs. 8 and 9 for a step-index and a graded-index.fiber. It is possible to see that the fringes are present on a region whose x-dimension does not depend on y for the step-index fiber, while

this is not true for the graded-index fiber. This fact can be
better appreciated by looking at an enlargement of the picture
referring to the graded-index fiber (see Fig. 10).

A simple calculation shows that the complex degree of coherence
is strictly related to the index-profile law. To this end let us
consider the final fiber-section. At each point $\underline{r}$ the emitted radiation completely fills a cone (see Fig. 11), whose aperture is given
by

$$\theta_o(r) = \{n^2(r) - n_c^2\}^{1/2} , \tag{32}$$

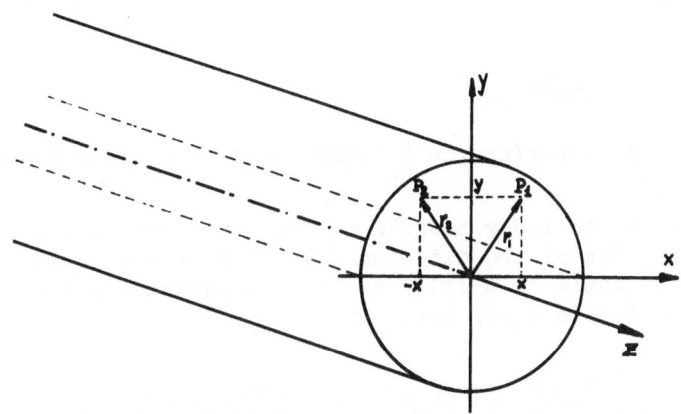

Fig. 6. Experimental Geometry for Measuring $\gamma(x,y,-x,y,z)$.

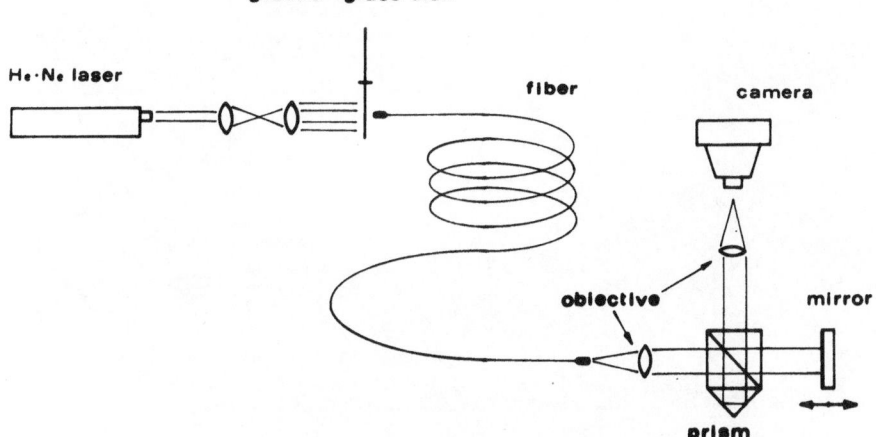

Fig. 7. Experimental Set-Up for Observing the Interference Fringes.

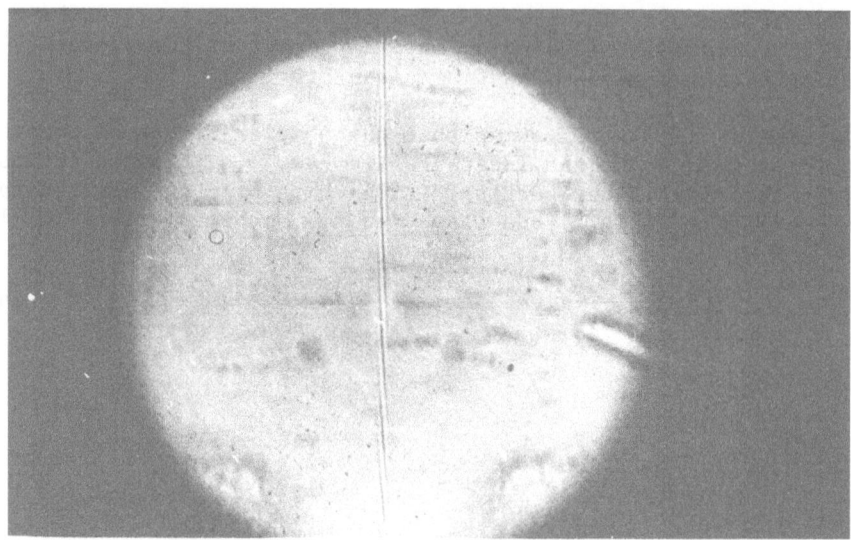

Fig. 8. Interference Fringes for a Step-Index Fiber.

where n(r) is the index profile of the fiber and n_c represents the refractive index of the cladding (7), so that indicating the associated solid angle by $\Omega_o(r)$, one has the radiance $B(\underline{r},\underline{s})$ in the direction $\underline{s}$ inside $\Omega_o(r)$ given by

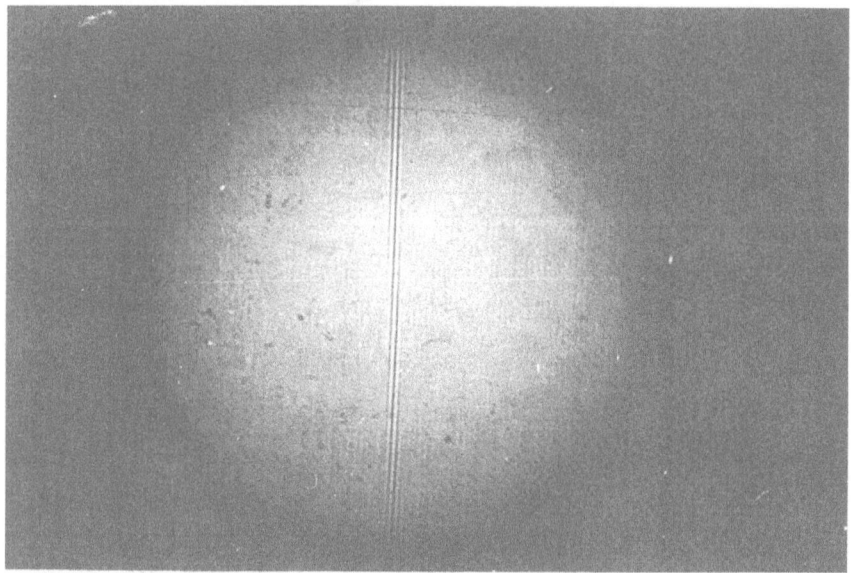

Fig. 9. Interference Fringes for a Graded-Index Fiber.

Fig. 10. Enlargement of the Picture Shown in Fig. 9.

$$B(r,\underline{s}) = \begin{cases} I(r)/\Omega_o(r) & \text{for } \underline{s} \text{ inside } \Omega_o(r) \\[2em] 0 & \text{for } \underline{s} \text{ outside } \Omega_o(r) \end{cases} \qquad (33)$$

having made use of the relation (8)

$$I(r) = \int B(r,\underline{s})\cos\theta(r)d\Omega \qquad . \qquad (34)$$

On the other hand, it can be demonstrated that, for quasi-isotropic sources, the radiance is connected with the Fourier-transform of the complex degree of coherence by the relation (9)

$$B(\tilde{r},\underline{s}) = k^2\cos\theta(\tilde{r}) \ I(\tilde{r})\hat{\gamma}(k\underline{s}') \quad , \qquad (35)$$

where $\tilde{r} = (\underline{r} + \underline{r}')/2$, $k = \omega/c$ and

$$\hat{\gamma}(k\underline{s}') = (1/2\pi)^2 \int \gamma(\underline{\rho})e^{-ik\underline{s}'\cdot\underline{\rho}}d\underline{\rho} \quad , \qquad (36)$$

and where $\underline{s}'$ is obtained by projecting $\underline{s}$ in a plane orthogonal to z and γ depends on the coordinate difference. This allows one to evaluate $\hat{\gamma}(k\underline{s}')$ in our geometry (see Fig. 11), thus obtaining

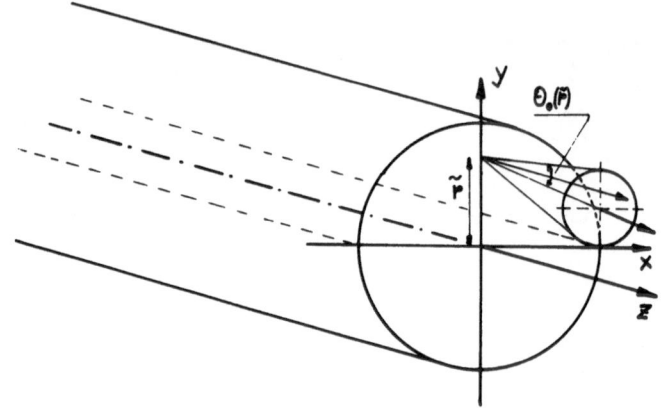

Fig. 11. Emitted Radiation Cone at $\tilde{r}$.

$$\hat{\gamma}(k\underline{s}') = \begin{cases} 1/\{k^2\Omega_o(\tilde{r})\} & \text{for } \underline{s} \text{ inside } \Omega_o(\tilde{r}) \\ \\ 0 & \text{for } \underline{s} \text{ outside } \Omega_o(\tilde{r}) \end{cases} \tag{37}$$

By Fourier-inverting this relation, one has

$$\gamma(2x,0)=\{1/k^2\Omega_o(y)\} \int_{\Omega_o(y)} e^{2iks'_xx} k^2 ds_x ds_y = 2J_1\{2kx\theta_o(y)\}/2kx\theta_o(y), \tag{38}$$

so that the locus of the point of constant complex degree of coherence is given by

$$x = \text{const}/2k\{n^2(y) - n_c^2\}^{1/2} , \tag{39}$$

showing that the width of the region in which the fringe visibility exceeds a fixed value is proportional to the inverse of the square root of the fiber index-profile.

REFERENCES

1. M. Born and E. Wolf, "Principles of Optics," Pergamon, Oxford, 1970.
2. D. Marcuse, "Theory of Dielectric Optical Waveguides," Academic, New York, 1974.
3. B. Crosignani, P. Di Porto, and C.H. Papas, J.O.S.A. 67, 1300 (1977).

4. B. Crosignani, B. Daino, and P. Di Porto, Appl. Phys. Letters 27, 237 (1975).
5. B. Daino, S. Piazzolla, and A. Sagnotti, "A New Method for Measuring the Index Profile of Optical Fibers," presented at the Fourth European Conference on Optical Communication, Geneva, September, 1978.
6. M. Carnevale and B. Daino, Optica Acta 24, 1099 (1977).
7. J.A. Arnaud and R.M. Derosier, Bell Syst. Tech. J. 55, 1489 (1976).
8. E.W. Marchand and E. Wolf, J.O.S.A. 64, 1219 (1974).
9. W.H. Carter and E. Wolf, J.O.S.A. 67, 785 (1977).

A COMPARISON OF SINGLE-MODE AND MULTIMODE FIBRES FOR LONG-

DISTANCE TELECOMMUNICATIONS

W. A. Gambling and H. Matsumura

Department of Electronics

University of Southampton, Southampton SO9 5NH, England

When research on optical fibre communications was first started in 1966[1] the main interest was in single-mode fibres since it was thought that the bandwidth of step-index fibres, the only type under consideration at the time, would be rather limited. Subsequently, with the development of SELFOC and other types of graded-index fibres, a high degree of equalisation of the group velocities of the fibre modes became possible so that the bandwidth available with multimode fibres was greatly increased. Interest then shifted away from single-mode fibres because of the difficulties of handling, launching, jointing and fabrication and also because the lifetimes of semiconductor lasers, the only small and relatively efficient sources possible, were not high. However the limits of bandwidth which are theoretically possible with an optimum refractive-index distribution are difficult to achieve in practice and the technology has now advanced to the stage where single-mode fibres may not be as difficult to incorporate into a practical system as was first thought. A renewed interest has therefore been shown in small-core fibres and a comparison of their properties with those of multimode fibres is presented here.

Attenuation

Transmission loss due to the fibre material itself is no longer a serious problem. Attenuations in both single-mode and multimode fibres have been reduced[2,3], in the laboratory at least, to levels below 1dB/km with minima of 0.5dB/km at a wavelength of 1.27μm. In single-mode fibres the loss in the region of cladding adjacent to the core is important since it carries an appreciable proportion of the transmitted power. A minimum thickness of high-quality cladding of about 7 times the core radius[4] is necessary to keep the

effect of a lossy jacket below 1dB/km. Nevertheless the total
volume of ultra-low loss material required is much less than in
the case of the multimode fibre. A disadvantage is that mode
conversion or mode scattering, which can be caused by inhomogeneities
or core/cladding interface imperfections, including microbending,
cause an additional power loss in the core of single-mode fibres
but has no effect on the bandwidth whereas the opposite may be the
case with multimode fibres depending on the degree of excitation.
Thus in multimode fibres which are underexcited, due to the
launching conditions or because of the loss of higher-order modes
during propagation, mode conversion is likely to produce additional
bound modes so that the total transmitted power is unchanged but
the bandwidth may be decreased. Careful choice of the numerical
aperture (NA) is necessary[5,6] with the single-mode fibre to
minimise such loss due to weak confinement and values of 0.1 to
0.12 are preferred. With larger values the core diameter becomes
too small to be practicable.

Fabrication

Both vapour deposition and double-crucible methods can be used
for the fabrication of either type of fibre although the former
method is much the more flexible. With multimode fibres chemical
vapour deposition can produce a desired refractive-index profile
so as to minimise the group velocity spread between modes. Either
method is suitable, in principle, for making single-mode fibres
which generally can be produced faster and more cheaply than the
graded-index multimode fibres. Particularly with vapour deposition
much longer lengths can be made for the same quantity of deposited
material. In the double-crucible process the whole of the cladding
must be of the highest quality and not simply that region nearest
the core so that the quantity of extremely pure glass required is
much greater.

Limiting core diameter for single-mode operation

Dielectric fibres can sustain one or more modes depending on
the ratio of core radius a to wavelength λ and the relative
refractive-index difference $\Delta = (n_1 - n_2)/n_2$ between core and
cladding. Single-mode operation requires[7], approximately, that:

$$(ka)\{2 \int_0^1 [n^2(R) - n_2^2]RdR\}^{\frac{1}{2}} < 2.405 \qquad \ldots (1)$$

where $k = 2\pi/\lambda$, $R = r/a$ is the normalized radial co-ordinate and
n(R) the radial variation of refractive index. With a constant
refractive index in the core this simplifies to

$$V = (ka)(n_1^2 - n_2^2)^{\frac{1}{2}} < 2.405 \qquad \ldots (2)$$

where V is the normalized radius or normalized frequency of the
fibre. During fabrication some diffusion of material occurs at

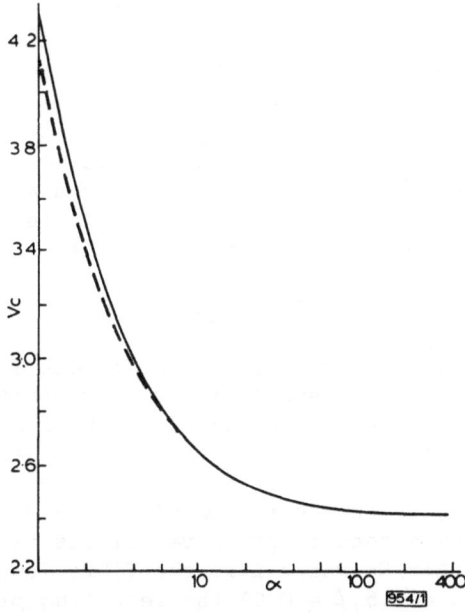

Fig. 1   Variation with profile parameter α of cut-off
frequency V_C in single-mode graded-index fibre

the core/cladding interface so that there is a radial variation of
refractive index and since the guidance factor of the fibre is
thereby decreased the normalized frequency is increased.   An
interesting class of profiles is represented by

$$n(R)^2 = n_o^2 (1 - 2\Delta R^{\alpha})   R < 1$$
$$= n_o^2 (1 - 2\Delta) = n_2^2   R > 1   \qquad \cdots (3)$$

In multimode fibres a suitable value of α is chosen so as to
minimise waveguide dispersion; usually $\alpha \simeq 2$.   The same equation
has been used[8] to estimate the effect on the normalized cut-off
frequency V_C of diffusion at the core/cladding interface in the
single-mode fibre, with the result shown in Fig.1.   The solid curve
gives the exact result while the dashed curve is obtained from the
approximate eqn(1).   It can be seen that as the profile becomes
more rounded V_c rises, being about 3.6 for a square-law profile.
In fibres made by vapour deposition an evaporation of some of the
constituents may take place during the preform collapsing stage
and this gives rise to a depression, or dip, in the refractive
index at the centre of the core.   Because the degree of guidance
is somewhat decreased, a rise in cut-off frequency again occurs[9]
but to a lesser extent than for a comparable loss of the same
constituent at the edge of the core.   The reason may lie in the fact

that the second-order (LP_{11}) mode has a zero of intensity at the
core centre.

Multimode fibres typically have core diameters in the range
50∼80μm and attempts are being made to standardise at about 60μm.
If microbending loss is to be restricted to an acceptable value
then the core diameter of a step-index single-mode fibre must not
be greater than about 4μm. However, with a graded profile and
operation at λ = 1.3μm the corresponding limit is 10μm or even more
so that the difference compared with multimode fibres is much reduced.

Bandwidth

As stated above, the bandwidth of multimode fibres is
primarily determined by the spread in group velocities in the
various propagating modes. This spread can be minimized by choice
of the appropriate refractive index distribution and for the so-
called α profile of eqn(3) the optimum value, in the absence of
material dispersion, is in the region of α = 2 depending on the
materials used. The spread in group velocities is then given
approximately by $n_o L \Delta^2 / 8c$ where L is the fibre length and c =
$3 \times 10^8 ms^{-1}$ and for n_o= 1.5, Δ = 0.01 the resulting predicted bandwidth
is a few tens of gigahertz over 1km. Unfortunately this bandwidth
has not been achieved in practice for a number of reasons. Firstly,
it is extremely difficult to realise the required refractive-index
distribution to the very high degree of accuracy[10] required.
Secondly, the radial variation in refractive index requires a

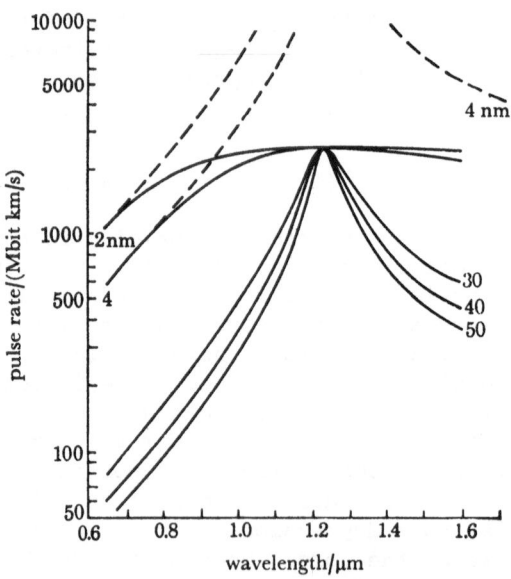

Fig. 2 Pulse rate for various source linewidths for
graded (solid) and single-mode (dashed) fibres

corresponding variation in composition which, in turn, implies a
variation in material dispersion. The result is that a degradation
in pulse dispersion is observed even though the radial variation of
material dispersion (i.e. profile dispersion) can be taken into
account[11]. In a single-mode fibre group delay dispersion is absent
and, to a first approximation, the refractive-index distribution is
immaterial. The limiting dispersion parameters are then mode
dispersion of the HE_{11} mode and material dispersion. With a monochromatic source bandwidths of hundreds of gigahertz over 1km become
possible although they have not yet been measured experimentally.
In practice the available bandwidth is somewhat reduced because of
the finite width of the semiconductor lasers available and there is
therefore great advantage to be gained by operation at a wavelength
where the material dispersion is zero[12]. For silica-based fibres
the optimum wavelength is near the range 1.25-1.30µm and the
advantage to be gained by operating in this region for both single-
mode and multimode fibres is illustrated by Fig.2.

A critical feature is, of course, the linewidth of the source
and a great improvement could be obtained at non-optimum wavelengths
for both types of fibre if semiconductor lasers could be produced
which operated stably in a single longitudinal mode, unaffected by
changes of driving current and temperature. If the problems posed
by material dispersion, and the resulting profile dispersion in
multimode fibres, can be overcome then waveguide dispersion becomes
dominant and in this respect single-mode fibres are clearly superior.
In fact single-mode fibres are essential if the maximum use is to
be made of the low attenuation which is now possible. For example,
if cables can be manufactured having a loss of 1dB/km then a

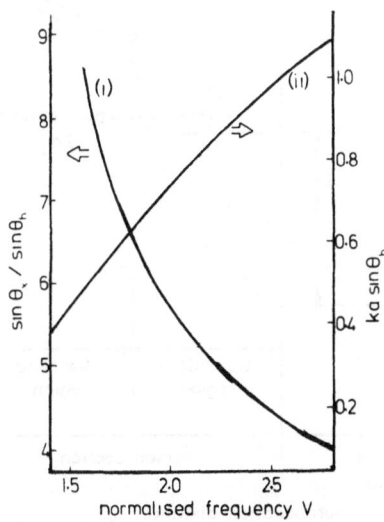

Fig. 3   Variation of θ_x and θ_h with V

repeater spacing of perhaps 50km becomes possible.   However with a
multimode fibre having a bandwidth x length product of 1GHz km,
which is still not easy to achieve in practice, then the total
system bandwidth is limited to $10^9/50$ = 20MHz.   In order to obtain
a system bandwidth of, say, 500MHz a bandwidth x length product of
25GHz is required and this can only be achieved with a single-mode
fibre.   It has not yet proved possible to measure the pulse
dispersion in single-mode fibres, but the best reported value[13,14]
so far of less than 0.1ns/km being limited by the time resolution
of the equipment used.

Characterisation

It is clear from the above discussion that two important
parameters of fibres are the refractive-index difference or profile,
from which the numerical aperture can be derived, and the core
diameter.   These can be obtained for multimode fibres by a number
of methods including interferometry[15], reflection measurements[16]
and near-field scanning[17].   With single-mode fibres the problems are
more severe since the degree of spatial resolution required is
comparable with the wavelength of measurement.

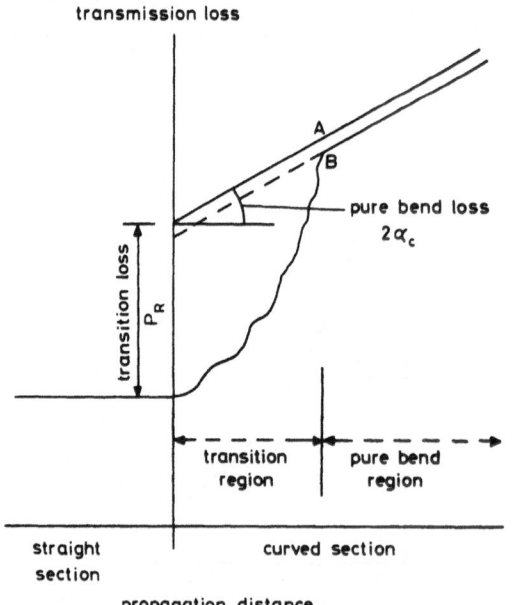

Fig.4 Loss near start of bend

One method has been suggested whereby the normalised cut-off frequency of the 2nd higher-order mode is measured by bending the fibre. However it has been shown[18] that this method gives an effective value of V which is much higher than the actual cut-off frequency. This is because microbending in the fibre can produce a high loss of the LP_{11} mode just above its cut-off frequency which increases the effective value of V. In practice we have observed single-mode operation in even short lengths of fibre having V = 2.8.

A simple method has been found[19] for determining a and Δ from the measurement of the far-field pattern at a single wavelength. It can be shown from theory, as well as experimentally, that in addition to the main bean the far-field pattern has several subsidiary lobes. The ratio $\sin\Theta_x/\sin\Theta_h$, where Θ_x is the angular width to the first minimum and Θ_h is the output angle at which the intensity has fallen to half of its central maximum, is an unambiguous function of V as is $k a \sin\Theta_h$. Therefore measurement of Θ_x and Θ_h gives V and from the next curve, $k a \sin\Theta_h$,the core radius a can be calculated as illustrated in Fig.3.

Another possibility with fibres fabricated by chemical vapour deposition is to make measurements on the preform and assume that

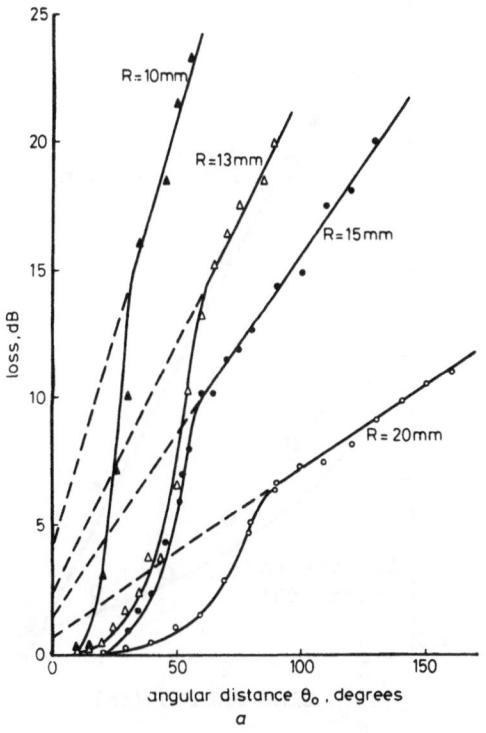

Fig.5 Measured total bend loss

only linear changes in geometry occur as the preform is drawn into
a fibre. This can only give an approximate result since some
diffusion between core and cladding can occur[19] during fibre drawing.
Other methods of characterising single-mode fibres reliably are
needed, particularly techniques for measuring the refractive index
profile.

Bending and Microbending Losses

        In general the radiation loss due to bends in a fibre arise
from two different physical mechanisms. One is the pure bend effect
whereby the velocity of the wave front at some distance from the
centre of curvature approaches that of an unguided wave and all
energy beyond that point is therefore lost[20]. The other mechanism
is the mode conversion which occurs at changes of curvature or,
more accurately, the inefficient coupling between the mode appropriate
to one degree of curvature to the equivalent mode at the following
curvature. At the junction between a straight and a curved fibre
the increase in transmission loss might be expected to follow curve
A in Fig.4. However the mode conversion or transition loss does

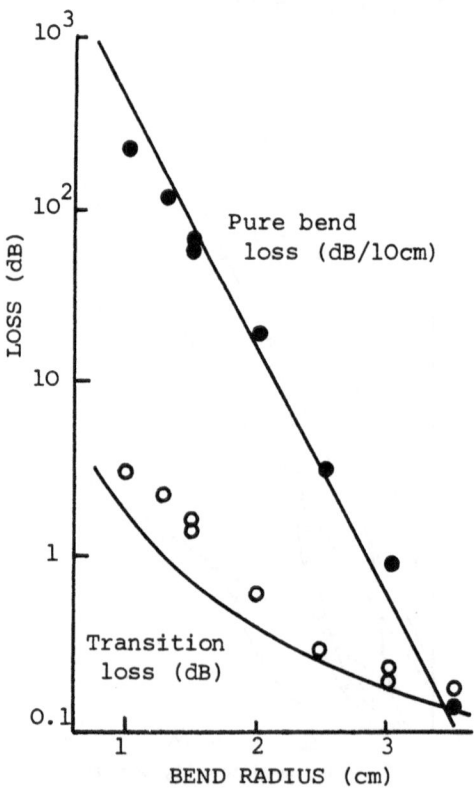

Fig.6   Measured and calculated
        bend losses

not take place instantaneously because the power coupled to the
radiation field leaks away gradually and in addition an abrupt
change of curvature is prevented by the mechanical stiffness of the
fibre.   Intuitively one might therefore expect[21] the increase in
loss with distance to take the form of curve B.   That this is indeed
the case is shown by the experimental measurements of Fig.5 which
were obtained[21] for a single-mode fibre of V = 2.4 and NA = 0.06.
The relative effects of pure bend loss[22] and transition loss[23,24]
are shown in Fig.6 as a function of radius of curvature.   It is clear
that in any practical situation both types of loss must be taken
into account.

    Microbending losses arise from very localized changes of
curvature and again both mode conversion and pure bend loss mechanisms
must be considered.   Representative microbend loss curves are given
in Fig.7 as a function of $2\pi a/\lambda$ for a multimode (dashed curve) and
a single-mode (solid curve) fibre.   Both curves are calculated for
$\Delta = 0.01$ and $n_2 = 1.457$.   In the case of the single-mode fibre the
microbend loss depends strongly on the index difference Δ and for
NA $\simeq$ 0.1 it is roughly of the same magnitude as with a multimode fibre.

Splice Losses

    A severe problem with single-mode fibres is the very tight
mechanical tolerance that must be maintained during splicing and

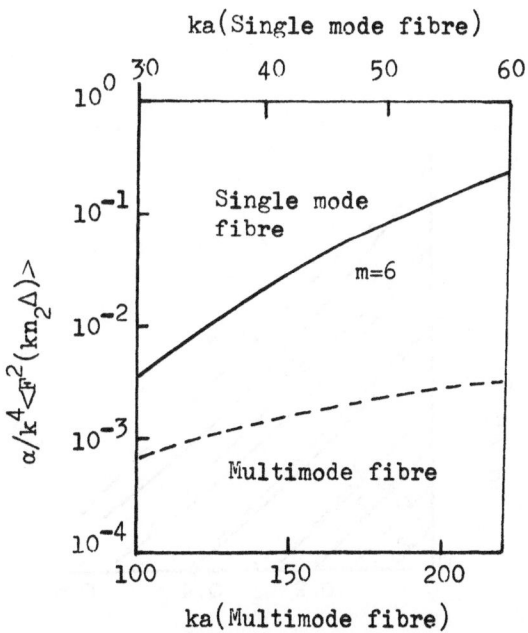

Fig.7   Normalized microbend losses

jointing.   For permanent connections fusion splices are feasible
and losses as low as 0.2dB per splice[25] have been reported. However
the design and implementation of demountable connectors is much more
difficult.   Transverse displacement d and angular misalignment α
need to be accurately controlled and contribute interdependently to
the joint loss α_t by the following equation[26]

$$\alpha_t = 2.6D^2 + 7.6(\alpha n/NA)^2 + 6.0(\alpha nD/NA)$$

where D = d/a.   Fig.8 illustrates the combinations of α and d that
produce various fixed values of α_t.   For a jointing efficiency of
90% an angular misalignment of less than 3° or an offset of less
than 1µm is required.   The corresponding figures for a multimode
fibre of core diameter 50µm are $\alpha = 3^o$ (i.e. the same as for a
single-mode fibre) and d = 5µm.

Sources and Detectors

Detectors, of course, are equally applicable to any type of
fibre and there is no particular problem to be solved.   As far as
sources are concerned a multimode fibre can accept radiation from
either a light-emitting diode or a laser.   The coupling efficiency
from a GaAs laser with an output beam width of 30° is approximately
30% into a fibre of NA = 0.15.   Coupling into a single-mode fibre
is more difficult to achieve but efficiencies of 40% or more may be
possible with microlens or other types of coupling.

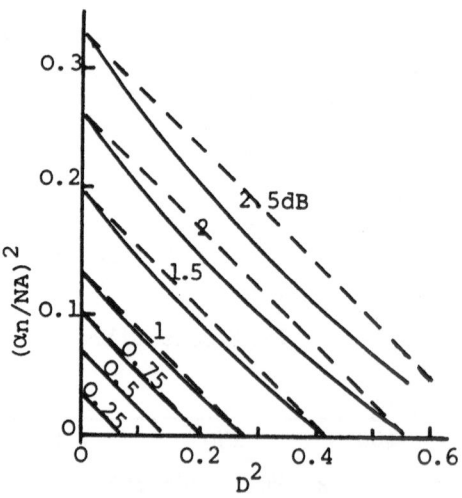

Fig.8   Lines of constant bend loss

References

1. K.C.Kao and G.A.Hockham, Proc.IEE 113, 1151-1158,1966.
2. M.Kawachi, A.Kawana and T.Miyashita, Electron.Lett. 13, 442-3,1977
3. M.Horiguchi and H.Osanai,Electron.Lett. 12, 310-312, 1976
4. M.H.Kuhn, Archiv Electron.Ubertragungstech 20, 201-204, 1975
5. R.Olshansky, Second European Conf on Optical Fibre Communication, 101-103,1976
6. K.Petermann, Electron.Lett. 12, 107-109, 1976.
7. W.A.Gambling, H.Matsumura and C.M.Ragdale,Opt. & Quantum Electron. 10, 301-309,1978
8. W.A.Gambling, D.N.Payne and H.Matsumura, Electron.Lett. 13,139-140, 1977
9. W.A.Gambling, D.N.Payne and H.Matsumura,Electron.Lett. 13,174-5 1977
10. R.Olshansky and D.B.Keck, Appl.Opt. 15, 483-491,1976
11. D.Gloge, I.P.Kaminow and H.M.Presby, Electron.Lett.11,469-471,1975
12. D.N.Payne and W.A.Gambling, Electron.Lett. 11,8-10, 1975
13. W.A.Gambling and D.N.Payne, First European Conference on Optical Fibre Communication, Proceedings 197-200, 1975
14. H.Matsumura, unpublished work
15. W.E.Martin, Appl. Opt. 13, 2112-2116, 1974
16. M.Ikeda, M.Tateda and H.Yoshikiyo, Appl.Opt. 14,814-815,1975
17. D.N.Payne, F.M.E.Sladen and M.J.Adams, First European Conf on Optical Fibre Communications, Proceedings 43-45,1975
18. W.A.Gambling, D.N.Payne, H.Matsumura and S.R.Norman, Electron. Lett. 13, 133-135, 1977
19. W.A.Gambling, D.N.Payne, H.Matsumura and R.B.Dyott, Microwaves, Optics and Acoustics 1, 13-17, 1976
20. D.Marcuse, Bell System Technical J. 55, 937-955, 1976
21. W.A.Gambling, H.Matsumura and C.M.Ragdale, Microwaves, Optics and Acoustics 3, 1978 (in the press)
22. E.F.Kuester and D.C.Chang, IEEE J. Quantum Electron. QE11, 903-907, 1975
23. M.Miyagi and G.L.Yip, Opt. and Quantum Electron. 8,335-341,1976
24. W.A.Gambling, H.Matsumura and C.M.Ragdale, Electron.Lett. 14, 130-132, 1978
25. H.Tsuchiya and I.Hatakeyama, Optical Fiber Transmission II, post-deadline paper, 1977
26. W.A.Gambling,H.Matsumura and C.M.Ragdale (to be published)

GEODESIC LENSES IN INTEGRATED OPTICS

V.Russo Checcacci

Istituto di Ricerca sulle Onde Elettromagnetiche, C.N.R.

Via Panciatichi 64 - Firenze, Italy

INTRODUCTION

In recent years various types of waveguide lenses have been developed for integrated optics, a field where they should play the same beam-control and optical data processing role that conventional lenses play in bulk optical systems. However the implementation of the integrated optical processors has been limited by the difficulty encountered in the fabrication of high quality lenses compatible with the other elements of the circuit. Special problems are the correction of aberrations and the minimization of scattering losses at the insertion region of the lens into the planar circuit.

Let us briefly recall that a lens is a phase transforming element. The most familiar property of lenses is their ability to form images or to focus a plane beam. The most remarkable and useful property of a converging lens is its inherent ability to perform Fourier Transformations. This operation, generally associated with bulky, complex and expensive electronic spectrum analizers, can be performed with extreme simplicity in a coherent optical system. The distribution of light amplitude across the back focal plane of a lens turns out to be the Fourier Transform of the amplitude distribution on the front focal plane. A lens which has to be used as a Fourier Transforming element must present higher performances than in conventional uses: the tolerated aberrations must be very small over a large field angle.

Let us also recall that any significant (> $\lambda/4$) departure from
the spherical shape of the wavefront emerging from the lens is an
"aberration". The effect of aberration on a focusing element is the
increasing of the focal spot size. The effect on the Fourier Transform operation is the lowering of the contrast of each spatial frequency, so that the resolving power and the cutoff frequency are
strongly reduced. Aberrations can have a severe effect on the fide-
lity of the system.

In bulk optics, for each specific operation, aberrations can
be easily compensated by a suitable combination of different types
of lenses. In Integrated Optics we choose to design perfect optical
elements, by taking advantage of the twodimensional geometry of a
waveguide.

For later purposes I shall briefly recall the different methods
of obtaining converging lenses in bulk optics and in microwave optics. Lenses for integrated optics, as well as other components, can
be designed by taking into account the various methods already investigated in both fields.

In optics, lenses with spherical boundaries, and n = cost > 1
are very easily to be made, but they suffer from aberrations. Aspherical lenses are perfect instruments but it is very difficult to
make them. Fresnel lenses, with a minimum cumbersome suffer from
aberrations, especially over large field angles.

In integrated optics, twodimensional lenses which are the counterparts of the ones above mentioned, have been already constructed
and tested. The first attempts, the so-called mode-index lenses,
[1] [2] had long focal length and problems of mode conversions due
to discontinuities; moreover they were affected by spherical aberration. The thin film Fresnel lenses, recently built [3], in addition
to the above mentioned extrassial aberrations, require a high accuracy in the fabrication technique.

At microwaves, converging lenses (although they are not very
much used) can be made with the technique of the guided propagation
between parallel metal plates of suitable shape (n < 1), or by means
of holographic techniques. However other types of perfect lenses,
twodimensional lenses, were obtained for a specific application
(rapid scanning) and their working principle is still of interest

today in integrated optics.

Let us consider lenses with spherical symmetry and inhomogeneous refractive index distribution. Maxwell found in 1884 the well known fish-eye lens where n(r) = 2/1 + r^2. Later Luneburg (1944) conside-red a different index distribution n(r) = $\sqrt{2 - r^2}$. In the first ca-se, all the rays from an arbitrary point O meet in a diametrically opposite point I. In the second case, every incident pencil of pa-rallel rays is brought to a sharp focus F.

A principle of equivalence [4], [5] was found between an inho-mogeneous refractive index distribution and a twodimensional Riemann space. The starting point for its application is represented by the Fermat principle:

$$1) \qquad \delta \int n(P)\, ds = 0$$

stating that the optical path is stationary along a ray.

Let us consider then a non-Euclidean space corresponding point by point to the physical (Euclidean) space and having the line ele-ment ds_r given by:

$$2) \qquad ds_r = n\, ds$$

This space will be called the Fermat space corresponding to the gi-ven distribution of refractive index. By substituting from (2), eq. 1 transforms into:

$$3) \qquad (3)\ \delta \int ds_r = 0$$

that is into the condition for a geodesic. Hence we can conclude that the light rays of the physical system characterized by a given distribution of refractive index n(P) correspond to the geodesics of the Fermat space characterized by (2).

The correspondence between the physical and the Fermat space is a conformal mapping so that the angles are preserved. A further pro-perty is that both the electric and the magnetic vectors of a light wave travelling along a geodesic of the Fermat space remain parallel to themselves.

Consider now a plane where the refractive index distribution
corresponds to one of the preceeding optical systems. It is readily
found that the twodimensional analog of the fish-eye is a spherical
surface. Here all great circles through a point intersect each other
at the diametrically opposite point. The twodimensional analog of
the Luneburg lens is the well known Rinehart lens. A twodimensional
Riemann space was easily achieved at microwaves by bending two parallel metal plates. The propagation occurred along the mean surface. Many of these configuration or geodesic lenses, with rotational symmetry were studied at microwaves with different shapes and
different performances.

In integrated optics, geodesic lenses and inhomogeneous index
distribution lenses can be obtained. A dielectric thin film of constant thickness formed on a curved substrate of different refractive
index, constitutes a twodimensional Riemann space for light waves.
The propagation is not substantially disturbed if the radii of curvature are everywhere large with respect to the wavelength. The rays
are the geodesics of the surface. The performances of geodesic lenses are independent of the substrate material and of the operating
wavelength.

The use of geodesic lenses in integrated optics was introduced
by our group in 1972.[6]

Also Luneburg lenses can be formed on a thin film, by depositing
a high index film through a circular mask with shaped edge. [7] [8]

It is to be noted that with substrates of low refractive index
Luneburg lenses are an alternative answer to the problem of high
quality lenses. With substrates having high refractive index, geodesic lenses represent the only feasible solution.

Geodesic lenses can be designed corrected from aberrations, or
perfect. They can be matched to the planar surface of the circuit
by a rough edge-rounding. Otherwise the rotation surface of the lens
has to be designed in order to match the planar circuit without discontinuity. The drawbacks due to the sharp variations of the waveguide thickness (or of the refractive index), that are present in the
other types of lenses, are thus overcame.

PERFECT ASPHERIC GEODESIC LENSES WITH SMOOTH JUNCTIONS

The simplest geodesic lens is constituted by a spherical surface (protrusion or depression). However only a quarter of the spherical surface behaves like a perfect focusing element [6] and it is
difficult to insert it in a planar circuit, (Fig.1). Any other fraction of the spherical surface focuses with strong spherical aberration [9] [10]. By combining a spherical depression with a mode index lens it is possible to reduce considerably aberrations [11].
Anyhow problems associated with mode index lenses are still present.
Such lenses have high f/number (focal length/linear aperture).

An alternative method to obtain corrected or perfect geodesic
lenses is to give an "aspherical shape" to the focusing rotation
surface. [12] [13]

In 1954, Toraldo and Kunz indipendently found a family of rotation surfaces which are twodimensional perfect optical instruments.
The Rinehart lens which is free from aberrations over the entire
aperture and which has plane input and output turned out to be a
member of this family. However it has the disadvantage of a strong
conflection of rays along the input and output line.

Also the oblate spheroidal lenses recently proposed in integrated optics have the same defect [14].

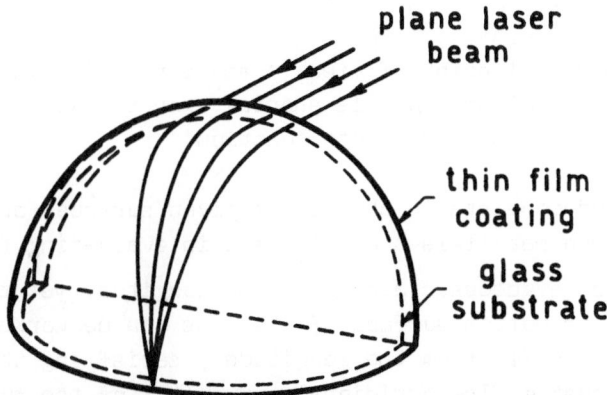

Fig.1 - Sketch of a geodesic lens constituted by a thin film formed
on a quarter of spherical surface.

In order to avoid the scattering losses along the conflection
lines the edges of the lens can be made round [15]. In this case
the focusing effect of this portion of surface has to be taken into
account as it has been done at microwaves [16].

A different approach to the problem of the sharp edges is that
of requiring the continuity of the surface in the design of the optical instrument.

Starting from the principle of equivalence between inhomogeneous index distribution and geodesic lenses, Kassai et alii [17] evaluated the profile of a new geodesic lens with focal points in the
planar region, by means of a rather complicated and approximated numerical calculation.

Using a complete different method, in 1957 Toraldo gave a complete solution to the problem of perfect lenses with no discontinuity, discussing a general problem on the geodesics of a surface of revolution [18].

The problem was approached by asking whether a surface of rotation might exist with the same properties of the Rinehart lens but
without having any parallel of conflection. It was found that such
a surface exists provided that the condition which guarantees the
perfect collimation in the Rinehart case:

4) $$\phi = \frac{\pi}{2} + \psi$$

be satisfied only for $\psi > \psi'$ (Fig.2).

From a physical point of view it means that Toraldo lens is perfect, but the useful aperture is shorter than the Rinehart lens's,
due to the presence of the smooth junction.

The method consists in choosing a given surface for the smooth
junction between parallels A and A_1, and in evaluating the longitude
ϕ_0 described by a geodesic along it. The longitude yet to be described along the bottom surface of the lens can be worked out as
the difference of $2\phi_0$ from the longitude ϕ satisfying the condition
of perfect focusing. The meridional curve $l(r)$ of the surface S is
derived from the longitude expression by means of rather long calculations, given in details in a paper in press.

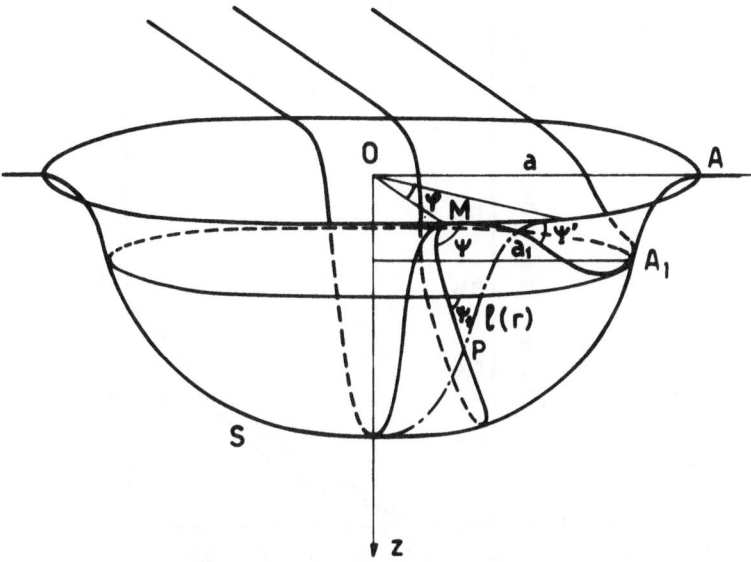

Fig.2 - Geometry of a Toraldo lens where:

ϕ = longitude of a general point P of the surface S with
           respect to the origin M.

r   = distance of the point P from the axis of revolution
           z.

$l(r)$ = meridional curve of the surface S.

a   = radius of the parallel A (input and output line).

a_1 = radius of the parallel A_1 limiting the junction.

ψ = angle made by a geodesic leaving M with the starting
           parallel A.

ψ' = value of ψ, for which a geodesic is tangent to the
           parallel A_1.

All the geodesics which intersect the parallel A_1 are perfectly collimated.

       The expression of $l(r)$ in cylindrical coordinates z, r can be
written as follows:

$$z(r) = \int_0^r \left\{ \left[-\left(1 + \frac{a^2 - r^2}{h^2}\right) + \frac{2}{\pi} \frac{(a_1^2 - r^2)^{1/2} (a^2 - a_1^2)^{1/2}}{h^2} + \right.\right.$$

$$\left.\left. + \frac{1}{\pi} \left(1 + 2\frac{a^2 - r^2}{h^2}\right) \arcsin\left(\frac{a_1^2 - r^2}{a^2 - r^2}\right)^{1/2} \right]^2 - 1 \right\}^{1/2} \quad dr \quad 0 \le r \le a_1$$

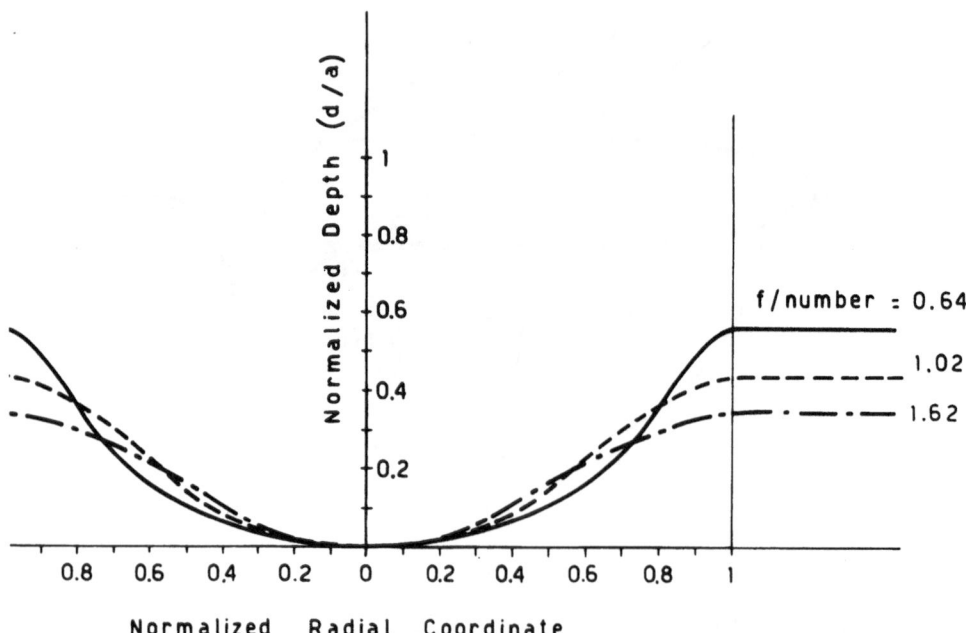

Fig.3 - Toraldo lens profiles for different values of the ratio
a/2a$_1$ (f/number).

$$z(r) = \int_0^r \left[(\frac{a^2 - r^2}{h^2})^2 + 2\ \frac{a^2 - r^2}{h^2} \right]^{1/2} dr\ \ a_1 \leq r \leq a$$

where:

$$h = a\ (\frac{\pi - 2\gamma - \sin 2\gamma}{\gamma})^{1/2}\ \ \ \ \sin \gamma = a_1/a$$

As one can choose the ratio a$_1$/a arbitrarily, the above equations
represent a family of perfect aspheric lenses, having different
apertures and different parameters for the junction.

   Fig.3 shows three meridional curves corresponding to three
different values of the ratio a/2a$_1$ (f/number).

   Various geodesic lenses of this family have been built and
tested having different sizes. The most recent one has the following specifications: focal length f = 4.5 mm and f/number = 0.64

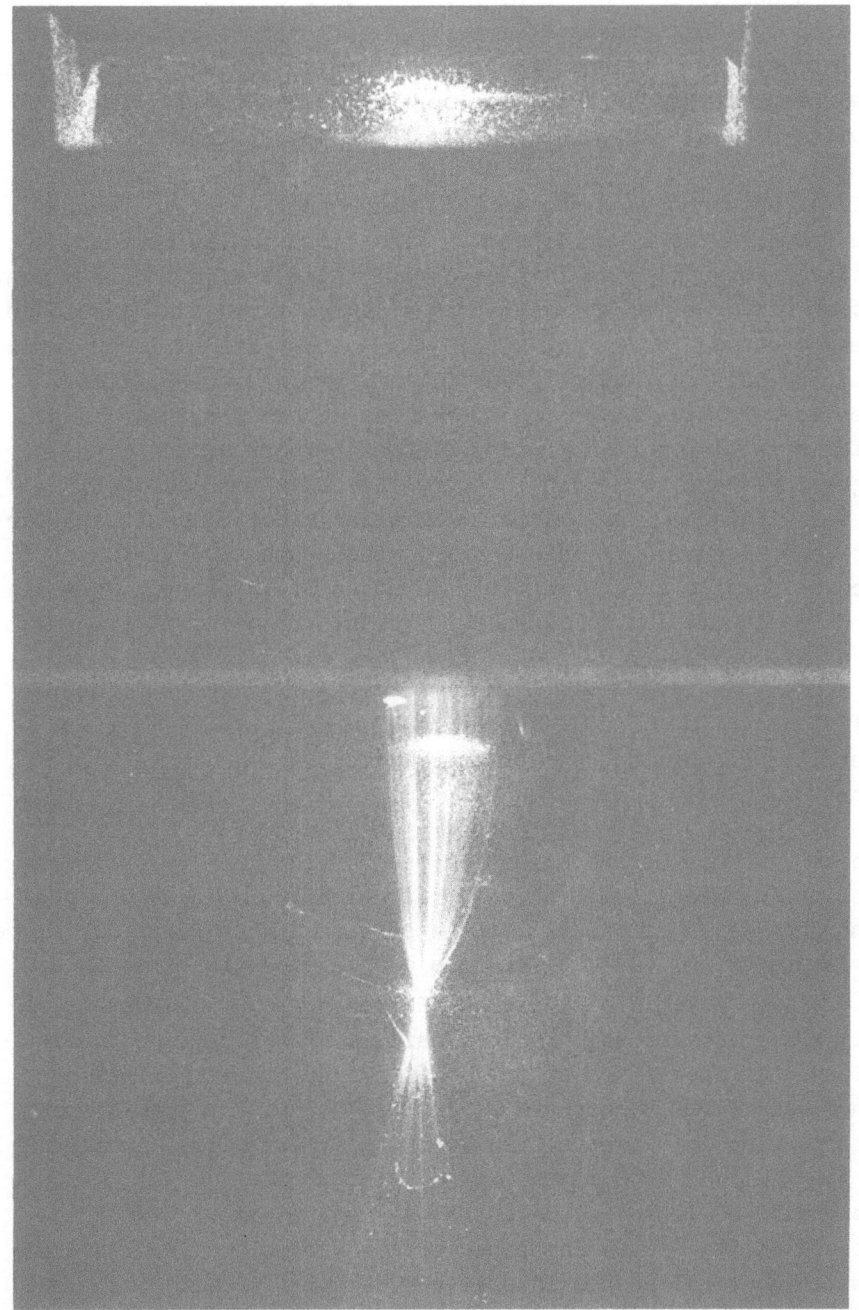

Fig.4 - Focusing of a 5 mm wide laser beam by a Toraldo Lens.
Focal length $f = a = 4.5$ mm, $a_1 = 3.5$ mm, f/number $= 0.64$, depth $= 2.5$ mm.

[19]. The lens has been built starting from a thin metal template
with the desired profile obtained by the photolithographic technique. Then a depression was mechanically grounded in a glass substrate by using the template as a comparison. The finest Syton polishing
powder has been used in the final step. A thin film epoxy resin with
uniform thickness has been spinned on the substrate. The film was
doped with Rodamine B to make visible the guiding beam.

Fig.4 shows the lens focusing a plane beam, 5 mm wide, prism
coupled from a He-Ne laser. The lens was theoretically perfect, but
a defective fabrication could have deteriorated its characteristics.
As a rough test, we looked at the focus region through a microscope:
no displacement was observed, within a precision of 50 micron, when
the entering beam was reduced from 5 mm to paraxial rays. A more
precise test was made by trying to separate three collinear beams
generated by a diffraction grating placed in front of the lens.
Three gratings were used with frequency of 20 linees/mm, 12 linees/
mm and 8.1 linees/mm. The angular separation of contiguous beams turned out to be 8 mrad, 4.7 mrad and 3.2 mrad respectively. These figures correspond to an arc length separation of 36 micron, 21.4 micron and 14 micron. The focal spots were well resolved in all the
three cases.

Fig.5 shows the case of three beams separated by 14 micron. A
more detailed discussion on these results will be given in a paper
in press.

For what concerns the ability of performing Fourier Transformations it was theoretically found that the above mentioned geodesic lens shows a phase distribution aberrated by the field curvature.
The curvature of the output line is a common characteristic to all
the lenses with rotational symmetry. On the other hand, in many cases only the intensity spectrum analysis is of real interest. Geodesic lenses of this type can be used for instance in an integrated convolver of the type sketched in Fig.6 [20]. This device can
find application in optical communications or for signal processing.

The family of lenses we have tested seems to be very suitable
for optical digital communications, because it is free from aberrations which could alter the intensity of spatial frequencies. In
addition, the arbitrary choice of the ratio a_1/a allows us to vary
the profile, the focal length and the f/number, in order to obtain
lenses with the best characteristic for the considered purpose.

Fig.5 - Resolution of three guided beams on the focal line of the
        lens of Fig.4. The angular separation of contiguous beams
        is 3.2 mrad which corresponds to an arc length separation
        of 14   micron. An optical fiber of 90 micron is shown for
        comparison.

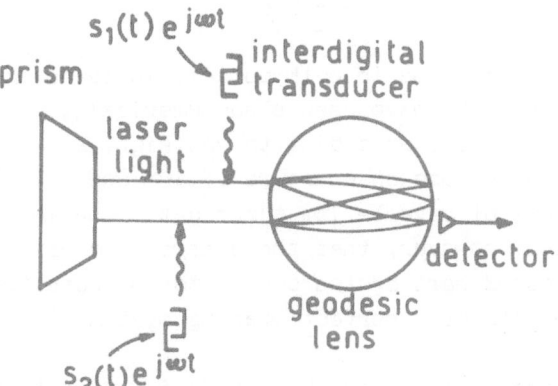

Fig.6 - Sketch of an integrated optical convolver employing a geo-
        desic lens.

HEMISPHERICAL GEODESIC CORRELATOR

Another interesting optical system is constituted by the hemispherical surface which is the twodimensional analog geodesic element of the Maxwell fish-eye. It shows the characteristic of imaging
without aberrations (Fig.7) and its symmetry properties allow us
to expect a suitable Fourier Transform locus [19]. Such an imaging
system,Fig.8a, can be compared with the bulk imaging system of Fig.
8b. In bulk optics, forming an image of the object is a sufficient
condition for producing the Fourier Transform on a plane (VANDER
LUGT,A. 1974). We investigated if analogous property is valid also
in the case of an unusual imaging system such as the hemispherical
surface. For symmetry reasons, in agreement with the focusing property of a quarter of spherical surface, the Fourier Transform locus
under investigation must be the great circle that divides the surface into two parts (Fig.8a).

To verify this assumption, one can evaluate the lack of parallelism on the Fourier Transform line of the beams diffracted by
point sources on the input line. The lack of parallelism or angular
aberration δ of the beam from S, with aperture α and field angle
Ω defined in Fig.8a, turns out to be

$$\delta = - \frac{1}{2} \Omega \alpha^2$$

Analogously the wave aberration turns out to be

$$W = \frac{1}{6} \Omega \alpha^3$$

It corresponds to the coma in bulk optics, while no axial aberration
exists. The aberrations have been also numerically evaluated. As an
example, if we consider, according to Rayleigh, a maximum wave aberration of $\lambda/4$, we can use an aperture of 10° and a field angle of
9°, while the lack of parallelism turns out to be less than 0.1°.
It is to be noted, however, that the aberrations are completely corrected by the second part of the correlator because the semispherical surface constitutes a perfect imaging system.

Numerical evaluations have been made in order to clarify the
effects of wave aberration on the complex amplitude of the Fourier
Transform and on the autocorrelation or cross correlation of binary
signals. It has been found that the amplitude difference with res-

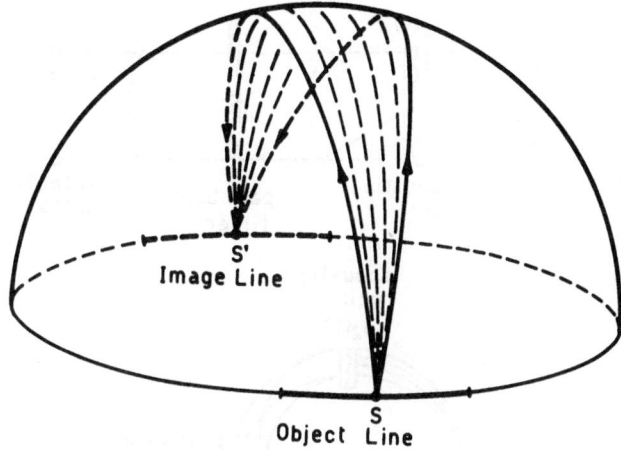

Fig.7 - Perfect imaging semispherical geodesic system.

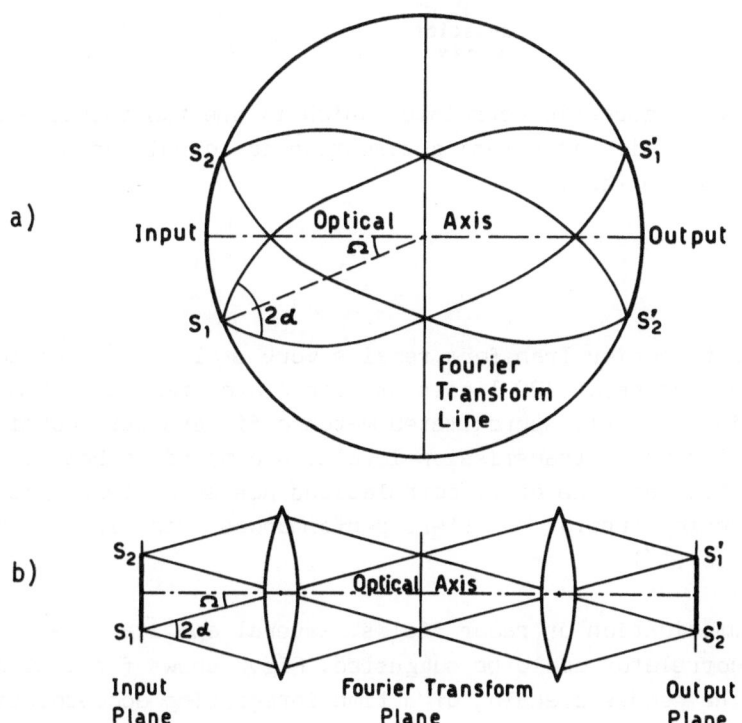

Fig.8 - (a) Top view of the hemispherical system compared with
        (b) a bulk double diffraction system.

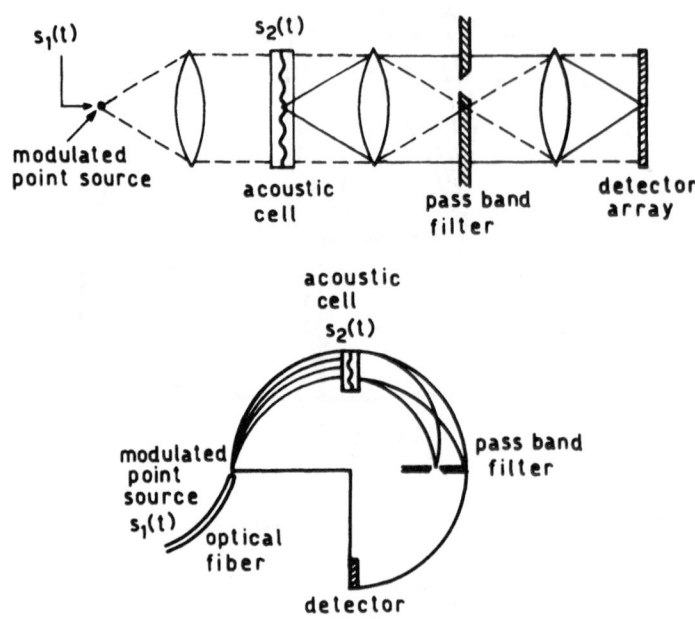

Fig.9 - Sketch of a geodesic correlator which is the twodimensional
        analog of a time integrating acoustooptic correlator sugges-
        ted in bulk optics.

pect to the exact Fourier Transform remains very small (10^{-5}) up to
the spatial frequencies of 800 linees/mm (correlator radius = 10 mm,
input signal width = 5mm). Approximated matched filters were considered with two and three transmission levels. A comparison between
the autocorrelation and the cross correlations has shown that this
particular geodesic correlators offers performances which are acceptable in practice [19].

     In the communication or radar fields, several applications of
the geodesic correlator could be suggested. Fig.9 shows for instance
the sketch of the geodesic analog of a time integrating acoustooptic
correlator suggested in bulk optics [21]. The comparison with the
bulk system shows that even rather complex optical systems can be
realized with spherical surface. This kind of correlator could be
particularly interesting for the possibility of coupling it to an optical fiber which brings the input signal.

CONCLUSIONS

The problem of building perfect waveguide lenses can be solved by using aspherical geodesic lenses with rotational symmetry. They are constituted by a uniform thickness thin film formed on the substrate, where a depression having the required shape has been previously grounded. Geodesic lenses can be suitably designed for a planar circuit, that is, matched without discontinuity to the plane, in order to avoid scattering, radiation losses and mode conversions. Their performances are independent of the substrate material and of the operating wavelength.

Experimental tests recently made on a geodesic lens belonging to the family of Toraldo lenses, having focal length = 4.5 mm and f/number = 0.64 have been reported. The lens was able to resolve in its focal region three plane beams of angular separation: 3.2 mrad. A lens of this type seems suitable to be used in integrated optical processes as for example spectrum analizers.

It is noticeable that the free choice of the parameters a_1 and a in the design of a Toraldo lens allows to easily match the lens characteristics to the particular application one has in mind.

As to the fabrication technique, it could be improved by using the ultrasonic impact grinding [10] or other techniques recently investigated in bulk optics for aspherical surfaces [22] [23]. It seems possible that, once appropriate grinding and polishing procedures have been used, mass-production of geodesic aspherical lenses can be also easier than an analogous production of other kind of thin film lenses.

The geodesic lenses investigated so far present the only problem of the field curvature. The Fourier Transform locus is a circle which lies in the light guiding plane. Such a characteristic, common to all revolution lenses, makes it very difficult to assemble two lenses in order to constitute a thin film correlator analogous to a bulk double diffraction system. Geodesic systems that reduce this disadvantage are under investigation.

It has also been shown that the hemispherical geodesic lens can constitute a thin film correlator and can perform analog operations analogously to the bulk double diffraction system. As an application, a time integrating acoustooptic correlator which employs the hemispherical geodesic system has been suggested.

REFERENCES

(1) R.Shubert and J.H.Harris, "Optical guided-wave and diffraction",
    J.Opt.Soc.Am., vol.61, pp.154 - 161, 1971.
(2) R.Ulrich and R.J.Martin, "Geometrical optics in thin film light
    guides", Appl.Opt., Vol.10, pp.2077-2085, 1971.
(3) P.R.Ashley and W.S.C.Chang, "Fresnel lens in thin film wavegui-
    des", Topical Meeting on Integrated and Guided Wave Optics,
    Tech.Digest, OSA, Washington D.C.,1978, paper MA3-1.
(4) K.S.Kunz, "Propagation of microwaves between a parallel pair
    of doubly curved conducting surfaces", J.Appl.Phys., Vol.25,5,
    pp.642-653, 1954.
(5) G.Toraldo di Francia, "A family of perfect configuration len-
    ses of revolution", Opt.Acta, Vol.1, 4 pp.157-163,1955.
(6) G.C.Righini, V.Russo, S.Sottini and Toraldo di Francia, "Thin
    film geodesic lenses", Appl.Opt., Vol.11, pp.1442-1443,1972.
(7) F.Zernike, "Luneburg lens for optical waveguide use", Opt.
    Commun. Vol.12, pp.379-381, 1974.
(8) S.K.Yao, D.B.Anderson, C.M.Oania, G.Kreismanis, "Mask synthesis
    for diffraction-limited waveguide Luneburg lenses", Topical Me-
    eting on integrated and guided wave optics, Tech.Digest, OSA
    Washington D.C., 1978.
(9) C.M.Verber, D.W.Vahey and Van E.Wood, "Focal properties of geo-
    desic waveguide lenses", Appl.Phys.Lett. Vol.28, pp.514-516,
    1976.
(10) Bor-Uei Chen, E.Maron, A.Les, "Geodesic lenses in single mode
    LiNbO$_3$ waveguides", Appl.Phys.Lett. Vol.31, pp.263-265, 1977.
(11) E.Spiller and J.S.Harper, "High resolution lenses for optical
    waveguides", Appl.Opt., Vol.13, pp.2105-2108, 1974.
(12) G.C.Righini, V.Russo, S.Sottini and G.Toraldo di Francia, "Geo-
    desic lenses for guided optical waves", Appl.Opt. Vol.12, pp.
    1477-1481, 1973.
(13) G.C.Righini, V.Russo, S.Sottini and G.Toraldo di Francia, "Geo-
    desic lenses for integrated optics", Proc.1973 Eur.Microwave
    Conf., G.Hoffman Ed. Brussels, 1973, Vol.1, paper B.5.5.
(14) D.W.Vahey and Van E.Wood, "Focal characteristic of spheroidal
    geodesic lenses for integrated optical processing", IEEE J.
    Quantum Electron., Vol. QE-13, pp.129-134, 1977.
(15) Van E.Wood, "Effects of edge rounding on geodesic lenses", Appl.
    Optics, Vol.15, pp.2817-2820, 1976.
(16) L.Ronchi, "Geometrical optics of toroidal junctions in configu-
    ration lenses", Opt.Acta, Vol.2, pp.64-80, 1955.

(17) D.Kassai, B.Chen, E.Maron, D.G.Ramer and M.K.Barnoski, "Aber-
      ration corrected geodesic lens for integrated optics circuits",
      Topical Meeting on Integrated and Guided Wave Optics, Tech.Di-
      gest, OSA, Washington D.C., 1978, paper MA2-1.
(18) G.Toraldo di Francia, "Un problema sulle geodetiche delle super-
      fici di rotazione che si presenta nella tecnica delle microonde"
      Atti Fondaz.Ronchi, Vol.12, pp.151-172, 1957, see also G.Toraldo
      di Francia, "A problem on the geodesics of a surface of revo-
      lution which is of interest in microwave optics", Technical Note
      n.9, April 1957 - A.R.D.C. Contract AF 61 (514) - 903.
(19) G.C.Righini, V.Russo, S.Sottini, "Thin film integrated signal
      processors", Optical Fibres, Integrated Optics and their mili-
      tary applications, AGARD Conference Proceedings n.219, London,
      16 may 1977.
(20) P.Das, D.Shumer, "Optical communications using surface acoustic
      waves", Intern.Conf.on Application of Holography and Optical
      Data Processing, Proceedings of Jerusalem 1976.
(21) Sprague,R.A., and C.L.Koliopoulos, 1976, "Time integrating acou-
      stooptic correlator", Appl.Opt., Vol.15, p.89.
(22) R.E.Parks, "Aspheric lens element and spline functions", Appl.
      Opt. Vol.12, pp.2541-2543, 1973.
(23) D.J.Bajuk, "Computer controlled generation of rotationally
      symmetric aspheric surfaces", Opt.Engin. Vol.15, pp.401-408,
      1976.

GUIDED OPTICAL WAVE ACTIVITY IN GERMANY

E. Voges

Fernuniversität

D-5800 Hagen 1, Postfach 940

The work on guided optical waves in Germany is reviewed in short.
New results of work on optical components for fibre communication
are given and field trials are described. The current status of
monomode and multimode integrated optical components and new
developments are reported.

1. OPTICAL FIBRE COMMUNICATION

The research on optical fibre communication in Germany comprises
fibre technology, fibre characterization and fibre cables, LED's
superluminescent diodes, lasers and photodiodes, as well as fast
repeaters and fibre system studies.
The progress in fibre technology aims at
- the achievement of high deposition rates for the economical
  production of large preforms
- the fabrication of graded-index fibres with optimum profiles
  for low dispersion
- the fabrication of graded-index fibres with low attenuation and
  high numerical aperture for fibre transmission with LED-sources
- the fabrication of monomode fibres in particular for wavelengths $\lambda \simeq 1,3 \ldots 1,6 \mu m$.
New silica fibres with fluorine doping are produced by a plasma
activated chemical deposition technique, that leads to very large
preforms and allows the drawing of fibres of 40 km length or more
/1/. A fluorine concentration of only 3,5% yields a numerical
aperture of 0,22 and attenuation of 5 dB/km at $\lambda = 1,05 \mu m$ and
fibres with low mechanical strain.

Graded index fibres with extremely smooth index profiles are

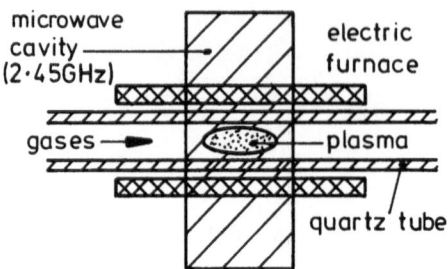

Fig. 1. RF-plasma activated CVD-process

fabricated by means of a RF-plasma activated CVD-process /2/. Here
the plasma reactor consists of a microwave cavity combined with
an electric furnace around the silica tube.(Fig. 1) High deposition
rates without soot formation are realized and the deposition
zone follows the moving cavity instantaneously. Therefore, the
cavity can be moved with high speed, resulting in a large number
of layers per unit time. In this way thousands of layers are
applied. Fibres drawn from GeO_2/B_2O_3 doped silica preforms have
a numerical aperture of 0,2 and a pulse widening of 150 ps/km at
$\lambda \simeq 0,9$ μm /3/.

In order to increase the numerical aperture of silica fibres,
multi-component doping of high concentration is applied in a
modified CVD-process. Fibres from this process reach beyond 0,35
in their numerical aperture with less than 1,5 ns/km pulse
widening and 4 dB/km attenuation at $\lambda = 0,82$ μm /4/.

The transmission losses in fibres approach the ultimate lower
limits, and signal dispersion determines the maximum repeater
spacing. Therefore, growing interest is given to monomode fibres
/5/ in particular for the wavelength region 1,3 ... 1,6 μm.

The experimental fibre characterization uses common techniques.
In connection with analytical studies, emphasis is given to the
degrading effects of profile inhomogeneities on signal dispersion /6/, and to the evaluation of optimum profiles /7/.

Fibre cables have been developed using modifications of conventional cable manufacturing. The reinforced cables contain 10
or more separate fibres loosly fitted into plastic tubes. They
withstand tensile forces up to 500 N and have a bending radius
down to 5 cm /8/. For the connection of fibres a variety of
self-adjusting rugged connectors have been developed, the
splicing of field cables, however, still imposes problems of
practability.

The development of light sources for fibre communication is

concentrating on GaAlAs laser diodes. The degradation of laser
diodes, and single mode operation at high modulation frequencies
are of primary importance. Laser lifetimes in excess of 10^4h have
been achieved. Thus, interest is turning towards buried channel
hetero-structure lasers, and laser diodes for wavelengths between
1,3 µm and 1,6 µm are envisaged. LED's are still of interest in
particular for analog signal transmission. High-speed LED's with
a 3 dB modulation bandwith in excess of 1 GHz and still 3,5W/srcm^2
of radiance with a spectral linewidth of 36 nm at λ = 0,89 µm have
been fabricated by liquid-phase epitaxy /9/. The active area was
limited to a diamter of 50 µm by proton bombardment. In addition,
high speed superluminescent diodes with linewidths of 10 nm and
narrow radiation lobes have been fabricated. These diodes provide
increased coupling into fibres and do not show resonant behaviour
during pulse modulation with risetimes less than 1 ns /10/.

Besides photodiodes with gain-bandwidth products in excess of
300 GHz /11,12/, transparent photodiodes have been developed /13/
(Fig. 2). These diodes show absorption losses down to 1 dB at
wavelength exceeding 0,8 µm, they have a sensitivity of 0,18 mA
per mW incident light power and showed a response time of about
1 ns in first measurements. Transparent photodiodes are inserted
directly into the transmission path, and can be used for monitoring or regulating purposes.

The prospect of Gbit/s transmission rates in fibre transmission
systems has stimulated the research on new repeater techniques.
Diode regenerators and multiplexers employing step-recovery
diodes have been built in thin-film technology and operate for
bit rates up to about 8 Gbit/s /14/. Extremely fast pulse generation and multiplexing is accomplished with series-gate sampling
lines using Schottky-diodes as gates /15/. For high speed processing of continuous signals, several sampling-lines operate in

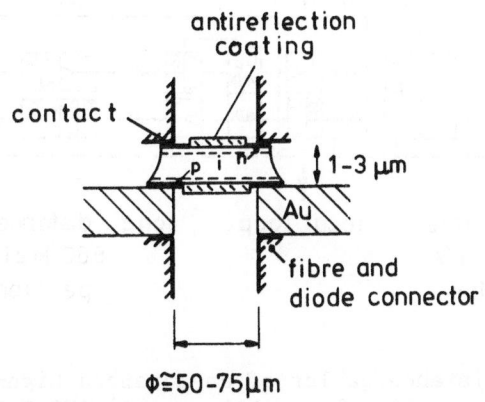

Fig. 2. Transparent silicon pin-photodiode

parallel. At present sampling-line circuits operate at bit rates
of 1,25 Gbit/s, but the technique allows bit rates as high as
10 Gbit/s.

System studies and field trials take into account fibre systems
with bit rates up to and beyond 1 Gbit/s, and consider in particular the use of optical fibre links in wideband integrated
communication networks which combine voice, data and video transmission. A 34 Mbit/s (PCM 480) field trial is being installed
at Berlin. The fibre experiments are conducted on an interoffice
trunk route /16/, which has been selected to involve all problems
and difficulties that are encountered under most adverse conditions. It is supported by the German Federal Post Office and
by the German Federal Ministry of Research and Technology. Four
industrial groups (AEG-Telefunken, Siemens, SEL, TEKADE-Felten &
Guilleaume) are participating at the field trial, and are each
contributing a fibre cable of 4,3 km length containing at least
6 graded-index fibres, and the receiving and transmitting units.
This field-trial aims at a competition of different systems under
realistic conditions.

An advanced "Experimental system of an integrated communication
network with glass fibres" /17/ is being installed at the
Heinrich-Hertz-Institut in Berlin. This large-scale field trial
is supported by the German Federal Ministry of Research and
Technology and various industrial groups contribute to this
project. The local cables (Fig. 3) contain 4 graded index fibres
each operating at 280 Mbit/s. The repeater distances are at least
3 km. The long distance cables contain 2 graded index fibres,
each, and operate at 560 Mbit/s. Altogether, about 150 km glass
fibres will be installed. The results of this large-scale field
test will provide criteria for the introduction of optical fibre

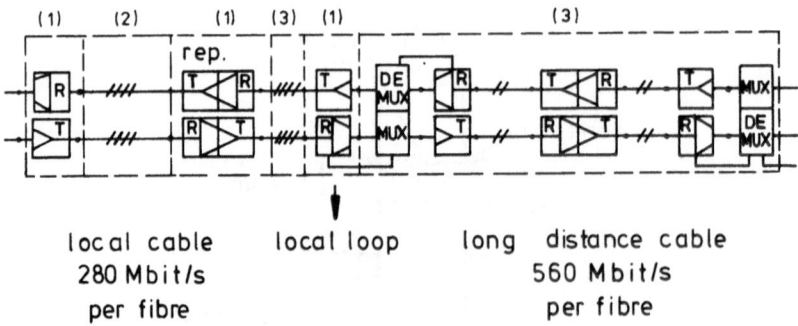

local cable   local loop   long distance cable
280 Mbit/s   560 Mbit/s
 per fibre   per fibre

Fig. 3
Local and long distance cables of an advanced high-speed
fibre transmission system. Contributors: (1) AEG-Telefunken,
(2) Schott, (3) Siemens.

links as components of advanced integrated systems into the
communication network of the German Federal Post Office.

2. INTEGRATED OPTICS

The rapid development of optical fibre communications has provided
the strongest motivation for research on integrated optics.
Therefore, most effort is directed towards this particular
application. One expects, however, that the concepts of integrated optics could well be applied to other fields. Consequently,
the term "integrated optics" should encompass all aspects of
monomode and multimode guided wave planar optics /18/, and may -
in this rather broad sence - find applications for example in the
general fields of signal processing, controlling and monitoring.
The search for applications, where the concepts of integrated
optics lead to superior solutions, is developing but is considered to be a difficult task.

Therefore, the research on integrated optics is still dominantly
connected with fibre transmission, and has so far mainly been
directed to monomode optical components with $LiNbO_3$ being the
preferred substrate material. In particular, electrooptic modulators are investigated by different groups. Fig. 4a shows an
electrooptic modulator of the Mach-Zehnder interferometer type,
which requires 6V modulation voltage for a phase shift of $\pi/2$
radians in each arm /19/. The switched Y-junction in Fig. 4b /20/
operates by controlling the superposition of the two modes E_{11}^x
and E_{21}^x /21/ in one arm, which are symmetric and antisymmetric
with respect to x = 0 and are, therefore, influenced by the
modulation voltage in a different way. Low levels of cross-talk
will be difficult to achieve in this structure, therefore,
somewhat different configurations are investigated in addition/21/.
The cut-off modulator in Fig. 4c /22/ provides direct intensity
modulation. In this new structure, the monomode waveguide is just
above cut-off. When applying a sufficiently high voltage to the
electrodes, the waveguide section between the electrodes gets
below cut-off, and the guided wave radiates into the substrate.
The length of the electrodes can be less than 1 mm, their capacitance, therefore, is small, and very fast switching is possible.
In preliminary measurements a modulation depth of 50% has been
achieved with a pulse amplitude of 18V at a rise time of about
2 ns.

The waveguides in the above modulators are fabricated by the
well-known technique of Ti-indiffusion /23/. Since the diffusion
coefficient along the c-axis is by a factor of roughly 3 larger
than the diffusion coefficient normal to the c-axis /24/, the
waveguides for electrooptic modulators are broadened during the
Ti-indiffusion, and the overlap between the modulating electric

field and the optical wave is decreased. Moreover, the alignment
between monomode waveguides and the electrodes is critical. Ion
implantation may provide a solution for both problems /25/. After
an homogeneous Ti-indiffusion, the waveguides are formed by
He$^+$-implantations through a metal mask (Fig. 4d). The He$^+$-implan-
tation decreases the refractive index by 1...3% at an ion flux
of $\emptyset \cong 10^{16} cm^{-2}$, with a depth of about 1,1 µm for an He$^+$-energy
of 350 keV. Rib-guides with relatively close lateral confinement
are fabricated in this way. The electrodes can be aligned in a
self-adjusting technique, when covering the substrate with another
metal, this is indicated by the dashed lines in Fig. 4d, and
applying selective etching.

Planar optical waveguides are attractive for nonlinear interactions (for example second harmonic generation or mixing). The
necessary high power density is easily achieved, and maintained
over large distances without limitations due to diffraction
spreading. By tailoring the waveguide index profile and/or by
using periodic structures, the conditions for phasematching are
noncritical, and can be implemented in a flexible manner. Second
harmonic generation with very high efficiencies was realized in
Ti/LiNbO$_3$ waveguides /26/. At 45 W peak fundamental power in
waveguides of 17 mm length an efficiency for TM$_o$ (λ=1,08µm) →
TE$_1$ (λ=0,54µm) conversion up to 25% was achieved in pulse mode
operation to avoid optical damage.

Until now relatively little work has been done on multimode
integrated optical components. Optical phenomena and calculations
are more complicated in multimode guides, the application of

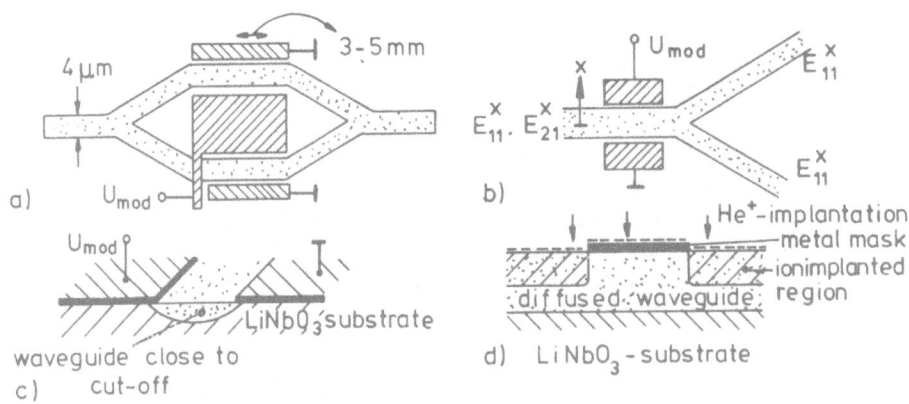

Fig. 4   Electrooptic modulators
         a) Mach-Zehnder interferometer   b) Switched Y-junction
         c) Cut-off-modulator d) Waveguide fabrication by
         combined diffusion and ion implantation

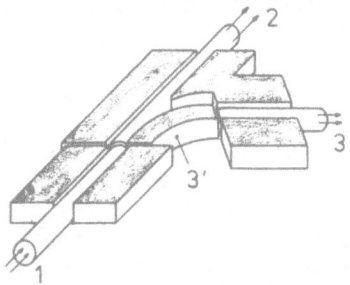

Fig. 5. Multimode branching and fibre holding structure. The light
is fed from fibre 1 to fibre 2 with a small portion coupled into
fibre 3 through the **strip-waveguide 3'**.

microwave principles is not obvious and the fabrication of thick
transparent planar waveguide structures imposes technical problems.
There is on the other hand a real need for multimode planar optical
components in particular at fibre terminals and repeaters or in
optical branching networks.

One technique for the fabrication of multimode planar waveguide
structures has been developed at the Siemens Research Laboratory
München. The principle is shown in Fig. 5 /27/. A glass substrate
is laminated with light-sensitive plastic sheets of higher refractive index up to a thickness of about 1oo μm. After suitable
exposure through a mask, multimode strip-waveguides are formed
and simultaneously deep grooves, into which optical fibres are
snapped in and butt-coupled to the strip-waveguides. The coupling
between waveguides is localized, and the coupled power can be
varied by slight lateral shifts of the guides as shown in Fig. 5.
This technique makes use of planar technologies, and assures
reproducibility and tight tolerances. The use of plastic waveguides

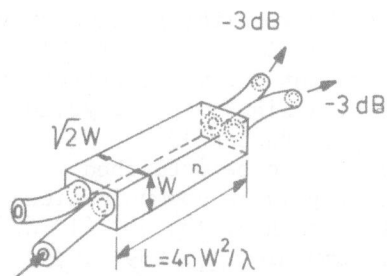

Fig. 6. Multimode 3 dB directional coupler with self-imaging
rectangular waveguide.

may, however, impose problems with respect to longterm and thermal stability.

With the aid of self-imaging waveguides /28/ a variety of multi-mode optical devices can be fabricated. Fig. 6 shows a self-imaging 3 dB directional coupler /29/. It consists of a homogeneous-index multimode waveguide of perfectly rectangular cross-section with an aspect ratio $1:\sqrt{2}$. By choosing a particular length $L = 4\ nW^2/\lambda$, the modes excited by the input fibre can be made to interfere constructively at the opposite end forming a double image of the input. This imaging device is a true multimode directional coupler, and does not suffer from packing-fraction losses. The losses mainly arise from imperfect imaging and are typically below 1 dB /18/.

REFERENCES

/ 1/   A. Mühlich et al.,3rd ECOC, München 1977
/ 2/   D. Küppers et al., J.Electrochem.Soc.123, 1079 (1976)
/ 3/   J.P. Hazan et al., Electron.Lett. 13, 540 (1977)
/ 4/   Schott Gen. Mainz, (1977)
/ 5/   M. Börner, S. Maslowski, Proc. IEE 123, 627 (1976)
/ 6/   K. Behm, AEÜ 31, 45(1977), K.Petermann, AEÜ 31, 201 (1977).
/ 7/   S. Geckeler, Electron. Lett. 13, 440 (1977)
/ 8/   U. Oestreich et al., 3rd ECOC, München 1977
/ 9/   J. Heinen et al.,Electron.Lett. 12, 553 (1976)
/10/   W. Harth, M.C. Amann, Electron.Lett. 13, 291 (1977)
/11/   K. Berchtold et al., Appl. Phys. Lett. 26, 585 (1975)
/12/   J. Müller, IEEE Trans. ED-25, 247 (1978)
/13/   J. Müller et al., 3rd ECOC, München 1977
/14/   U. Barabas et al., Electron. Lett. 14, 62 (1978)
/15/   R. Schwarte et al., 7th EuMC, Copenhagen (1977)
/16/   J. Feldmann et al., NTZ 29, 235 (1976)
/17/   K. Fußgänger et al., Frequenz June 1978
/18/   R. Ulrich, 3rd ECOC, München 1977
/19/   F. Auracher, Research Laboratory Siemens AG München, private
       communication
/20/   U. Langmann, Ruhr-Universität Bochum, private communication
/21/   M Papuchon et al., Appl. Phys. Lett. 31, 266 (1977)
/22/   W. Sohler, Universität Dortmund, private communication
/23/   R.V. Schmidt, I.P. Kaminow, Appl.Phys.Lett.25, 458 (1974)
/24/   P. Baues, 3rd ECOC, München 1977
/25/   J. Heibei, E. Voges, to be published
/26/   W.Sohler, H.Suche, private communication
/27/   H.H. Witte, Opt. Communic. 18, 559 (1976)
/28/   R. Ulrich, Opt. Communic, 13, 259 (1975)
/29/   A. Simon, R.Ulrich, Appl.Phys.Lett. 12, 600 (1976)

RECENT WORK ON OPTICAL FIBRE SYSTEMS AND COMPONENTS IN THE UK

J.E. Midwinter

Post Office Research Centre
Martlesham Heath
Ipswich, Suffolk, UK

INTRODUCTION

The United Kingdom played a major role in starting the present
worldwide activity on optical fibre communications with the publication in 1966 of the paper by Kao and Hockham (1) proposing dielectric waveguides made of glass for carrying optical signals.
This marked the beginning of a growing activity by Standard Telecommunications Laboratories (STL) and the British Post Office (BPO)
which continues to this day. Since then many other have become
involved, as will be apparent from this summary which covers work
carried out over the last two or three years only. The work is
divided by technical topic, starting with the components and leading
to complete system demonstrations and studies.

OPTICAL FIBRE

Four fibre pulling processes have been studied in some depth
in the UK. Pilkington Glass use a preform process (2) in which a
step index preform is produced by pulling from a two component melt,
cladding glass floating on core glass, so that a rod pulled upwards
is formed with a skin of cladding glass surrounding the core. This
preform is then converted in a preform fibre puller to fibre.
Fibres made by this technique have been produced and sold in quantity and typically have losses of under 50 dB/km, NA of 0.48 and
core diameter of 120 or 200 microns. More recently, Pilkington have
been developing the Phasil process first described by P.B. Macedo (3)
and this has led to production fibres with losses of under 20 dB/km.

Fibres have been produced by chemical vapour deposition (CVD)

using the inside tube process at a number of laboratories. Southampton University (4) pioneered the use of phospho-silicate glass
formed by the oxidation of $POCl_3$ and has continued to study the CVD
process in great depth, publishing extensively (5-9) on the properties of the materials so produced, particularly in the longer wavelength region around 1.3 micron. STL (10) have used the CVD process,
mainly with GeO_2 as the core dopant, to produce both graded index
and single mode fibers while GEC has also made graded fibres of very
low loss by the CVD technique. All of these laboratories are
regularly producing fibres having losses below 3 dB/km at 840 nm
and losses below 1 dB/km at 1.1 micron have been achieved.

      The BPO (11) has developed the double crucible process using
sodium borosilicate glasses and has achieved the lowest losses ever
reported by this method (12), 3.4 dB/km at 840 nm. The fibres can
be graded by using a diffusible dopant (13) in the core glass so that
the core cladding boundary inter-diffuses while the glass is molten
within the crucible structure, and in this way, low mode dispersion
of 2 ns/km (all modes excited) has been achieved. The process has
great attraction as a mass production method for the manufacture
of very low cost intermediate performance fibre for use in applications such as local distribution or CATV and this usefully
complements the CVD process.

FIBRE PROPAGATION

      Recent studies have centred on both graded index and monomode
fibres. A major study of graded index fibre has been carried out
by the BPO in association with an experimental system installation
(14). Twenty-one kilometres of fibre cable were involved, each
cable containing two Corning fibres. These were measured in the
laboratory to obtain their profile dispersion and attenuation.
Detailed studies were then made of the links' dispersion as the
fibres were jointed together in steps of 1 km. It had been expected
that the rms pulse width, due to mode dispersion, would follow a
law of the general form:

$$\sigma_n = \left(\sum_{i=o}^{n} \sigma_i^b \right)^{1/b}$$

where $\frac{1}{2} \leq b \leq 1$ (corresponding at the extremes to the cases of
heavy mode coupling and zero mode coupling respectively) and n is
the number of sections. In practice, much more complex effects
were observed which could be described by the following relation
(15):

$$\sigma_{mode}^2 = \sum_{i=o}^{n} \sigma_i^2 + \sum_{i=o}^{n} \sum_{j=o}^{n} c_{ij}\, \sigma_i\, \sigma_j$$
$$i \neq j$$

where c_{ij} is a correlation coefficient and takes values in the range $-1 \leq c_{ij} \leq 1$. This surprising behavior arises because of the absence of mode coupling in the cables and the interaction between pairs of graded index fibres. If successive fibres are alternately over and under compensated, then ray paths or modes in one that has relatively slow group velocity will have relatively fast group velocity in the second and vice-versa, so that a pulse that spreads out in the first fibre can, by mode dispersion, be compressed in the following fibre. Such a situation corresponds to $c_{ij} < 0$. Pairs of like fibres lead to values of $c_{ij} > 0$ while heavy mode coupling corresponds to $c_{ij} = 0$.

This generalised model has been used to explain the detailed behaviour of the whole link of 40 km of fibre and excellent agreement between theory and practice has been observed (16). Other effects were observed involving materials dispersion, which was complicated by the fact that the lasers chirped (their wavelength swept during the drive pulse) and this greatly complicated the interpretation of some of the data (17).

Southampton University (18-27) has been heavily engaged in the study of propagation in monomode fibre. Both theoretical and experimental analysis has been carried out using fibres made by the CVD process. Of particular interest have been their studies of curved monomode fibres, the shift in the mode position and the steady state and transient losses associated with the bends. Other more general studies have considered the characterisation of such fibres and the losses at joints. A theme here has been the development of simple approximations to predict the observed effect to allow system design to proceed more readily without the complexity of an exact model.

The losses from small V-value fibres through tunnelling have been studied by J. Arnold (28) at Queen Mary College, London, as part of a continuing study of fibres that are intermediate in size to the monomode fibre at one extreme and the typical multimode fibre that carries some 500 to 1000 modes at the other. Arnold (29) has also made a study of the group delay dispersion in graded fibres of small V-value and noted that very rapid variation of rms pulse width with profile parameter α, typical of 500 mode fibres, does not occur; nor does the minimum dispersion occur for the same α value, since the effect of the cladding on the guided modes is no longer negligible.

FIBRE MEASUREMENT TECHNIQUES

A very wide range of measurement techniques has been developed and deployed in the various active laboratories. Two techniques for studying the refractive index profile of a fibre have been des-

cribed. The first, by Southampton University (30) is the now widely
used near field profile technique in which the near field intensity
on the end of a short length of fibre is scanned when it is excited
by diffuse illumination. The technique works well but requires
processing of the experimental data with some simple correction
factors to remove the effect of tunnelling leaky modes (31). A
modified technique which avoids the use of correction factors, re-
lying only on refracting rays, was described by W.J. Stewart (32)
and has been further developed by the BPO (33).

A powerful facility for the study of the wavelength dependence
of fibre parameters has been built by Southampton University and
consists of a flashlamp pumped dye laser which is used to pump an
optical parametric oscillator. This system can deliver fast pulses
over a very wide wavelength range, 400 nm to at least 1.6 microns,
and by the use of an electro-optic modulator to chop the pulses, can
generate impulses as short as 0.5 ns. This equipment provides an
almost unique facility for the study of fibres in and around the
zero material dispersion region at 1.3 microns.

W.J. Stewart (34) of Plessey Research Laboratories has also
developed some elegant methods for the study of the mode structure
and properties of multimode step or graded fibres. In general,
these involve the use of prism couplers, of the type used for the
excitation of planar integrated optical guides, to excite pure
single modes in the fibre or to detect the energy associated with
such a mode.

At the BPO (35), a special apparatus was designed some years
ago for the calorimetric measurement of fibre absorption loss and
this has been used for routine analysis of fibres ever since. This
is backed by all the usual facilities for the measurement of total
insertion loss, bandwidth, pulse echo, scatter loss, wavelength
variation of attenuation, etc. Two test vehicles have been equipped
for use as mobile fibre measurement laboratories and they can
operate independently or as a pair to make insertion loss, echo or
bandwidth measurements on installed cables in operational duct
situations (36).

Cables have been described in a number of papers (37-40) and
have opened up new fields for the study of fibres in a very complex
environment. A selection of these papers illustrates the range of
measurement problems met.

FIBRE JOINTING, SPLICING AND CONNECTORS

A number of field splicing techniques have been developed by
the BPO, BICC and GEC using variants of the V-groove principle in
which the fibres to be connected are aligned in a miniature V block,

usually formed in copper by impression. The fibres are index
matched by epoxy resin or a liquid and held in place by either the
resin or a clamp plate. A simple jig for making such joints in the
field has been described by the BPO (41) together with tools for end
preparation and these have been tested in the BPO field installation
exercise (42,43).

A separate approach to splicing fibres in the field has been
developed by STL (44) who use a low melting point glass sleeve into
which the fibres are sealed. One fibre end is placed in the sleeve
to its mid-point and that end of the sleeve is heated and allowed
to shrink down onto the fibre. The second fibre end is then placed
in the free sleeve end and either glued or sealed in place. The
completed assembly is mounted within a longer metal tube sleeve for
mechanical protection.

Fibre connector designs have been published by a number of
manufacturers and the reader is referred to the literature for
further details of these (45,46).

SOURCES

The development work on sources for fibre systems has been
concentrated in three centres – the BPO, Plessey and STL. At the
BPO, short cavity proton isolated stripe GaAlAs lasers have been
studied and devices with attractive performance have been made.
In particular, low threshold current and "kink-free" output devices
have been sought (47-49).

Plessey has devoted most effort to the development of high
radiance LEDs and offers a range of products in both the 900 nm
wavelength region using GaAs and the 1.1 micron region based upon
GaInAs. Devices with integral lens construction to increase the
efficiency of power coupling between a very small area device and
a multimode fibre are also offered.

STL has developed a number of different GaAlAs lasers, some
of which are offered as products. Their current work is centred
on oxide isolated narrow stripe lasers for fibre systems and such
lasers have been successfully used in a number of system demonstrations. The device fabrication work is backed by a substantial
exploratory programme as evidenced by their published work (50-59).

Studies of the failure mechanisms (60-62) of GaAlAs lasers have
been carried out by the BPO and STL and have resulted in a number
of publications concerned with the methods of testing and the interpretation of the results so obtained, in addition to showing that
devices with usefully long lives are now becoming available.

SYSTEM DESIGN

Considerable attention has been applied to the design of both
transmitter and receiver in the fibre transmission system. The
major problem addressed in transmitter design has been the strategy
for controlling the laser electrically to give the desired optical
output to the fibre. In all cases, a monitor photodiode has been
assumed, but a number of control strategies have been analysed and
tested. STL has described the use of a simple mean power feedback
control (63) and a more sophisticated control circuit (64) which
samples the laser 0 and 1 levels only when a long sequence of
continuous 0s or 1s, respectively, is present. In both cases, low
bandwidth control circuitry is used and hence the power consumption
and cost of the components is low. In the former case, a single
signal is derived which can be used to control either the pulse
drive or the bias current; in the latter case two control signals
are derived so that both variables can be controlled.

The BPO has also described the use of simple mean power feedback circuits but has also studied three other strategies (65). The
first uses a fast detector to monitor the peak and trough power
levels directly, but this is rather extravagant with electrical
power. The second uses the switch-on delay of the laser to provide
a control signal for the laser bias and a separate mean power feedback circuit to control the pulse drive current, while the third
design monitors the position of the device threshold by applying a
low frequency sinusoidal modulation on the bias current and monitors
the optical signal during the 0 transmit period, servo controlling
the bias to the threshold point at which there is a rapid change
in slope in the light output/drive current curve. This has proved
to be a low power consumption design (66). All these designs have
been tested in system trials.

Optical receivers have been studied in detail by the BPO. A
simplified version of the Personick theory (67) led to detailed consideration of the effects of extinction ratio on error rate (68) and
of the relative attractions of avalanche photodiodes (APD) versus
pin photodiodes (69). Following from this study has been the
development of high performance pin-fet hybrid receiver circuits
with comparable performance to APDs but without the complications
of the latter, either in fabrication or operation (70).

SYSTEMS EXPERIMENTS

A number of experimental systems have been constructed and
tested, some in conditions that closely resemble those of a fully
operational system. These are listed in Table 1 in order of bit
rate, all having been binary digital transmission systems. The
major activity has been at 8 and 140 Mbit/s levels with application

TABLE 1

2 Mbit/s - 30 Channel Systems

repeater spacing	route length	cable/ fibre	source	estab- lishment	location
6 km	1 km	GEC/ TCL	Plessey LED	GEC/TCL	British Rail Wilmslow
10 km	10 km	BICC/ PO	ELED	PO	Maidstone/ West Malling
15 km	10 km	BICC/ PO	PO laser	PO	(planning)

8 Mbit/s - 120 Channel Systems

repeater spacing	route length	cable/ fibre	source	estab- lishment	location
13 km	13 km	BICC/ Corning	PO laser	PO	Martlesham/ Ipswich
6/7 km	13 km	BICC/ Corning	BNR LED	PO	Martlesham/ Ipswich
6 km	12 km	BICC/ Corning	Plessey LED	Plessey/ BICC	Slough/ Maidenhead
9 km	9 km	STC	STL laser		Stevenage/ Hitchin
17 km	6 km	TCL/GEC	Plessey LED	GEC	Uxbridge/ Ruislip

140 Mbit/s - 1920 Channel Systems

repeater spacing	route length	cable/ fibre	source	estab- lishment	location
8 km	6 km	BICC/ Corning	STL laser	PO	Martlesham/ Kesgrave
3 km	9 km	STC	STL laser	STC	Stevenage/ Hitchin
11 km	-	GEC	laser	GEC	laboratory

in the junction and trunk networks respectively in mind.

   Two of the 8 Mbit/s demonstrations are of particular interest.
The BPO link (71) between the Research Centre at Martlesham and the
group switching centre in Ipswich, 13 km away, has been operating
for over a year. It can operate without an intermediate repeater
when laser sources are used, or with one intermediate repeater
when LED sources are used. The receivers in this system achieve
10^{-9} error rate with a mean receive power of -63 dBm when a laser
source is used. An LED source gives rise to a penalty of some
2 dB for the same error rate because of the material dispersion
(over the shorter section length, 6 to 7 km versus 13 km). Several
transmitters have been tested on the system, an LED unit employing
a Bell Northern device, and two laser units, one with an STL laser
operating at 160 mA threshold current and 150 mA bias and one with
a BPO laser operating at 50 mA threshold and zero bias current.

   The GEC link shown in the table is of interest because,
although it has not yet been tested in the field, in the laboratory
it has shown satisfactory operation over 17 km of fibre cable when
operated with an LED source. This performance was possible because
the GEC fibre used had a loss in cable of less than 3 dB/km. It
is thus an interesting pointer to things to come.

   At the 140 Mbit/s level, the STC system is interesting as being
the most fully engineered repeater system yet shown (72,73). It
operates over a 9 km route with two intermediate repeaters. These
are housed in footway boxes at the side of a main road and are
remotely powered and operated. The terminal equipment is housed
in BPO exchange buildings and numerous tests have been carried out
over the completed link, including the transmission of two 70 Mbit/s
digital television channels.

   The BPO 140 Mbit/s system (74) has been operated over 6 and 8 km
of fibre cable running from the Research Centre to a nearby telephone exchange building at Kesgrave. As with the STC system, the
cable was installed in ducts (75) already used by existing telephone
cables and the terminal equipment was housed in buildings. No
intermediate repeaters were needed. The receiver performance was
checked over the 6 and 8 km routes and mean receiver powers of -47
dBm and -43.5 dBm were found with 10^{-9} error, the difference being
caused by the degradation of the signal over the additional 2 km
of fibre in cable. Detailed studies of the factors controlling the
system performance have been made for both the BPO systems and the
results compared with theoretical predictions (76).

   Finally, we note another experimental system, tested only in
the laboratory by GEC (77), in which operation over 11 km of fibre
has been demonstrated at 140 Mbit/s. As with their 8 Mbit/s system,
the long repeater section achieved reflects both low fibre loss

(under 3 dB/km) and good profile control in the graded index fibre used. A laser source was used for this system, since material dispersion would rule out the use of an LED.

These experiments have demonstrated that fibre systems can be installed and operated successfully and without any major departure from current installation practice. In addition, the performances that have been demonstrated, namely repeater spacings to about 10 km at the 140 Mbit/s level and about 15 km at the 8 Mbit/s level, appear most attractive to network planners and it seems likely that systems of this general form will begin to find their way into the transmission network in a traffic carrying capacity over the next few years.

Longer term studies in the UK cover the development of longer wavelength devices at a number of laboratories and integrated optics at Glasgow University and University College, London. The results of this work do not seem likely to appear until the second generation of fibre systems, the first generation being firmly based on graded index fibres, sources fabricated from GaAs and GaAlAs with silicon detectors.

REFERENCES

1   Kao, K.C., and Hockham, G.A., Proc IEE, 113, 1151 (1966).
2   Reid, A.M., Harper, D.W., and Forbes, A., British Patent
    50549 (1969).
3   Macedo, P.B., et al., US Patent 3,938,974 (1976).
4   Gambling, W.A., Payne, D.N., Hammond, C.R., and Norman, S.R.,
    Proc IEE, 123, 570 (1976).
5   Payne, D.N., and Gambling, W.A., Electron Letts., 11, 176
    (1975).
6   Payne, D.N., and Hartog, A.H., Electron Letts., 13, 627 (1977).
7   Adams, M.J., Payne, D.N., Sladen, F.M.E., and Hartog, A.H.,
    Electron Letts., 14, 64 (1978).
8   Sladen, F.M.E., Payne, D.N., and Adams, M.J., Electron Letts.,
    13, 212 (1977).
9   Hammond, C.R., Opt. Quant. Electron, 10, 163 (1978).
10 Black, P.W., Irven, J., Byron, K., Few, I.S., and Worthington,
    R., Electron Letts., 10, 239 (1974).
11 Beales, K.J., Day, C.R., Duncan, W.J., Midwinter, J.E., and
    Newns, G.R., Proc. IEE, 123, 591 (1976).
12 Beales, K.J., Day, C.R., Duncan, W.J., and Newns, G.R., Electron
    Letts., 13, 755 (1977).
13 Beales, K.J., Day, C.R., Duncan, W.J., and Newns, G.R., Third
    European Conference on Optical Communications, Munich,
    September, 1977, Pub: VDE, Berlin (1977).
14 Eve, M., Hensel, P.C., Malyon, D.J., Nelson, B.P., Stern, J.R.,
    Wright, J.V., and Midwinter, J.E., Opt. Quant. Electron, 10,
    253 (1978).

15   Eve, M., Opt. Quant. Electron, $\underline{10}$, 41 (1978).
16   Midwinter, J.E., and Stern, J.R., IEEE Trans. Comm., to be
      published July, 1978
17   Wright, J.V., and Nelson, B.P., Electron Letts., $\underline{13}$, 361 (1977).
18   Gambling, W.A., Payne, D.N., and Matsumura, H., IEE Trans. MOA,
      $\underline{1}$, 13 (1976).
19   Gambling, W.A., Payne, D.N., Matsumura, H., and Dyott, R.B.,
      Electron Letts., $\underline{12}$, 546 (1976).
20   Gambling, W.A., Payne, D.N., and Matsumura, H., Electron
      Letts., $\underline{12}$, 567 (1976).
21   Gambling, W.A., Payne, D.N., Matsumura, H., and Norman, S.R.,
      Electron Letts., $\underline{13}$, 133 (1977).
22   Gambling, W.A., Payne, D.N., and Matsumura, H., Electron
      Letts., $\underline{13}$, 139 (1977).
23   Gambling, W.A., Matsumura, H., and Sammut, R.A., Electron
      Letts., $\underline{13}$, 695 (1977).
24   Gambling, W.A., and Matsumura, H., Electron Letts., $\underline{13}$, 691
      (1977).
25   Gambling, W.A., and Matsumura, H., Electron Letts., $\underline{13}$, 532
      (1977).
26   Gambling, W.A., Matsumura, H., and Cowley, A.G., Electron
      Letts., $\underline{14}$, 54 (1978).
27   Gambling, W.A., Matsumura, H., and Ragdale, C.M., Electron
      Letts., $\underline{14}$, 130 (1978).
28   Arnold, J.M., IEE Trans. MOA, $\underline{1}$, 93 (1977).
29   Arnold, J.M., IEE Trans. MOA, $\underline{1}$, 203 (1977).
30   Sladen, F.M.E., Payne, D.N., and Adams, M.J., Appl. Phys.
      Letts., $\underline{28}$, 257 (1976).
31   Sladen, F.M.E., Payne, D.N., and Adams, M.J., Electron Letts.,
      $\underline{12}$, 282 (1976).
32   Stewart, W.J., Conference on Integrated Optics and Optical
      Communications, Kyoto, July, 1977.
33   White, K.I., to be published.
34   Stewart, W.J., First European Conference on Optical Communica-
      tions, London, September, 1975, Pub: IEE, London (1975).
35   White, K.I., Opt. Quant. Electron, $\underline{8}$, 73 (1976).
36   Eve, M., Hensel, P.C., Malyon, D.J., Nelson, B.P., and Wright,
      J.V., POEEJ, to be published June, 1978.
37   Jackson, L.A., Reeve, M.H., and Dunn, A.G., Opt. Quant.
      Electron, $\underline{9}$, 493 (1977).
38   Reeve, M.H., Electron Letts., $\underline{14}$, 47 (1978).
39   Lees, J., Proc. IEE, $\underline{123}$, 597 (1976).
40   Basket, R.E.J., and Foord, S.G., Electrical Comm., $\underline{52}$, 49
      (1977).
41   Hensel, P.C., North, J.C., and Stewart, J.H., Electron and
      Power, February, 1977.
42   Hensel, P.C., Electron Letts., $\underline{11}$, 581 (1975).
43   Hensel, P.C., Electron Letts., $\underline{13}$, 603 (1977).
44   Dalgoutte, D.G., Third European Conference on Optical Communi-
      cations, Munich, September, 1977, Pub: VED, Berlin (1977).

45   Hensel, P.C., Electron Letts., 13, 734 (1977).
46   Archer, J.D., Inter Nepcon 76, Brighton, October, 1976,
      Pub: Kiver Comm. (1976).
47   Steventon, A.G., Fiddyment, P.J., and Newman, D.H., Opt. Quant.
      Electron, 9, 519 (1977).
48   Newman, D.H., Bond, D.J., and Stefani, J., IEE Trans. Solid
      State and Electron Devices, 2, 41 (1978).
49   Matthews, M.R., and Steventon, A.G., Electron Letts., to be
      published.
50   Thompson, G.H.B., and Kirkby, P.A., IEEE J. Quant. Electron,
      QE-9, 311 (1973).
51   Goodwin, A.R., Peters, J.R., Pion, M., Thompson, G.H.B., and
      Whiteaway, J.E.A., J. Appl. Phys., 46, 3126 (1975).
52   Kirkby, P.A., and Thompson, G.H.B., J. Appl. Phys., 47, 4578
      (1976).
53   Selway, P.R., Proc. IEE, 123, 609 (1976).
54   Selway, P.R., and Goodwin, A.R., Electron Letts., 12, 25
      (1976).
55   Thompson, G.H.B., Henshall, G.D., Whiteaway, J.E.A., and
      Kirkby, P.A., J. Appl. Phys., 47, 1501 (1976).
56   Whiteaway, J.E.A., and Thompson, G.H.B., IEE Trans. Solid State
      and Electron Devices, 1, 81 (1977).
57   Thompson, G.H.B., Lovelace, D.F., and Turley, S.E.H., IEE
      Trans. Solid State and Electron Devices, 2, 12 (1978).
58   Henshall, G.D., and Hinton, R.E.P., IEE Trans. Solid State and
      Electron Devices, 2, 31 (1978).
59   Selway, P.R., Kirkby, P.A., Goodwin, A.R., and Thompson, G.H.B.,
      IEE Trans. Solid State and Electron Devices, 2, 38 (1978).
60   Ritchie, S., Godfrey, R.F., Wakefield, B., and Newman, D.H.,
      J. Appl. Phys., to be published 1978.
61   O'Hara, S., J. Phys. D, Appl. Phys., 10, 409 (1977).
62   Goodwin, A.R., Kirkby, P.A., Pion, M., and Baulcomb, R.S.,
      IEEE J. Quant Electron, QE-13, 696 (1977).
63   Ramsey, M.M., Horsley, A.W., and Epworth, R.E., Proc. IEE, 123,
      633 (1976).
64   Epworth, R.E., and Allsop, B.E., Colloquium on Optical Communi-
      cation Systems, May, 1978, IEE, London.
65   Salter, S.R., Smith, D.R., White, B.R., and Webb, R.P., Third
      European Conference on Optical Communications, Munich,
      September, 1977, Pub: VDE, Berlin (1977).
66   Hodgkinson, T.G., Hooper, R.C., Smith, D.W., and White, B.R.,
      Colloquium on Optical Communication Systems, May, 1978, IEE,
      London.
67   Smith, D.R., and Garrett, I., Opt. Quant. Electron, 10, 211
      (1978).
68   Hooper, R.C., and White, B.R., Opt. Quant. Electron, 10, 279
      (1978).
69   Smith, D.R., Hooper, R.C., and Garrett, I., Opt. Quant. Elec-
      tron, 10, 293 (1978).
70   Smith, D.R., Hooper, R.C., and Garrett, I., Colloquium on
      Optical Communication Systems, May, 1978, IEE, London.

71   Brace, D.J., and Ravenscroft, I.A., POEEJ, 70, 146 (1977).
72   Hill, D.R., Howard, P.J., and Weston, J.D., Colloquium on
      Optical Communication Systems, May, 1978, IEE, London.
73   Chown, M., Electrical Comm., 52, 170 (1977).
74   Berry, R.W., and Ravenscroft, I.A., POEEJ, 70, 261 (1978).
75   Brace, D.J., and Hensel, P.C., POEEJ, 71, 54 (1978).
76   Berry, R.W., Brace, D.J., and Ravenscroft, I.A., IEEE Trans.
      Comm., to be published July, 1978.
77   Collins, P.V., Colloquium on Optical Communication Systems,
      May, 1978, IEE, London.

FIBER AND INTEGRATED OPTICS IN FRANCE

J. M. Decaudin

DRET
26, Boulevard Victor
75996 Paris, France

The purpose of this paper is to present the activities in fiber and integrated optics in French universities and industry. This presentation is not exhaustive but does include the main teams involved in this field.

The first part describes middle- and long-term research on integrated optics, while the second part is devoted to short term developments in optical fibers, especially in the components range.

INTEGRATED OPTICS

Both universities and industry are concerned with theoretical and basic research essential for developing components and systems.

The largest and most advanced group is the Thomson-CSF Research group in L.C.R. (Laboratoire Central de Recherches). Investigations started there in 1971 and major problems in the fabrication of microstructures have been solved using techniques of thin layer deposit, masking, and etching, as well as diffusion, ion exchange, and implantation. Most of the guides are diffused (titanium into lithium niobate), and they are checked either by the deposit of a fluorescent layer on the guide or by observation of the mode structure at the edge.

One important problem in the development of devices has been the connecting of various elements like lasers, monomode fibers, and microguides. Permanent soldering between fibers or between fiber and guide have been made with a photopolymer material. A large effort has been devoted to integrated lasers and photodetectors.

Among the leading active components soon to be developed, we may mention:

- a phase modulator (800 MHz bandwidth and 0.3 V/rad driving voltage);

- an amplitude modulator (rise time < 700 ps, modulation ratio of 95% for a driving voltage of 1.5 Volt);

- a digital fast switch COBRA (rise time < 500 ps, crosstalk < 34 dB and driving voltage between 2 and 6 Volts);

- an optical active bifurcation BOA (crosstalk < 18 dB and driving voltage of 8 V and -18 V).

The state of development of these components is advanced enough for them to be joined together in more complex devices (for signal processing, for instance).

The second well known French team in this field is the Laboratoire d'Electronique et de Technologie de l'Informatique in Grenoble. The LETI undertook investigation of monolithic integrated optics in zinc telluride five years ago. This material has very interesting properties in the visible and near infrared range for electrooptic modulation and detection. Various guides have been manufactured (mainly by ion implantation) and the best results obtained using the confining effect induced by a dielectric microstrip above or on each side of the guide. Three dimensional guidance was observed on planar guides with boron implantation.

With regard to modulation, research has been made of high resistivity substrates having good crystalline quality. A non-optimized polarization modulator with a frequency of more than 50 MHz was manufactured and tested. Photodetection in ZnTe has been obtained with oxygen doping and the last problem to solve is the fabrication of lasers in ZnTe to develop a monolithic device.

Some smaller teams with less resources have advanced our knowledge of optical waveguides. The CNET (Centre National d'Etude des Télécommunications) has done some interesting experiments on modulators and two way commutators in lithium tantalate. Five research workers are involved. We may also mention the electro-optic laboratory at Nice University, where investigators are studying the modulation of optical guided wares at 10.6 μ by free carriers. The same team is trying to join integrated optics and the Josephson effect, and examining the Raman effect for fiber lasers. Some theoretical work is being done at the University of Marseille on the local distortion of the surface of optical waveguides.

This survey is not exhaustive, but it does attempt to give an idea of the main teams involved and some other, smaller teams partly involved in work with integrated optics in France.

OPTICAL FIBERS

Industry is much more concerned with optical fibers because it has the ability to make fibers, cables, connectors, systems, etc., which require money, time, and people. Moreover, without slighting theoretical and scientific problems met in optical fiber communications, technology seems for the moment to limit their applications. University groups are therefore mainly concerned with materials and characterizations.

Investigators at the University of Rennes have developed fluoride glasses from ZrF_4, BaF_2, and ThF_4 with very good characteristics (less than 20 dB on the preform) in spite of the lack of careful fabrication methods. These glasses have a melting point of about 600°C and a low foreseeable cost. The Institute of Optics in Orsay and the University of Limoges have achieved a set-up for measuring the index profile of optical fibers.

Between university and industrial activities, CNET plays a leading role. Aside from the well known test methods for fibers using microcalorimetry, CNET has made progress with the CVD method, the corresponding knowledge being ready to transfer to industry.

Research in fibers has been directed towards very large bandwidth fibers (gradient index or monomode fibers). Cable structure with a large number of fibers has been defined. Lasers manufactured by French or foreign industry have been characterized and some experiments made on CW lasers in Bagneux. Optical transmission at 1.3 μ and components for this wavelength have been considered.

The need for a link (34 Mbits, 7 km, 60 fibers) engaged a large part of the investigations, with only a few experiments for 140 Mbit links being carried out.

C.G.E. Group

This group has been working on optical fiber links for 5 years, now with 90 technicians and engineers, all of them in CLTO, which has been created in the regrouping of the activities of Filotex and Cables de Lyon and supported by Laboratoire de Marcoussis. Systems are designed by CINTRA and CIT Alcatel.

Research and developments cover the whole range: fibers, cables, connectors, emitters and receivers, and systems (30% short distance and 60% long distance).

For fibers, two production methods have been chosen:

  - the CVD method for low loss and large bandwidth fibers (3
    dB/km, > 400 MHz are usually obtained);

  - multicomponent glasses like sodocalcic or borosilicate are
    elaborated (20 dB/km, Na > 0.4) for plastic or glass cladded
    fibers.

Various connectors for bundles or single fibers, with the
necessary cutting, striping, and soldering tools are available.
With a portable blowlamp, fibers were soldered with less than
0.2 dB loss.

For the terminal components, CGE has provided GaAlAs double hetero-structure light-emitting diodes and a CW laser with a power output of 5 to 10 mW for a threshold current of 150 mA with a lifetime
of more than 5000 hours. Avalanche and PIN photodiodes are also
manufactured.

Thomson-CSF Group

Like the CGE group, Thomson is concerned with all the components and devices for fiber optic links, but activities are not
centralized; a lot of branches of the group are involved.

Basic research on fibers and LED CW lasers are the responsibility
of the Laboratoire Central de Recherche. Two types of fibers have
been developed:

  - a doped silica fiber for long range telecommunications (last
    results are 3 dB/km at 0.85 μ and 700 MHz bandwidth);

  - a fiber with phase separated glasses; the attenuation is now
    30 dB/km and the aim is to obtain a gradient.

To improve the lifetime of the lasers (currently more than 5000
hours) accelerated degrading tests at 70°C are performed on a large
number of lasers.

An avalanche photodiode using InP for 1.3 μ transmission has
been obtained and tested in the laboratory.

Fiber pulling and cabling take place in the Cables Isolés Division and in the new subsidiary, LTT.

LCR is in charge of connectors for the studies and SOCAPEX for
their development.

St. Gobain Group

     Quartz et Silice and Sovis are the subsidiaries of St. Gobain
involved in work on optical fibers.

     Quartz et Silice uses an inductive plasma blowlamp for the
growing of silica preforms. This high-purity silica has very low
metallic ion content. Rods heavier than 70 kg are usually pulled
(they provide 500 km fibers with a 200 μ core diameter and less
than 5 dB/km). The 200 μ to 1 mm core diameter fibers are coated
with silicon and tefzel, which gives good mechanical characteristics.
Quartz et Silice cooperates closely with ATI for connectors and
emitter or receiver modules.

     Sovis has turned its attention to the flexible bundles used as
light conductors and to coherent fiber optic plates.

FORT

     This small society, very well-known in endoscopy, has some
interest in optical communications. They have gotten preforms from
an English group and are now trying to make these by themselves to
provide silica core fibers with silicone cladding for graded index
fibers. They also make cables available.

     In addition to these French societies, subsidiaries of ITT and
Philips have some activities in France, with technical help of the
whole group.

     LCT, from the ITT group, has realized point-to-point or data
BUS links and specific components (a star coupler with a tube mixer).

     LEP and RTC, from the Philips group, are involved in different
ways. RTC has been developing terminal components with very good
results:

     - fast PIN photodetectors (0.4 A/W and risetime of < 1 ns);

     - low power emitting diodes (1 to 5 W cm^{-2} Sr^{-1}) and soon high
       power emitting diodes (50 W cm^{-2} Sr^{-1}).

LEP tries to integrate these components into modules with particular
efforts to improve the coupling.

     One important thing that was not mentioned but is often a limit
for optical fiber development is the connection. In addition to the
previous groups, nearly all the connector manufacturers (Deutch,
Radial, Souriau, AMP, etc.) have optical connectors for sale, but
these are often extrapolations from electrical connectors and are

not optimized (maybe because cables and interfaces are still not well standardized).

To bring this to a conclusion, the French government is very interested in fiber and integrated optics activities and supports a large part of the research through services like CNET, DRET, and DGRST. It will try in the future to get some standardization for all devices developed in France.

A NATIONAL REVIEW ON FIBER AND INTEGRATED OPTICS

ACTIVITY IN ITALY

A. M. Scheggi

Istituto di Ricerca sulle Onde Elettromagnetiche
Consiglio Nazionale delle Ricerche
Firenze, Italia

The research on optical fibers for communications has had, as is well known, an explosive growth since the beginning of the 1970's, especially in the U.S.A., Japan, and in Europe, the U.K., Germany, France, and The Netherlands. Today the industrial stage has already been achieved in some areas. In Italy also, research activity in this field has grown rapidly in several research laboratories. More recently several industries have also entered the field.

The main activities have been carried out at the following laboratories:

Centro Studi e Laboratori Telecomunicazioni S.p.A. (C.S.E.L.T.), Torino
Fondazione Ugo Bordoni (F.U.B.), Roma
Centro Informazioni Studi Esperienze (C.I.S.E.), Milano
Centro Elettrotecnico Sperimentale Italiano (C.E.S.I.), Milano
Istituto di Ricerca sulle Onde Elettromagnetiche del Consiglio Nazionale delle Ricerche (I.R.O.E.-C.N.R.), Firenze
Fondazione Guglielmo Marconi (F.G.M.), Pontecchio Bologna
Istituto di Elettronica dell'Università (I.E.BO), Bologna
Istituto di Elettronica ed Elettrotecnica dell'Università, Padova

The industries which have become involved include:

Pirelli, Milano	FACE Standard, Pomezia (Roma)
SGS-ATES, Milano	Elettronica S.p.A., Roma
SIT-Siemens, Milano	Telettra, Vimercate

C.S.E.L.T. belongs to the STET-IRI Group (a group of telecommunication equipment manufacturers and operating companies) and has
been involved in research on optical fiber communication since 1970;
in particular, C.S.E.L.T. participates in a research program and
telecommunication experiment in collaboration with Pirelli (in the
framework of an agreement with Corning Glass Works), SIP (the Italian
telephone company) and SIRTI (Società Italiana Reti Telefoniche
Interurbane) as cable installer.

F.U.B., a non-profit, private organization working in cooperation with the Post and Telecommunication Department, has been carrying out a research program on optical fiber communications since
1972. F.G.M. and I.E.BO collaborate in the framework of a joint
venture. C.I.S.E. and C.E.S.I. are research organizations supported
mainly by industries with a broad field of interest. As concerns
optical fibers, they have worked mainly on short distance data
transmission systems for special purposes.

The main research subjects can be identified as: 1) fibers and
cables; 2) sources and detectors; 3) measurement methods; and 4)
systems.

1.   FIBERS AND CABLES

In Italy, optical fibers have been fabricated only at C.S.E.L.T.
and I.R.O.E. The purpose of this production was mainly that of ac-
quiring knowledge from direct experience in this field. The first
fibers produced at C.S.E.L.T. are of the graded index type obtained
by CVD technique. They give an attenuation of $\sim$ 4 dB/km and bandwidths up to 1 Ghz·km (Fig. 1) [1-3].

At I.R.O.E. a technology has been developed for fabricating
silica core silicone claded fibers which are of interest for applications in short distance communications and constitute in this
field a good compromise between cost and performance. These fibers
have been obtained by a technique which in principle is not new;
however, some noticeable modifications have been introduced such as
the use of R.F. induction heated furnaces with an inert gas laminar
flow for protecting the silica rod surface, thus avoiding the contamination of the core surface, which in this type of fiber is an
important cause of loss [4]. The silicone cladding is operated on
line at the exit from the furnace by dipping the fiber into the
resin and curing it in an oven placed in cascade with the resin
vessel. A second plastic protection coating is applied by using a
similar line in cascade with the previous one. Fibers have been
produced having a core of Suprasil 1 and W1 (Heraeus), cladding of
Sylgard 182 (Dow Corning), and plastic coating of Kynar 7201 (Penn-
walt), with an external diameter of about 300 μm and a core diameter

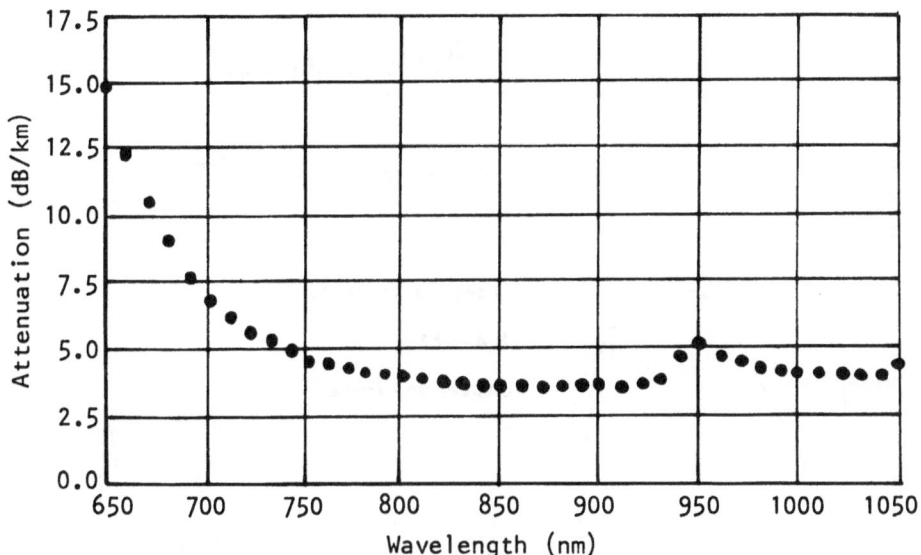

Fig. 1. Spectral attenuation curve of C.S.E.L.T. fiber.

of 140 μm. The numerical aperture is 0.2 and the bandwidth is about
140 Mhz·km. Fig. 2 shows the typical behavior of the spectral attenuation of these fibers with a minimum of about 10 dB/km. Mechanical tests performed on 0.5 m long samples give a fracture strength
with a mean value of 4.93 kg with a standard deviation of 1.52 kg [5].
These fibers have been supplied to F.U.B. and I.E.BO-F.G.M. for
setting up particular measurement methods which have been developed
there [6].

     As for cable, the Research and Development Laboratories of
Pirelli Industries are directing their activity mainly towards such
fiber cabling problems as fiber jacketing, stranding of jacketed
fibers, application of end protection, and behavior of the cable
during the laying operation. For the last aspect, Pirelli collabo-
rates with SIRTI. The main characteristic of the cable constructed
by Pirelli for the COS 2 experiment whose details will be given in
the sequel, is the choice of the loose jacketing of the fibers in
polypropylene tubes having a 1 mm internal diameter. The cable
structure is suitable for incorporating up to 8 fibers in one layer
around a plastic jacketed central steel strength member on which the
tubes containing the fibers are stranded. The task of the metallic
member is that of absorbing construction and laying stresses. The
cable has a length of 1 km, an outer diameter of 11 mm, and a weight
of 100 g/m. For reasons of cost, only 3 graded index Corning fibers

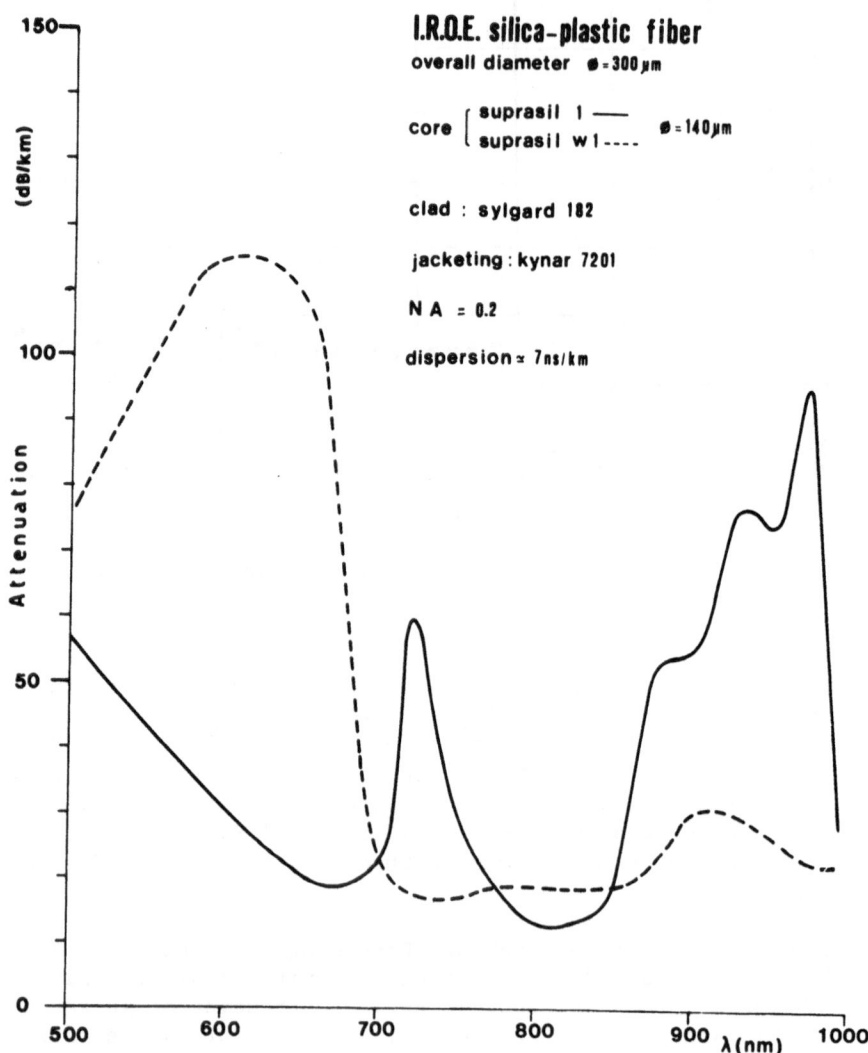

Fig. 2.   Spectral attenuation curve of I.R.O.E. fiber.

were inserted into the cable, with attenuation of 3.8 dB/km and a
bandwidth of about 700 MHz [2,3].   For the same experiment, Pirelli
constructed another cable (1 km long) with seven fibers: four fibers
supplied by Corning and three by C.S.E.L.T.   In the framework of a
collaboration with Elettronica S.p.A. and F.G.M. a cable for short
distance applications has been manufactured with six Pilkington
fibers, utilizing the same structure as that of the long distance
cable.

1.1. Propagation in Optical Fibers

It is worthwhile to mention the theoretical studies of propagation in optical fibers which have recently been performed by the different research groups, in particular by C.S.E.L.T., I.E.BO-F.G.M., I.R.O.E., and F.U.B. Such studies range from the e.m. analysis of propagation in fibers with any index and any loss distribution to the investigation of the problem of power launching and of coupling efficiency between multimode optical fibers. Investigations have also been performed on the dispersion characteristics of fibers with different refractive index distributions (of practical interest) as well as on the coherence propagation properties in multimode fibers, also in view of its relationship to mode coupling. However, for the details, reference can be made to the bibliography [7-21].

2. SOURCES AND DETECTORS

After initial technological activity developed at C.I.S.E., there is at present only one group (a collaboration between C.S.E.L.T. and SGS-ATES) operating at SGS-ATES (Castelletto, Milano). This group has the task of setting up sources and detectors suitable for applications in optical fiber communication systems. High radiation LEDs of the Burrus type have been constructed. The characteristics of these LEDs are shown in Table I for two types of Zn sources used in the diffusion process [22]. More recently [23] high radiation LEDs have been fabricated by utilizing an original diffusion process of Zn in GaAs which has allowed the reduction of commutation times to 1/4-1/5 of the value of those with extended diffusion ($\sim$ 5 ns). These LEDs are suitable for 50 Mbit/sec communication system. Experimental investigations have also been carried out on the coupling efficiency between a high radiation LED and the fiber.

Table I

	Source	
	ZnAs$_2$	As/Zn/As
Diffusion Temperature	700°C	850°C
Junction Depth	12 μm	13 μm
Typical Rise Time	5 ns	30 ns
Typical Radiation @ 300 mA	3 W/sr cm^2	35 W/sr cm^2
Emission Wavelength	903 nm	903 nm
Spectral Emission	30 nm	30 nm

In the field of AlGaAs lasers, the activity is mainly dedicated to the development of a laser emitting at 840 nm. To this purpose, the liquid phase epitaxy system has been set up, the first double heterojunction structure has been grown, and the first lasers are now being tested.

PIN detectors with advanced characteristics have been constructed with a high responsivity in the 600–950 nm range (Fig. 3) [22]. Silicon avalanche detectors are being developed for which the most promising technology seems to be the planar one using ion implantation.

Some research on source characterization is being carried out at F.U.B. Spatial and chromatic delay of the radiation emitted by LEDs have been examined as well as the physical models which can explain these effects [24,25].

3. MEASUREMENT METHODS

Typical measurements (such as attenuation, pulse spreading, dispersion, numerical aperture, index profiles, and so on) are performed at the different laboratories in order to check the different components of communications systems. However, it is to be noted

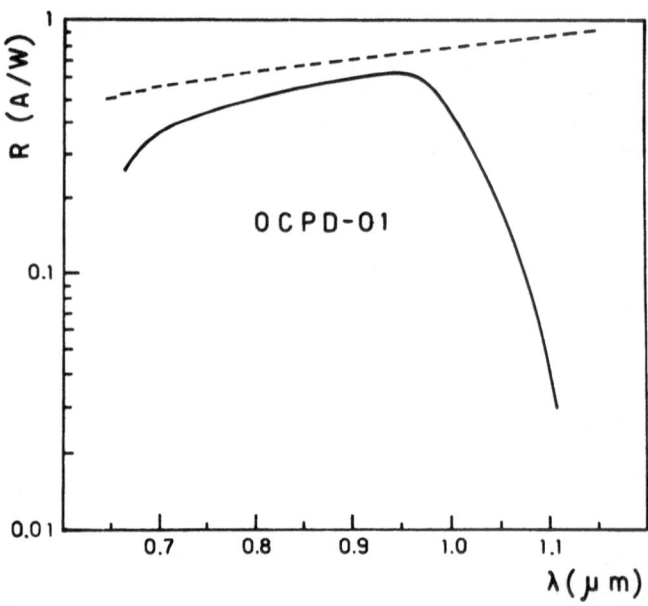

Fig. 3. Responsivity of the S.G.S.–ATES OCPD–01 detector. The dashed line represents the theoretical limit.

that some characteristic parameters of optical fibers are not yet
completely defined and that the results of the measurements can in
some cases depend on the conditions in which they are performed.
For instance, the dependence of the determination of attenuation,
dispersion and index profile on the fiber length is strictly related
to the illumination conditions. This seems to indicate that there
are still open questions in the research on measurement methods.
Research of this type is carried out in Italy at C.S.E.L.T., I.E.BO-
F.G.M., and F.U.B.

At I.E.BO-F.G.M. a nondestructive method has been developed
for spectral attenuation measurement in the 700-1100 nm range. The
method utilizes light scattered by the outer surface of the fiber
and collected by two detectors placed along the fiber in suitable
positions. With its high sensitivity, this method is suitable for
the fast determination of the lengths at which the losses tend to
the stationary value as a function of the employed optical source
[26]. Fig. 4 shows an example of spectral curves measured on a
particular fiber coated with a thin plastic layer, by placing the
first detector at 10, 20, and 50 m from the illuminated end face of
the fiber and keeping the other one at 1 km distance from the source.
The details will be presented at the IV European Conference on Optical Communications (ECOC) in Geneva (September, 1978). Another
parameter of interest for fiber characterization is the microbending

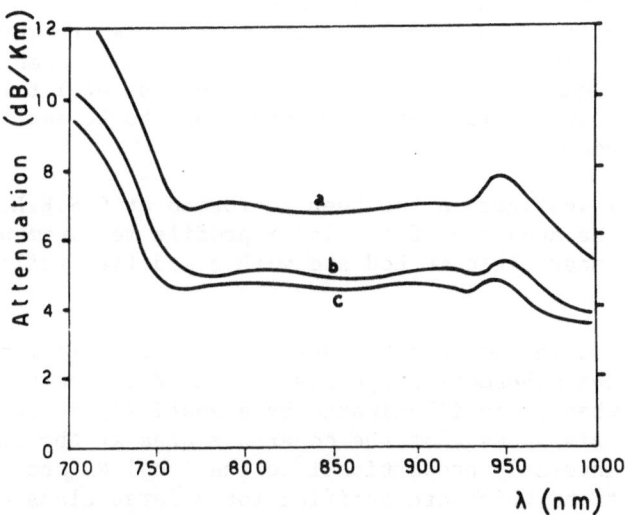

Fig. 4.   Spectral attenuation of an AWA 168 fiber measured at
          I.E.BO-F.G.M. Curves a, b, and c correspond to three
          different positions (10 m, 20 m, and 50 m, respectively,
          from the source) of the first lateral detector.

attenuation which occurs when the fiber is pressed on a rough surface. This problem has been tackled at I.E.BO-F.G.M. from both the
theoretical and experimental points of view for silicone cladding
fibers. The theoretical results have already been published [27].
They indicate a proportionality between microbending losses and the
square root of the cladding attenuation and an increase of such
losses when the wavelength increases. The experimental tests carried out on I.R.O.E. fibers gave results in good agreement with the
theoretical predictions and will also be presented at the IV ECOC
in Geneva.

A modified backscattering technique [28,29] has been developed
at both C.S.E.L.T. and F.U.B. for nondestructive attenuation measurements. The method consists in sending a short optical pulse
into the fiber and analyzing the behavior of the light backscattered
from the same input face of the fiber versus time. A difficulty of
this type of measurement lies in the weak power level of the backscattered signal which requires either elimination of the reflection
at the fiber input face or operation of a suitable processing on
the return signal. The experimental apparatus set up at F.U.B. is
shown in Fig. 5 (upper part). The elimination of the first reflection is obtained by electronic gating on the photodetector. In the
apparatus set up at C.S.E.L.T. the beam splitter is placed in an
index matching cell where the refractive indexes are so accurately
matched as to eliminate all the spurious reflections (Fig. 5, lower
part). This method has the advantage of launching and detecting
the light from the same end face; it provides detailed information
on attenuation characteristics along the fiber such as defects,
faults, discontinuities, and joints. Research is in progress at
F.U.B. for the analysis of joint characteristics by means of such a
method. Experimental tests are also in progress at I.E.BO-F.G.M.
on fiber-to-fiber and fiber-to-LED connectors to be emplyed in short
distance systems.

Another investigation has been performed at C.S.E.L.T. for
ascertaining the accuracy of the index profile measurements attain-
able with the near field method and with a modified reflection
method [30].

At F.U.B. a method is being developed which relies on the observation of the coherence properties of the field on the end face
of the fiber when it is illuminated by a spatially incoherent source.
A simple analysis shows that the coherence area at the end face of
the fiber is inversely proportional to the local NA and hence, under
certain conditions which are verified for a large class of fibers,
inversely proportional to the refractive index profile at the considered point. A visualization of the refractive index can be obtained at the end face of the fiber by imaging it through a magni-

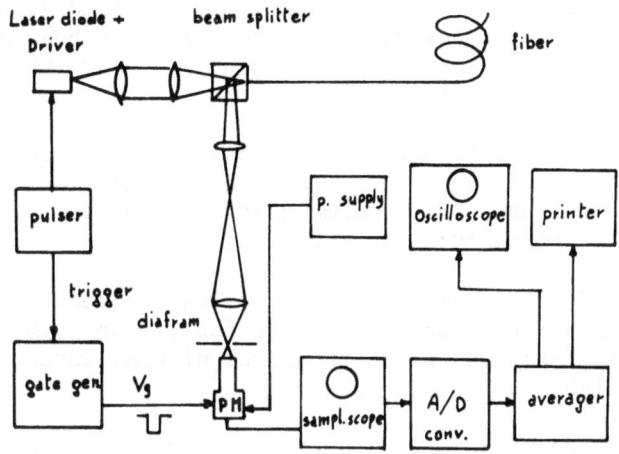

Experimental apparatus (FUB)

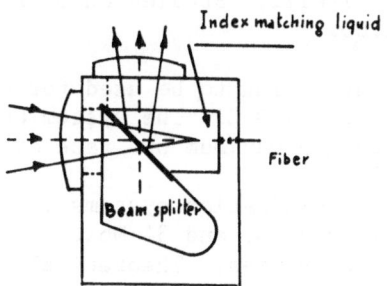

Beam splitter cell (CSELT)

Fig. 5. Experimental apparatus for the attenuation measurement with the modified backscattering technique.

fying lens and a reversing front interferometer on a photographic plate. Detailed results will be presented at the IV ECOC [31].

4. SYSTEMS

Applications presently developed in Italy can be divided into two main categories: 1) systems for telemetering and remote control purposes; and 2) telecommunication systems.

The development of commercial systems for the first category (essentially suitable for short distance data transmissions with velocities of the order of 1 Mbit/sec) is interesting because of

the variety of fields where such systems can find either military or
civil applications. Activity in this field has been developed at
Elettronica S.p.A. (jointly with F.G.M.), at FACE Standard, C.I.S.E.,
C.E.S.I., I.E.BO, and I.R.O.E.

At C.I.S.E., a telethermometer has been developed for geother-
mal wells which makes use of a passive optical transducer and of an
optical fiber link. This thermometer can operate up to 500°C with
a sensitivity of 0.1°C down to 2 km of depth.

At C.E.S.I., a number of f.m. analogic telemetering systems over
lengths of ∿20 m have been developed for use on high voltage or high
power testing stations where the level of e.m. interference is particularly high.

FACE Standard developed a data transmission system (2 Mbit/sec)
to be used over distances of up to 500 m. A system having nearly the
same performance for bit rate and distance has been developed at
Elettronica S.p.A. with the support of F.G.M. and using a cable
manufactured by Pirelli. Studies on short distance systems are in
progress at I.E.BO.

An experimental link to be used for connecting two computers
has been tested at I.R.O.E. The system uses 100 m of I.R.O.E.
fiber and can reach a maximum bit rate of ∿50 Mbit/sec.

As for telecommunication systems (that is for distances longer
than 3 or 4 km with 2, 8, and 34 Mbit/sec transmission rates) there
are a number of activities. Theoretical studies for system optimization (with particular regard to the receiver project) are performed
at C.S.E.L.T., F.U.B., SIP, and the University of Padova. Activity
dedicated to industrial realization of systems (with particular
contributions to the terminal apparatus) is carried out at SIT-Sie-
mens (2, 8, and 34 Mbit/sec), FACE Standard (8 Mbit/sec) and Telettra
(8 and 34 Mbit/sec). F.G.M. has a collaboration contract with
Telettra for this activity. It is worthwhile to mention the activities in this field developed at C.S.E.L.T. on a 140 Mbit/sec
system which is already at the development stage and will soon be
transferred to industry. Study of a 560 Mbit/sec system is still
at the research stage.

The only field test developed up to now in Italy is that performed at C.S.E.L.T. This experiment was developed in two phases:
in March, 1976, a first experimental optical cable (experiment COS 1)
of ∿1 km length manufactured by Pirelli with Corning fibers was laid
by S.I.R.T.I. in a trench with protective concrete boxes and a reinforced concrete cover at the C.S.E.L.T. site [32]. The experiment
consisted mainly of attenuation and dispersion measurements and environmental tests.

A second more complex experiment (COS 2) [1,2,4], in progress
since August, 1977 in collaboration with SIP, consists of laying
about 5 km of Pirelli cable (whose characteristics have been described above) in telephone ducts connecting the "Stampalia" and
"Lucento" exchanges in the Turin local telephone network. The total
length of the main cable was about 4 km and it was possible to pull
up to 1 km lengths of optical cable in the ducts to reduce the number
of joints; in addition, a second optical 1 km cable (with 7 fibers,
3 of which are C.S.E.L.T. fibers) was laid in loop with both ends
inside the Stampalia exchange. The purpose of this loop cable was
to make possible the jointing to the main cable of one or more
1 km lengths in order to get links of different lengths. For
jointing the cable a particular splicing technique developed at
C.S.E.L.T. was employed (the *spring groove splice*) [3]. Attenuation
and dispersion measurements were also performed on the cable in this
case. But the important experiment is that of the 140 Mbit/sec
system developed at C.S.E.L.T. and with which it was possible to
obtain a link of about 9 km with a total attenuation of about 34
dB without repeaters and with an error rate of $< 10^{-9}$.

Aside from the above mentioned fields of application, it is
worthwhile to mention a recently renewed interest in fibers with low
attenuation either for scientific uses or for laser surgery and
therapy. In this line, a system for laser surgery has been developed
at I.R.O.E. which utilizes an I.R.O.E. fiber for transporting the
radiation of an Argon laser with 2 W power. The fiber is terminated
with a diopter obtained by CO_2 laser fusion [33]. Preliminary tests
have been performed on animals by the neurosurgery staff of the
S. Maria Nuova Hospital in Florence. The results will be utilized
for setting up a more definite apparatus with higher power lasers
for systematic experiments.

5. INTEGRATED OPTICS

Another subject of research activity in Italy which is strictly
related to optical fibers, is integrated optics. Research groups
in this field are working at the following laboratories:

Centro Studi Propagazione ed Antenne (C.E.S.P.A.) c/o Istituto
        di Elettronica del Politecnico, Torino
I.R.O.E.-C.N.R., Firenze
Istituto Elettrotecnico dell'Università, Bari
Istituto di Elettronica dell'Università, Palermo

The main activities carried out by these groups are listed below.

At C.E.S.P.A., activity has been devoted mainly to the investigation of propagation characteristics in guiding structures for in-

tegrated optics as well as to the fabrication and characterization
of dielectric wave guides [34-37].

At I.R.O.E., thin film geodesic lenses have been extensively
investigated since their first introduction in guided optics in
1972 [38,39]. At present, particular attention is devoted to the
design and implementation of one-dimensional signal processors [40,
41,42]. As for coupling devices, high frequency gratings obtained
in photoresist by laser interference techniques have been tested
as air-waveguide couplers via their evanescent diffracted waves
[43].

At the University of Palermo the following research subjects
have been considered:

- *thin film waveguide technology* - apart from the usual techniques, flash evaporation of compounds with a pulsed discharge
  CO_2 laser have been used for producing thin film waveguides
  [44];

- *electrooptical effects* - a simple method of fabrication of
  low loss junctions between dielectric and metal clad dielectric waveguides has been applied to realize an integrated
  optical device for detection of electrooptical nonlinearities
  in thin polymer films [45];

- *thin-film detectors* - a thin film detector for integrated
  optical systems which utilizes a photovoltaic effect in GaAs
  thin films and which can be fabricated on any dielectric substrate, has been fabricated and tested [46].

The research group of the University of Bari has been concerned
mainly with theoretical activity; in particular, propagation studies
have been performed in diffused anisotropic multilayer waveguides
for the design of electrooptical modulators. Mode conversion devices have also been investigated for the design of mode converters,
directional couplers, and electrooptical modulators [47-50].

REFERENCES

(1) G. COCITO, S. LONGONI, L. MICHETTI, L. SILVESTRI, D. TIBONE
F. TOSCO, A. VAGO : First Report on COS 2 Experiment Results -
3° Convegno Europeo sulle Communicazioni Ottiche, Munich, F.D.R.
(Sept. 1977) (p. deadline paper).

(2) Attività di Enti e Industrie : Il primo collegamento a fibre
ottiche in Italia - Alta Frequenza XLVII, n.3, 48 N (March 1978)

(3) G. COCITO, B. COSTA, S. LONGONI, L. MICHETTI, L. SILVESTRI,
D. TIBONE, F. TOSCO : COS 2 Experiment in Turin : Field Tests on
an Optical Cable in Ducts - IEEE Trans. on Communications (July
1978)

(4) P.F. CHECCACCI, A.M. SCHEGGI, M. BRENCI : R.F. Induction
Furnace for Silica Fiber Drawing - Electronics Letters 12, 265
(1976)

(5) M. BRENCI, P.F. CHECCACCI, F. COSI, R. FALCIAI, A.M. SCHEGGI :
Progressi nella tecnologia delle fibre ottiche in silice-plastica
realissate presso l'I.R.O.E.-C.N.R. - Alta Frequenza XLVII, 122
(1978)

(6) P.F. CHECCACCI, A.M. SCHEGGI, B. DAINO, G. SOMEDA : Caratteris-
tiche di fibre ottiche a basse perdite ottenute da dielettrici
disponibili commercialmente - LXXVIII Riunione Annuale AEI, Como
(Sept. 1977)

(7) P. DI VITA, R. VANNUCCI : Multimode Optical Waveguides with
Graded Refractive Index : Theory of Power Launching - Applied
Optics 15, 2765 (1976)

(8) P. DI VITA, R. VANNUCCI : Loss Mechanism of Leaky Skew Rays in
Optical Fibres - Optical and Quantum Electronics 9, 177 (1977)

(9) P. DI VITA, V. ROSSI : Effects of the Presence of a Cladding
of Finite Extent in Multimode Optical Fibre - Presented at the
International Workshop on Optical Waveguide Theory, Reisensburg
(Sept. 1977)

(10) P. DI VITA, V. ROSSI : Evaluation of Coupling Efficiency in
Joints between Optical Fibre - C.S.E.L.T. Rapporti Tecnici, Volume
V, n.5 (Dec. 1977)

(11) E. BIANCIARDI, V. RIZZOLI : Propagation in Graded-Core Fibres:
a Unified Numerical Description - Optical and Quantum Electronics 9,
121 (1977)

(12) E. BIANCIARDI, V. RIZZOLI, C.G. SOMEDA : Spatial Correlation of Field Intensity in Incoherently Illuminated Multimode Fibres - Electronics Letters 13, 25 (1977)

(13) E. BIANCIARDI, V. RIZZOLI : Propagation in Lossy Fibers and the Leaky-Mode Concept - 3rd ECOC, Munich, F.D.R., Sept. 1977 - Proc. of the Conference, p. 66 (1977)

(14) A. CONSORTINI, P. MAGI, L. RONCHI : Trasmissione di informazione attraverso fibre ottiche - Atti XXIII Congresso Internazionale di Elettronica - Sistemi di Communicazione su Fibre Ottiche, Roma (March 1976) p. 358

(15) A.M. SCHEGGI, P.F. CHECCACCI, R. FALCIAI : Dispersion Evaluation in Multimode Fibres by Numerical Technique : Application to Ring Shaped and Graded Index with a Central Dip - AGARD-CP-219 p. 32 (1977)

(16) P.F. CHECCACCI, R. FALCIAI, A.M. SCHEGGI : Effetto della dispersione del materiale in fibre ottiche con differenti distribuzioni di indice di rifrazione - Alta Frequenza XLVII, 135 (1978)

(17) P.F. CHECCACCI, R. FALCIAI, A.M. SCHEGGI : Modal Caustics and Dispersion Mechanism in Optical Fibers - Optics Letters 2, 25 (1978)

(18) B. CROSIGNANI, B. DAINO, P. DI PORTO : Statistical Coupled Equations in Loss Less Optical Fibers - IEEE Trans. MTT-23, 416 (1975)

(19) B. CROSIGNANI, B. DAINO, P. DI PORTO : Smoothing of the Mode Power Fluctuations in Loss Less Optical Fibers due to the Source Spectral Bandwidth - Optics Communications 18, 551 (1978)

(20) B. CROSIGNANI, B. DAINO, P. DI PORTO : Speckle Pattern Visibility of Light Transmitted through a Multimode Optical Fiber - J. Opt. Soc. America 66, 1312 (1976)

(21) B. CROSIGNANI, P. DI PORTO, C.H. PAPAS : Theory of Time Dependent Propagation in Multimode Light Guides - J. Opt. Soc. America 67, 1300 (1973)

(22) M. CONTI, G. CORDA, R. ROCAK : Componenti optoelettronici avanzati per sistemi in fibre ottiche - Report Presented at the 4-th MIEI, Torino (Sept. 1977)

(23) C.P. BASOLA, U. MICHI, G. RANDONE, P. ROCAK : LED ad alta radianza ottenuti con diffusione planare : tecnologia a prestazioni - Alta Frequenza XLVII, 208 (1978)

(24) B. DAINO, S. PIAZZOLLA, P. SPANO, M. BERTOLOTTI : Spatial
Dependence of Time Delay in Light-Emitting Diodes - J. Opt. Soc.
America 47, 2773 (1976)

(25) P. SPANO : Chromatic Delay in Light Emitting Diodes, Measurements and Theory - 3rd ECOC, Munich, F.D.R., Sept. 1977 - Proc. of
the Conference, p. 751 (1977)

(26) G. CANCELLIERI : Misure di attenuazione nelle fibre ottiche -
Alta Frequenza XLVII, 151 (1978)

(27) C.G. SOMEDA : Analisi delle microcurvature nelle fibre ottiche
con nucleo in silice e mantello in resina siliconica - Alta
Frequenza XLVII, 126 (1978)

(28) B. COSTA, B. SORDO : Experimental Study of Optical Fibers
Attenuation by a Modified Back Scattering Technique - 3rd ECOC,
Munich, F.D.R., Sept. 1977 - Proc. of the Conference, p. 69 (1977)

(29) B. DAINO, D. SETTE : The Measurements of the Transmission
Properties of Optical Cables - EUROCON 77, Venezia (May 1977)

(30) B. COSTA, B. SORDO : Measurements of the Refractive Index
Profile in Optical Fibres : Comparison between Different Techniques
2nd ECOC, Paris, Sept. 1976 - Proc. of the Conference, p. 81 (1976)

(31) B. DAINO, S. PIAZZOLLA, A.SAGNOTTI : A New Method for Measuring the Index Profile of Optical Fibres - Private Communication to
be presented at the 4th ECOC, Genova (Sept. 1978)

(32) B. CATANIA, L. MICHETTI, F. TOSCO, E. OCCHINI, L. SILVESTRI :
First Italian Experiment with Buried Optical Cable - 2nd ECOC,
Paris, Sept. 1976 - Proc. of the Conference, p. 315 (1976)

(33) G.C. RIGHINI, V. RUSSO, S. SOTTINI : Le fibre ottiche in medi-
cina - Alta Frequenza XLVII, n. 3, 165 (March 1978)

(34) V. DANIELE, I. MONTROSSET, R. ZICH : Accopiamento modale nello
spettro continuo per strutture guidanti ottico integrate - Atti
della I Riunione Nazionale di Elettromagnetismo Applicato, l'Aquila,
p. 141 (1976)

(35) C. BARBERO, G.P. BAVA, M. MOSCA : Misure di indice di rifrazione su film sottili per ottica integrata - Atti della I Riunione
Nazionale di Elettromagnetismo Applicato, L'Aquila, p. 169 (1976)

(36) C. BARBERO, G.P. BAVA, M. MOSCA : Modifiche del metodo di
Abeles per misure di indice di rifrazione su film sottili per
ottica integrata - Alta Frequenza 46, 239 (1977)

(37) C. BARBERO, R. ORTA : Caratterizzazione di materiali diffusi per ottica integrate - LXXVIII Riunione Annuale AEI, Como (1977)

(38) G.C. RIGHINI, V. RUSSO, S. SOTTINI, G. TORALDO DI FRANCIA : Thin Film Geodesic Lenses - Appl. Opt. 11, 1442 (1972)

(39) G.C. RIGHINI, V. RUSSO, S. SOTTINI, G. TORALDO DI FRANCIA : Geodesic Lenses for Guided Optical Waves - Appl. Opt. 12, 1477 (1973)

(40) G.C. RIGHINI, V. RUSSO, S. SOTTINI : An Unusual Correlator for Guided Waves - Colloque sur l'optique des ondes guidées, Paris, p. 11-7 (1975)

(41) G.C. RIGHINI, V. RUSSO, S. SOTTINI : Some Passive Components for Integrated Optics - Atti del 23° Congresso Internazionale per l'Elettronica, p. 233 (1976)

(42) G.C. RIGHINI, V. RUSSO, S. SOTTINI : Thin Film Integrated Signal Processors - AGARD Conference Proceedings n. 219, p. 25-1 (May 1977)

(43) G.C. RIGHINI, V. RUSSO, S. SOTTINI : KOR Photoresist in Integrated Optics - Optical and Quantum Electronics 7, 447 (1975)

(44) C. CALI, V. DANEU, A. ORIOLI, S. RIVA SANSEVERINO : Flash Evaporation of Compounds with a Pulsed-Discharge CO_2 Lasers - Appl. Opt. 15, 1327 (1976)

(45) S. AGLIERI, C. CALI, V. DANEU, S. RIVA SANSEVERINO : Integrated Optical Technique for Detection of Electrooptical Non-Linear Effect in Thin Films - Proc. of the 6th European Microwave Conference, p. 428, Roma (1976)

(46) C. CALI, V. DANEU, S. RIVA SANSEVERINO : Thin Film Detector for Integrated Optical Systems - Presented at the Topical Meeting on Integrated and Guided Wave Optics, Salt Lake City (January 1978) Paper MC3

(47) M.N. ARMENISE, M. DE SARIO : Propagazione in guide ottiche a molti strati per la progettazione di modulatori elettro-ottici - Alta Frequenza XLVII, 159 (1978)

(48) M.N. ARMENISE, M. DE SARIO : Modes in Diffused Bidimensional Multilayer Waveguide for Electrooptical Modulator Design - To be presented at the VI Colloquium on Microwave Communications, Budapest (Sept. 1978)

(49) M.N. ARMENISE, M. DE SARIO : Modal Characteristics of Diffused
Rectangular Anisotropic Waveguides to Design Integrated Optic
Devices - To be presented at VIII European Microwave Conference,
Paris (Sept. 1978)

(50) M.N. ARMENISE, M. DE SARIO : Progettazione di modulatori
elettro-ottici implicanti materiali dispersivi - To be presented
at the II Riunione Elettromagnetismo Applicato, Pavia (October
1978)

ECONOMIC ASPECTS OF FIBRE SYSTEMS

Giorgio PELLEGRINI

CSELT - Centro Studi e Laboratori Telecomunicazioni S.p.A.

Via G. Reiss Romoli, 274 - 10148 Torino (Italy)

1. Introduction

Optical fibre transmission systems are going to be widely used, in a near future, in telecommunication networks to transmit digital signals, at repetition rates ranging from few Mbit/s to hundreds of Mbit/s.

One of the main points to be considered in the introduction into the telecommunication network of a new transmission system is its economical aspect.

The aim of this paper is to discuss the economic aspects of fibre system starting with a general introduction to the so-called «cost-comparison» studies. Afterwards, a definition will be given of basic cost factors for fibre system; some informations about reliability and cost of each basic component of the system, and hence of the whole transmission system itself, will be provided as well. Finally a comparison between optical and coaxial cable transmission systems, from the point of view of costs, will be done. That comparison will concern digital systems at hierarchical rates, as standardized by international organizations, based on primary PCM group, of 30 voice channels, at a bit rate of 2048 kbit/s.

2. Cost-comparison studies

Cost-comparison studies are made to solve several types of economic decisions [1]; for instance:

Growth of services may require extension of existing facilities or addition to increase capacity;

Excessive operation costs may justify complete replacement of existing plant. An engineering economy study will show wheter it is more economical to continue present operation or replace with new plant;

Determination of the optimum size of plant increments and the proper time interval between additions. To this end periods of utilization should also be taken into account [1], [2]. Table I lists some typical service lifes as suggested by the CCITT.

All the cost factors involved in cost-comparison studies can be grouped in two broad classes:

TYPICAL SERVICE LIFE (YEARS)
AERIAL CABLE. 25
BURIED CABLE. 30
CARRIER SYSTEMS. 15
MICROWAVE EQUIPMENT 15
ANTENNAE AND OTHER ELECTRONIC EQUIPMENT. 15
BUILDING AND TOWERS. 40

Table I

a) One-time expenditures, even called «initial investment costs» (e.g. costs of material and equipment, installation costs, initial training of personnel, and so forth);
b) Continuous expenditures for operation and maintenance (O&M) even called «operation costs» (e.g. costs related to material and labour associated with the upkeep, costs of labour associated with the day-to-day operation of the plant, supervision costs, etc.).

These factors must be taken into account in economy planning when choices are needed to arrive at good decisions between two or more alternatives and to determine which will be the most economical and practicable in the long run.

The choices will generally refer to the following items:

i) Type of system: for instance microwave or coaxial cable or optical fibre cable; and moreover digital or analog transmission techniques. The comparison will be meaningful if all the factors affecting the cost of each system are taken into account. To this end, reference [1] provides useful suggestions by listing the basic cost factors and giving guidance on the specific cost contributions which should be taken into account for each type of transmission system.
However, it must be pointed out that factors common to all plans to be compared may be omitted, as these costs would only add the same amount to each plan.
ii) Optimum system capacity: in general the higher the capacity of a system, the lower the cost per circuit. But that is true only if the system is full utilized, as unused circuits give no returns.
Then an optimum should be attained between the capital cost per circuit provided and the rate of taking these circuits into service. Of course, such an optimization requires the knowledge of the statistics of the network links capacity, of the link lengths and the forecast of the circuits increasing rate.
iii) Type of carrier: for instance carrier size. This choice can't be completely free as several limitations are imposed by the manufacturers and by the standardizations. However when developing a new transmission medium some economic considerations could be done during the project. For example the carrier cost, generally increases with decreasing attenuation while the repeater cost generally decreases. Then total repeater plus medium cost can be optimized.
It must also be noted, when comparing different types of carriers, that installation costs can often exceed the cost of the carrier itself; therefore the installation cost can become very important and should be taken into account in the comparisons among different types of carriers.
iv) Maintenance and reliability: the connection existing between the reliability of a transmission system and its maintenance requirements can be easily understood

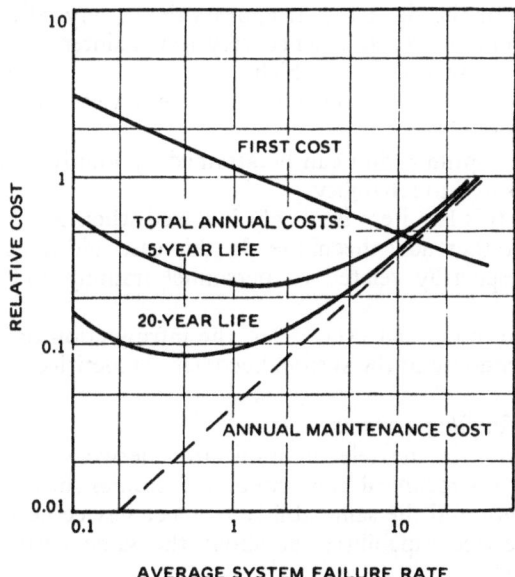

Fig. 1 - *Relationship between first cost of the system, annual maintenance cost and total annual cost*

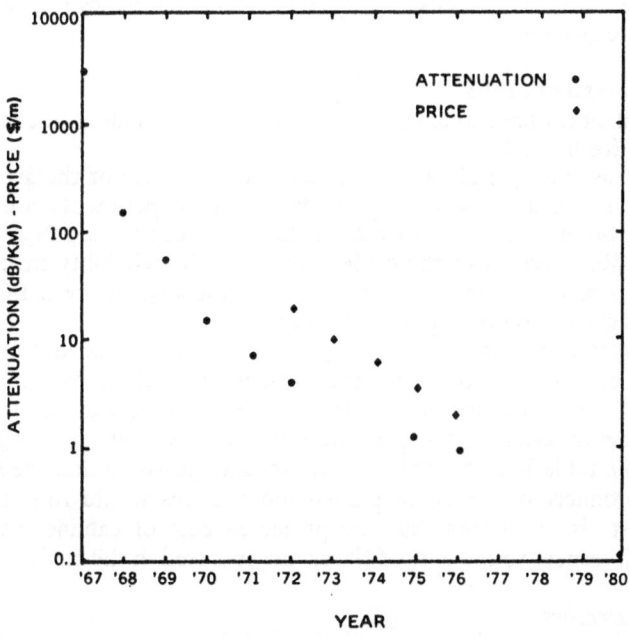

Fig. 2 - *Progress in optical fibre characteristics*

by the inspection of the fig. 1 [3]. It appears that a very reliable system requires practically no maintenance, and hence very low maintenance costs; but, on the other hand, its first cost will be very high.

Moreover a less expensive system is generally not very reliable, and hence the maintenance costs shall rise.

As a consequence, optimization can be attained by jointly considering the O&M costs and those related to reliability.

v) Innovation costs: this last item is mainly involved when one among the different alternatives is a rather new thecnique. In this case additional costs can occur; these costs are especially related to personnel training and to new operating procedures.

Such «innovation» costs can even delay the introduction of the new technique until the global economy of the system becomes competitive.

3. Basic cost factors for fibre systems

The introduction in the telecommunication network of a new transmission system can be planned if technical and economical studies show that it offers either the same service quality and the same ability to meet service needs at lower cost, or potentially greater service capabilities at about the same costs as the alternative systems already in use.

Optical fibre transmission systems seem to meet both the above requirement as consequence of the main characteristics of the fibres:
- large available bandwidth and low attenuation;
- low projected costs of the transmission medium (optical cable).

However economical analysis of optical systems must concern other constituents as well; to this end cost and reliability characteristics of the various system constituents are given in the following.

3.1 Optical fibres and cables

Optical fibres have undergone continuous improvement since their beginning, as can be seen from fig. 2.

As far as the optical cable is concerned, the state of the art, as it is today, appears to be satisfactory, at least from the cabling experiments conducted in Italy [4]. From the point of view of the cable reliability it can be said that this factor gives no problems. Moreover, after the cable is installed, its reliability mainly depends on fortuite breaks produced mechanically, and is independent from the fact that the cable contains glass fibres or copper conductors.

For what concerns the fibre cost, it must be pointed out that this cannot be evaluated on the basis of either the present production volume or the present high cost of the processes involved in fibre fabrication. As a consequence an attempt has been made to evaluate the projected cost in a sistuation of regular industrial production (see table II of ref [5]). It appears then justified that the selling price for a sufficient production volume of graded index fibres would run around 0.1 $/m.

It must also be noted that the projected cost of cabling seems always predominant with respect to the cost of fibres incorporated in the cable.

3.2 Optical detectors

No reliability problems exist for optical detectors (p-i-n or avalanche photo-diodes).

As far as prices are concerned, at present the price of an avalanche photodiode

(APD) is about one order of magnitude higher than that of a p-i-n, this latter being in the range of 35 $. However, it must be pointed out that, if the market demand, and hence the production volume will increase, the unit selling price of both components should fall below 10 $.

3.3 Optical sources

Optical sources suitable for fibre communication systems present today some limitations on their reliability. These limitations are stronger for the lasers than for the high radiance light emitting diodes (LED) whose life is, today, in the order of 10^5 hrs. Lasers available today have exhibited life periods in excess of 10^4 hrs at room temperature.

However, extrapolated «half-power» life of semiconductor lasers is in the range of 100,000 hrs. These figures would be in line with those of the expected life requirement on the complete repeater, as said in the preceeding paragraph, and would be therefore acceptable to the Operating Companies.

For what concerns the cost of the abovementioned optical sources it must be said that laser diodes are priced today at about twice the price of high radiance LED. This ratio is not likely to be reduced in the future, but perhaps be increased, due to the fact that the production volume of laser diodes is likely to be much lower than that of high radiance LED.

Even in the case of optical sources, the selling price will strongly depend on the production volume. A reduction of one order of magnitude in unit price with respect to the present values could be hypotized if the production volume could be higher than 100,000 pieces/year.

3.4 System cost sensitivity

An important point to be considered in economy studies is the «cost sensitivity». Cost evaluation are often based on a set of assumptions concerning the costs of the system constituents. It is generally very useful to know what is the incidence of evaluation errors on the final result.

Two examples of these calculations are reported in figs. 3 a) and b). They refer respectively to cost sensitivity of line digital systems at 8,448 kbit/s and 139,264 kbit/s.

From the inspection of the two diagrams it can be seen that there are some factors, among other, whose evaluation must be very carefully done; as a matter of the fact that errors in cost forecasts can give a large variation in the cost of the whole system and hence result in wrong economy planning and cost-comparisons; on the other hand, other factors can be allowed to have a large margin of uncertainty as a consequence of their low incidence on the total cost of the system. These are the cases, respectively, of the installation cost and of the line terminal cost in an optical line system. The former is always predominant in determining the cost of the line system, the latter being, on the contrary, the least important.

It can be interesting to observe, by comparing the diagrams of figs. 3 a) and 3 b), that the influence of the cabling and of the repeater costs are changed each other, because the higher the bit rate, the higher the cost of the repeater.

4. Cost comparison with copper cable systems

According to the methodology adopted in [5], a cost comparison analysis has been made between optical and copper cable digital transmission systems, operating at the hierarchycal rates based upon the 2,048 kbit/s PCM systems carrying 30 voice channels.

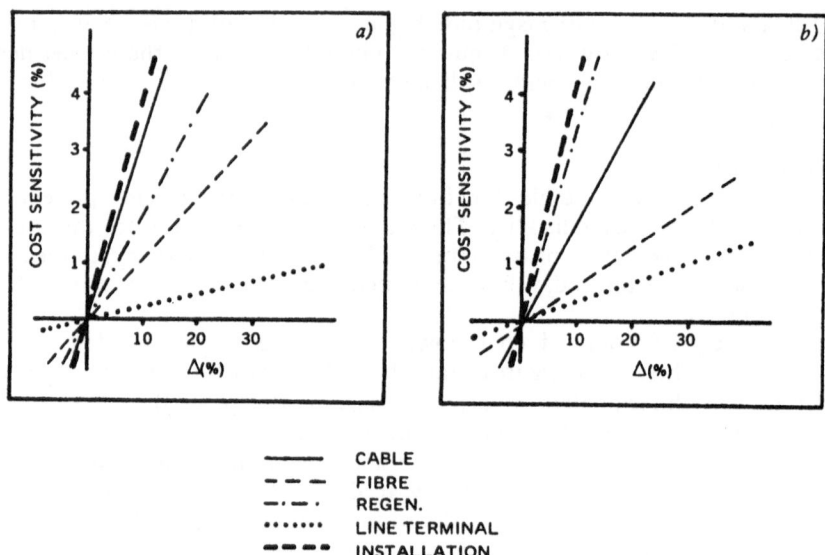

CABLE
FIBRE
REGEN.
LINE TERMINAL
INSTALLATION

Fig. 3 - Cost sensitivity of optical digital systems at 8,448 kbit/s a) and 139,264 kbit/s b)

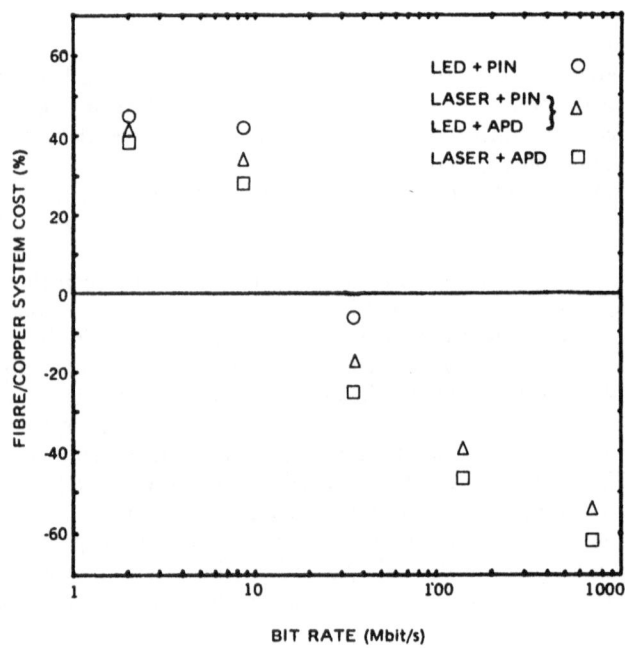

Fig. 4 - Comparison between cost of optical fibre systems and cost of copper cable systems for different bit rates

The fig. 4 shows the results of the computations performed taking into account the cost of the cable (including installation (*)), the cost of the regenerators and of the line terminal equipment including power feeding.

The cost of regenerators is assumed as being approximately the same for the two systems compared because the optical fibre system, on one hand, bears the additional costs of the O/E and E/O transducers, but, on the other hand, needs much simpler equalizers and easier clock extraction circuits.

The cost of the higher bit rate regenerators has been estimated to increase with the square root of the bit rate [5].

From the inspection of the diagram reported in fig. 4 it appears that optical fibre systems at bit rates higher than 8 Mbit/s will be probably more convenient than copper cable systems from the cost standpoint.

It may also be interesting to observe that the use of laser and APD is always advantageous (as the number of regeneration in the link is reduced). This conclusion is only valid, however, under the assumption that the cost of both optoelectronic devices will not constituite a significant amount of the complete regenerator cost.

5. Conclusions

From what precedes it can be concluded that economical evaluations are of primary importance in the planning of a telecommunication networks. Economy studies require careful evaluation of many factors effecting the cost of transmission systems to give a serious estimate of the cost characteristics of the alternatives under consideration.

As far as digital transmission is concerned, the optical fibre systems seem to be very attractive and to show several useful aspects which allow to expect such systems to become competitive with conventional transmission media when digital systems at high bit rates will be introduced into the telecommunication network.

(*) The cost of installation has been kept constant for the optical cable, but variable for the copper cable as it actually happens.

Bibliography
[1] C.C.I.T.T., C.C.I.R.: «Economical and technical aspects of the choice of transmission systems» - U.I.T. 1969.
[2] F. Bigi:«Aspetti economici nella scelta dei sistemi di trasmissione» - SIP-DG, Roma Internal Report, 1975.
[3] L.N. St. James: «Reliability and Military Economics» - Bell Laboratories Record, May 1965.
[4] G. Cocito et alii: «COS 2 Experiment in Turin: Field Test on Optical Cable in Ducts» - IEEE tr on Communications, to be published.
[5] C. Colavito, B. Catania, G. Pellegrini: «Reliability and cost evaluation in view of the intrduction of optical fibres in communication networks» - 2nd European Conference on Optical Fibre Communication, Paris, Sept. 1976.

INDEX

415